Security and Privacy for 6G Massive IoT

Security and Privacy for 6G Massive IoT

Edited by

Georgios Mantas
Instituto de Telecomunicações, Aveiro, Portugal
Faculty of Engineering and Science
University of Greenwich, Chatham Maritime, UK

Firooz B. Saghezchi
Chair for Distributed Signal Processing
RWTH Aachen University, Aachen, Germany

Jonathan Rodriguez
Faculty of Computing, Engineering, and
Science, University of South Wales, Pontypridd, UK

Victor Sucasas
Cryptography Research Center
Technology Innovation Institute, Abu Dhabi, UAE

WILEY

This edition first published 2025

Registered Offices
John Wiley & Sons, Inc., 111 River Street, Hoboken, NJ 07030, USA
John Wiley & Sons Ltd, New Era House, 8 Oldlands Way, Bognor Regis, West Sussex, PO22 9NQ, UK

For details of our global editorial offices, customer services, and more information about Wiley products visit us at www.wiley.com.

Wiley also publishes its books in a variety of electronic formats and by print-on-demand. Some content that appears in standard print versions of this book may not be available in other formats.

Library of Congress Cataloging-in-Publication Data

Names: Mantas, Georgios, editor. | Saghezchi, Firooz, editor. | Rodriguez, Jonathan, editor. | Sucasas, Victor, editor.
Title: Security and privacy for 6G massive IoT / edited by Georgios Mantas, Firooz Saghezchi, Jonathan Rodriguez, Victor Sucasas.
Description: Hoboken, NJ : Wiley, 2025. | Includes index.
Identifiers: LCCN 2024037423 (print) | LCCN 2024037424 (ebook) | ISBN 9781119987970 (hardback) | ISBN 9781119987987 (adobe pdf) | ISBN 9781119987994 (epub)
Subjects: LCSH: 6G mobile communication systems–Security measures. | Internet of things–Security measures.
Classification: LCC TK5103.252 .S42 2024 (print) | LCC TK5103.252 (ebook) | DDC 005.8–dc23/eng/20241030
LC record available at https://lccn.loc.gov/2024037423
LC ebook record available at https://lccn.loc.gov/2024037424

Cover Design: Wiley
Cover Image: © Just_Super/Getty Images

Set in 9.5/12.5pt STIXTwoText by Straive, Chennai, India

Contents

List of Contributors *xiii*
Acknowledgements *xix*
Introduction *xxi*

1 Threat Landscape for 6G-Enabled Massive IoT *1*
Maria Papaioannou, Georgios Mantas, Firooz B. Saghezchi, Georgios Kambourakis, Felipe Gil-Castiñeira, Raúl Santos de la Cámara, and Jonathan Rodriguez
1.1 Introduction *1*
1.2 6G Vision and Core Values *2*
1.3 Emerging Massive IoT Applications Enabled by 6G *3*
1.3.1 Enabling Sustainability *4*
1.3.1.1 E-health for All *5*
1.3.1.2 Institutional Coverage *6*
1.3.1.3 Earth Monitor *6*
1.3.1.4 Autonomous Supply Chains *6*
1.3.1.5 Sustainable Food Production *7*
1.3.1.6 Network Trade-Offs for Minimized Environmental Impact *7*
1.3.1.7 Network Functionality for Crisis Resilience *8*
1.3.2 Massive Twinning *8*
1.3.2.1 Digital Twins for Manufacturing *9*
1.3.2.2 Immersive Smart Cities *10*
1.3.2.3 Internet of Tags *11*
1.3.3 Telepresence *12*
1.3.3.1 Fully Merged Cyber–Physical Worlds *12*
1.3.3.2 Mixed Reality Co-design *14*
1.3.3.3 Immersive Sport Event *14*
1.3.3.4 Merged Reality Game/Work *15*
1.3.4 From Robots to Cobots *15*
1.3.4.1 Consumer Robots *15*
1.3.4.2 AI Partners *16*
1.3.4.3 Interacting and Cooperative Mobile Robots *16*
1.3.4.4 Flexible Manufacturing *17*
1.3.4.5 Situation-Aware Device Reconfiguration *17*
1.3.5 Trusted Embedded Networks *18*

1.3.5.1 Human-Centric Communications 18
1.3.5.2 Infrastructure-less Network Extensions and Embedded Networks 19
1.3.5.3 Local Coverage for Temporary Usage 20
1.3.5.4 Small Coverage, Low-Power Micro-Network in Networks for Production and Manufacturing 20
1.3.6 Hyperconnected Resilient Network Infrastructures 20
1.3.6.1 Sensor Infrastructure Web 21
1.3.6.2 AI-Assisted Vehicle-to-Everything (V2X) 21
1.3.6.3 Interconnected IoT Micro-Networks 22
1.3.6.4 Enhanced Public Protection 23
1.4 Overview of a 6G Network Architecture to Enable Massive IoT 23
1.4.1 Space Network 24
1.4.2 Air Network 25
1.4.3 Ground Network 25
1.4.4 Sea/Underwater Network 26
1.4.4.1 Surface Network 26
1.4.4.2 Underwater Network 26
1.5 Security Objectives in Massive IoT in 6G 26
1.5.1 Confidentiality 26
1.5.2 Integrity 27
1.5.3 Availability 27
1.5.4 Authentication 28
1.5.5 Authorization 28
1.6 Security Threats in Massive IoT in 6G 28
1.6.1 Security Threats to Data Confidentiality 29
1.6.2 Security Threats to Data Integrity 29
1.6.3 Security Threats to Availability 30
1.6.4 Security Threats to Authentication 31
1.6.5 Security Threats to Authorization 31
1.7 Conclusion 32
References 33

2 Secure Edge Intelligence in the 6G Era 35
Tanesh Kumar, Juha Partala, Tri Nguyen, Lalita Agrawal, Ayan Mondal, Abhishek Kumar, Ijaz Ahmad, Ella Peltonen, Susanna Pirttikangas, and Erkki Harjula
2.1 Introduction 35
2.2 Background 36
2.2.1 Edge Computing and Its Importance 36
2.2.2 Emergence of Edge Intelligence 38
2.2.3 Fusion of Edge Intelligence and 6G 39
2.2.4 The Need for Secure Edge Intelligence 40
2.3 Security Challenges in 6G EI 40
2.3.1 Computational Offloading 40
2.3.2 Security of Machine Learning 41

2.3.3 Post-quantum Cryptography *42*
2.4 Privacy Challenges in 6G EI *42*
2.5 Trust Challenges in 6G EI *44*
2.6 Security Standardization for EI and 6G *46*
2.7 Conclusion *47*
References *47*

3 Privacy-Preserving Machine Learning for Massive IoT Deployments *53*
Najwa Aaraj, Abdelrahaman Aly, Alvaro Garcia-Banda, Chiara Marcolla, Victor Sucasas, and Ajith Suresh
3.1 Introduction *53*
3.2 PPML for IoT *54*
3.3 Secure Multiparty Computation *56*
3.3.1 MPC: Security Models and Setups *60*
3.3.2 MPC: Privacy and Output Guarantees *61*
3.3.3 MPC: Arithmetic *61*
3.3.4 MPC: Preprocessing *63*
3.4 Fully Homomorphic Encryption *63*
3.4.1 FHE: Bottlenecks and Advantages *64*
3.4.2 Comparison with MPC *65*
3.4.3 Generations of FHE Schemes *66*
3.5 Oblivious Neural Networks (ONN) *67*
3.5.1 Two-Party (2PC) *67*
3.5.2 Two-Party with Helper ($2PC^+$) *67*
3.5.3 Three-Party (3PC) and Four-Party (4PC) *69*
3.5.4 n-Party (MPC) *69*
3.5.5 A Closer Look at PPML Frameworks *69*
3.5.5.1 CryptoNets *69*
3.5.5.2 SecureML *70*
3.5.5.3 MiniONN *70*
3.5.5.4 DeepSecure *72*
3.5.5.5 Gazelle *73*
3.5.5.6 Delphi *74*
3.5.5.7 ABY^3 *74*
3.5.5.8 Fantastic Four *77*
3.5.5.9 Crypten *77*
3.5.5.10 FANNG–MPC *78*
3.6 Decision Trees *78*
3.6.1 Structure of a Decision Tree *79*
3.6.2 Training *79*
3.6.3 Computational Setups *80*
3.6.4 Inference *81*
3.6.4.1 The Three Stages Model *81*
3.6.5 Relevant Contributions *83*

3.6.5.1 On Passive Security *83*
3.6.5.2 On Active Security *84*
3.7 Software and Frameworks *85*
3.8 Lightweight FHE and MPC for IoT *86*
3.8.1 MPC Outsourcing *86*
3.8.2 Hybrid-FHE *87*
3.9 Alternative Solutions *90*
3.10 Conclusions *90*
References *91*

4 Federated Learning-Based Intrusion Detection Systems for Massive IoT *101*
Filippos Pelekoudas-Oikonomou, Parya H. Mirzaee, Waleed Hathal, Georgios Mantas, Jonathan Rodriguez, Haitham Cruickshank, and Zhili Sun
4.1 Introduction *101*
4.2 Intrusion Detection Systems (IDSs) in IoT *103*
4.2.1 Fundamentals of IDSs *103*
4.2.2 Machine Learning Techniques for IDSs in IoT *104*
4.2.3 Limitations of ML-Based IDSs in IoT *106*
4.3 Federated Learning: A Decentralized ML Approach *107*
4.3.1 Definition of Federated Learning *107*
4.3.2 Categories of Federated Learning *108*
4.3.3 Federated Learning Architectures *108*
4.3.3.1 Horizontal Federated Learning (HFL) *110*
4.3.3.2 Vertical Federated Learning (VFL) *111*
4.3.3.3 Federated Transfer Learning (FTL) *111*
4.4 Federated Learning-Based IDSs for IoT *113*
4.4.1 FL-Based IDSs in IoT Use Cases *113*
4.4.1.1 FL-Based IDSs in Smart Homes *113*
4.4.1.2 FL-Based IDSs in Industrial IoT *114*
4.4.1.3 FL-Based IDSs in Agricultural IoT *115*
4.4.1.4 FL-Based IDS in Vehicular IoT Networks *116*
4.4.2 Cross-layer FL for Lightweight IoT Privacy-Preserving IDS *116*
4.5 Model Aggregation Approaches and Algorithms in FL *117*
4.5.1 Model Aggregation Approaches in FL *117*
4.5.2 Model Aggregation Algorithms in FL *119*
4.6 Challenges and Future Directions in FL-Based IDS for IoT *121*
4.6.1 Validation and Standardization *121*
4.6.2 Data Heterogeneity and Non-IID Data *121*
4.6.3 Security and Privacy Enhancements *121*
4.6.4 Communication Efficiency and Scalability *122*
4.6.5 Explainability and Interpretability *122*
4.7 Conclusion *122*
List of Abbreviations *123*
References *124*

5 Securing Massive IoT Using Network Slicing and Blockchain *129*
Shihan Bao, Zhili Sun, and Haitham Cruickshank
5.1 Introduction *129*
5.2 Background *131*
5.2.1 Massive IoT *131*
5.2.2 Blockchain *132*
5.2.2.1 Blockchain Applications *132*
5.2.2.2 Consensus Mechanisms *133*
5.3 Challenges on Massive IoT *135*
5.3.1 Security Requirements in Massive IoT *135*
5.3.2 Privacy Requirements in Massive IoT *136*
5.3.3 Location Privacy Challenges *138*
5.4 Securing IoT Using Network Slicing and Blockchain *138*
5.4.1 Network Slicing Standardization *138*
5.4.2 Network Slicing Security, Privacy, and Trust Threats for 5G and Beyond *142*
5.4.3 Blockchain-Enabled Secure Network Slicing *144*
5.4.3.1 Integration of Distributed Ledger Technology (DLT) with Network Slicing *144*
5.4.3.2 Blockchain-Powered Network Slicing in Verticals *144*
5.4.3.3 Multiple Participant Coordination and Trust Management by Blockchain for Network Slicing *145*
5.5 Open Challenges *146*
5.5.1 Edge Intelligence in Network Slicing *147*
5.5.2 Post-quantum Security *147*
5.5.3 Blockchain Scalability *147*
5.5.4 RAN Slicing *148*
5.6 Conclusion *148*
References *148*

6 Physical Layer Security for RF-Based Massive IoT *155*
Marcus de Ree, Seda Dogan-Tusha, Elmehdi Illi, Marwa Qaraqe, Saud Althunibat, Georgios Mantas, and Jonathan Rodriguez
6.1 Introduction *155*
6.2 Physical Layer-Based Key Establishment *156*
6.2.1 Introduction *156*
6.2.2 Channel Reciprocity-Based Key Establishment *157*
6.2.2.1 Principles and Assumptions *157*
6.2.2.2 Evaluation Metrics *159*
6.2.2.3 Key Establishment Model *159*
6.2.3 Signal Source Indistinguishability-Based Key Establishment *161*
6.2.3.1 Principles and Assumptions *161*
6.2.3.2 Evaluation Metrics *162*
6.2.3.3 Key Establishment Model *162*
6.2.3.4 Security and Performance Evaluation *163*
6.2.4 Final Remarks *164*
6.3 Physical Layer-Based Node Authentication *164*

6.3.1 Introduction 164
6.3.2 Key-Based Physical Layer Authentication 165
6.3.2.1 Tag-Embedding Authentication 166
6.3.3 Channel-Based Keyless Physical Layer Authentication 166
6.3.3.1 RSS-Based Authentication 167
6.3.3.2 Frequency-Based Authentication 167
6.3.3.3 CIR-Based Authentication 167
6.3.3.4 Pilot-Based Authentication 168
6.3.3.5 Machine Learning-Based Authentication 168
6.3.4 Device-Based Keyless Physical Layer Authentication 168
6.3.4.1 Radio Frequency Fingerprinting 168
6.3.4.2 Physically Unclonable Function 169
6.3.5 Final Remarks 170
6.4 Physical Layer-Based Data Confidentiality 171
6.4.1 Introduction 171
6.4.2 Key-Based Data Confidentiality – Physical Layer Encryption 171
6.4.2.1 Channel Coding-Based Physical Layer Encryption 172
6.4.2.2 Signal Modulation-Based Physical Layer Encryption 172
6.4.3 Keyless Data Confidentiality 173
6.4.3.1 Security Metrics 174
6.4.3.2 Channel Coding-Based Data Confidentiality 174
6.4.3.3 Artificial Noise-Based Extensions 175
6.4.3.4 Directional Modulation-Based Extensions 176
6.4.4 Final Remarks 176
6.5 Physical Layer-Based Detection of Malicious Nodes 177
6.5.1 Introduction 177
6.5.2 Data Injection Attack Detection 177
6.5.3 Sybil Attack Detection 178
6.5.4 Sleep Deprivation Attack Detection 179
6.5.5 Final Remarks 180
6.6 Conclusion 180
List of Abbreviations 181
References 182

7 Quantum Security for the Tactile Internet *193*
Shima Hassanpour, Riccardo Bassoli, Janis Nötzel, Frank H. P. Fitzek, Holger Boche, and Thorsten Strufe
7.1 Introduction 193
7.2 Preliminaries – Quantum Mechanics 197
7.2.1 Finite Dimensional Quantum Systems 197
7.2.2 Quantum Gates 197
7.2.3 Quantum Measurement 198
7.2.4 Fock Space 199
7.2.5 Continuous Variable Quantum States 199
7.2.6 Gaussian States 200
7.2.7 Conjugate Coding 201

7.2.8 Quantum Entanglement *201*
7.2.9 Entanglement Swapping *202*
7.3 Preliminaries – Security *203*
7.3.1 Formal Cryptographic Frameworks *204*
7.3.2 Practical Source and Security *204*
7.4 Quantum Key Distribution *205*
7.4.1 Discrete Variable Protocols *205*
7.4.1.1 The Example of BB84 *207*
7.4.1.2 Security Argument *209*
7.4.2 Continuous Variable Protocols *211*
7.5 Integrated Physical Layer Security *211*
7.5.1 Conventional Approaches *212*
7.5.2 Quantum Approaches *213*
7.5.2.1 Notation *213*
7.5.2.2 Quantum Physical Layer Security *214*
7.6 Challenges and Known Attacks on Quantum Security *214*
7.7 Oblivious Transfer (OT) *217*
7.7.1 Probabilistic Formulation of OT *218*
7.7.2 Relevance of OT *219*
7.7.3 Bit Commitment *219*
7.7.4 Why Bit Commitment Is Important to Oblivious Transfer *219*
7.8 Conclusion *220*
Acknowledgment *221*
References *221*

8 Physical Layer Security for Terahertz Communications in Massive IoT *229*
Vitaly Petrov, Hichem Guerboukha, Zhambyl Shaikhanov, Edward W. Knightly, Daniel M. Mittleman, and Josep M. Jornet
8.1 Introduction *229*
8.2 Information Theoretic Analysis of Eavesdropping *230*
8.3 Eavesdropping THz Links *232*
8.3.1 Security of Directional THz Links *232*
8.3.2 Metasurface-in-the-Middle Attack *233*
8.4 Multi-path THz Communications *236*
8.4.1 Multi-path THz Communications to Improve Security Against Eavesdropping *236*
8.4.2 Evaluation Framework *237*
8.4.3 Security vs. Capacity Trade-Off *239*
8.5 Absolute Security *241*
8.6 Jamming THz Links *243*
8.7 Summary and the Road Ahead *246*
References *246*

Index *251*

List of Contributors

Najwa Aaraj
Cryptography Research Center
Technology Innovation Institute
Abu Dhabi
UAE

Lalita Agrawal
Department of Computer Science and Engineering
Indian Institute of Technology Indore
Indore
India

Ijaz Ahmad
VTT Technical Research Centre of Finland
Espoo
Finland

Saud Althunibat
Faculty of Engineering
Al-Hussein Bin Talal University
Ma'an
Jordan

Abdelrahaman Aly
Cryptography Research Center
Technology Innovation Institute
Abu Dhabi
UAE

Shihan Bao
State Key Laboratory of Wireless Mobile Communications
China Academy of Telecommunications Technology
Beijing
China

Riccardo Bassoli
Deutsche Telekom Chair of Communication Networks
Technische Universität Dresden
Dresden
Germany

and

Centre for Tactile Internet with Human-in-the-Loop (CeTI)
Cluster of Excellence
Dresden
Germany

and

Quantum Communication Networks Research Group
Fakultät Elektrotechnik und Informationstechnik
Institut für Nachrichtentechnik
Technische Universität Dresden
Dresden
Germany

Holger Boche
Lehrstuhl für Theoretische Informationstechnik
TUM School of Computation, Information and Technology
Technische Universität München
Karlsruher Institut für Technologie
München
Germany

Raúl Santos de la Cámara
R&D Department
HI Iberia Ingeniería y Proyectos, S.L.
Madrid
Spain

Haitham Cruickshank
Institute for Communication Systems (ICS), 5G&6G Innovation Centre (5G&6GIC)
University of Surrey
Guildford
UK

Seda Dogan-Tusha
Division of Information and Computing Technology
College of Science and Engineering
Hamad Bin Khalifa University
Qatar Foundation
Doha
Qatar

and

Wireless Institute, College of Engineering
University of Notre Dame
South Bend, IN
USA

Frank H. P. Fitzek
Deutsche Telekom Chair of Communication Networks
Technische Universität Dresden
Dresden
Germany

and

Centre for Tactile Internet with Human-in-the-Loop (CeTI)
Cluster of Excellence
Dresden
Germany

Alvaro Garcia-Banda
Cryptography Research Center
Technology Innovation Institute
Abu Dhabi
UAE

Felipe Gil-Castiñeira
Telematics Engineering Department
University of Vigo
Vigo
Spain

Hichem Guerboukha
School of Science and Engineering
University of Missouri – Kansas City
Kansas, MO
USA

Erkki Harjula
Center for Wireless Communications
University of Oulu
Oulu
Finland

Shima Hassanpour
Chair of Privacy and Network Security
Technische Universität Dresden
Dresden
Germany

Waleed Hathal
Institute for Communication Systems (ICS), 5G&6G Innovation Centre (5G&6GIC)
University of Surrey
Guildford
UK

Elmehdi Illi
Division of Information and Computing Technology
College of Science and Engineering
Hamad Bin Khalifa University
Qatar Foundation
Doha
Qatar

Josep M. Jornet
Institute for the Wireless Internet of Things
Northeastern University
Boston, MA
USA

Georgios Kambourakis
Department of Information and Communication Systems Engineering, School of Engineering
University of the Aegean
Samos
Greece

Edward W. Knightly
Department of Electrical and Computer Engineering
Rice University
Houston, TX
USA

Abhishek Kumar
Center for Ubiquitous Computing
University of Oulu
Oulu, Oulun Yliopisto
Finland

Tanesh Kumar
Department of Information and Communications Engineering
Aalto University
Espoo
Finland

Georgios Mantas
Instituto de Telecomunicações
Aveiro
Portugal

and

Faculty of Engineering and Science
University of Greenwich
Chatham Maritime
UK

Chiara Marcolla
Cryptography Research Center
Technology Innovation Institute
Abu Dhabi
UAE

Parya H. Mirzaee
Institute for Communication Systems (ICS), 5G&6G Innovation Centre (5G&6GIC)
University of Surrey
Guildford
UK

Daniel M. Mittleman
School of Engineering
Brown University
Providence, RI
USA

Ayan Mondal
Department of Computer Science and Engineering
Indian Institute of Technology Indore
Indore
India

Tri Nguyen
Center for Ubiquitous Computing
University of Oulu
Oulu, Oulun Yliopisto
Finland

Janis Nötzel
Lehrstuhl für Theoretische Informationstechnik
TUM School of Computation, Information and Technology
Technische Universität München
Karlsruher Institut für Technologie
München
Germany

Maria Papaioannou
Faculty of Engineering and Science
University of Greenwich
Chatham Maritime
UK

Juha Partala
Center for Machine Vision and Signal Analysis
University of Oulu
Oulu
Finland

Filippos Pelekoudas-Oikonomou
Faculty of Engineering and Science
University of Greenwich
Chatham Maritime
UK

Ella Peltonen
M3S Research Unit
University of Oulu
Oulu
Finland

Vitaly Petrov
Division of Communication Systems
KTH Royal Institute of Technology
Stockholm
Sweden

Susanna Pirttikangas
Center for Ubiquitous Computing
University of Oulu
Oulu, Oulun Yliopisto
Finland

Marwa Qaraqe
Division of Information and Computing Technology
College of Science and Engineering
Hamad Bin Khalifa University
Qatar Foundation
Doha
Qatar

Marcus de Ree
Instituto de Telecomunicações
Universidade de Aveiro
Aveiro
Portugal

Jonathan Rodriguez
Faculty of Computing, Engineering and Science
University of South Wales
Pontypridd
UK

Firooz B. Saghezchi
Chair for Distributed Signal Processing
RWTH Aachen University
Aachen
Germany

Zhambyl Shaikhanov
Department of Electrical and Computer Engineering
Rice University
Houston, TX
USA

Thorsten Strufe
Chair of Privacy and Network Security
Technische Universität Dresden
Dresden
Germany

and

Centre for Tactile Internet with Human-in-the-Loop (CeTI)
Cluster of Excellence
Dresden
Germany

and

Chair of IT-Security
Karlsruher Institut für Technologie
Karlsruhe
Germany

Victor Sucasas
Cryptography Research Center
Technology Innovation Institute
Abu Dhabi
UAE

Zhili Sun
Institute for Communication Systems (ICS), 5G&6G Innovation Centre (5G&6GIC)
University of Surrey
Guildford
UK

Ajith Suresh
Cryptography Research Center
Technology Innovation Institute
Abu Dhabi
UAE

Acknowledgments

We thank the 6G-XR project[1] and the MATRIS project[2] for their contribution toward Chapter 1, and the PHYSEC project[3] for its contribution toward Chapter 6. Furthermore, we extend our deepest gratitude to our families, whose patience, cooperation, encouragement, and support have been the cornerstone of this book.

1 6G-XR project has received funding from the Smart Networks and Services Joint Undertaking (SNS JU) under the European Union's Horizon Europe research and innovation programme under Grant Agreement No 101096838 and from the Swiss State Secretariat for Education, Research and Innovation (SERI).
2 MATRIS project has received funding from the FCT/PTDC through national funds under grant 2022.07313.PTDC.
3 PHYSEC project has received funding from the NATO Science for Peace and Security Programme under grant SPS G5797.

Introduction

As the fifth-generation (5G) mobile networks are entering the deployment and commercialization phase across the world, the next-generation mobile communications systems (6G) have already come into the discussion, driven by the strict network performance requirements of the emerging disruptive massive IoT applications and the limitations of the existing 5G networks to meet 6G IoT-enabled use cases such as holographic telepresence, fully autonomous driving, and five-sense communications. In fact, the mismatch between the capabilities of the existing 5G systems and the stringent requirements of the future advanced applications leveraging massive IoT is one of the drivers behind the efforts toward the realization of the 6G systems by 2030. Therefore, the 6G systems are expected to extend the capabilities of their predecessors (5G) to a considerably higher level in order to enable the upcoming massive IoT applications that are highly demanding in terms of data rates, latency, reliability, energy efficiency, coverage, and connectivity.

However, the exponential growth in the number and diversity of IoT devices, along with the unprecedented increase in the volume and types of connections in the upcoming 6G-enabled massive IoT ecosystems are factors that are expected to pose many security and privacy challenges in future massive IoT applications. At the same time, attackers are becoming more and more powerful to carry out new and more sophisticated attacks on the massive IoT applications. Therefore, novel security solutions are essential to protect the 6G-enabled massive IoT applications, before they gain the trust of all involved stakeholders and fully harness their potential in the 6G era.

Toward this direction, this book intends to (i) give in-depth insights into the potential security and privacy challenges of the future 6G-enabled massive IoT applications and develop a comprehensive understanding of how security and privacy attacks against such applications can be carried out and how to counter them in the future; and (ii) provide a foundation for organizing research efforts to design and develop proper security and privacy-preserving solutions, based on the key 6G-enabling technologies and for these 6G-enabling technologies, in order to protect massive IoT applications in the 6G era, by building on existing 5G security research in a methodical way. To achieve both objectives, the following two key elements of the sixth-generation (6G) mobile communications systems are taken into consideration: (i) the new envisioned four-tier network architecture of 6G (i.e. space-air-ground-underwater/sea network architecture), enhanced by edge computing (EC) and edge intelligence (EI); and (ii) the main key 6G-enabling technologies including blockchain, distributed machine learning (ML), quantum computing, and technologies closely coupled with physical layer security (PLS) such as Terahertz (THz) communication.

Chapter 1 gives a comprehensive overview of representative massive IoT applications envisioned to be deployed in the 6G era to improve people's lives and identifies the threat landscape of these applications. The aim is to shed light on their potential key threats and identify the security and privacy issues that should be addressed before the 6G-enabled massive IoT applications build trust among all relevant stakeholders and realize their full potential.

Chapter 2 mainly explores potential security, privacy, and trust issues for 6G EI systems, such as threats related to computational offloading and distributed service provisioning on the edge, vulnerabilities due to the complex integration of various enabling technologies, and trust issues when multiple devices and other stakeholders (e.g. service providers and network operators) share information and resources through a common platform. In addition to the identified security, privacy, and trust issues, the work also briefly presents some of the potential solutions to these issues. Moreover, a brief discussion about ongoing security-related standardization activities for edge intelligence and 6G systems is provided.

Chapter 3 examines two novel techniques within the realm of privacy-enhancing technologies (PETs) that address privacy concerns in IoT and that can scale to massive IoT. Specifically, it covers secure multi-party computation (MPC) and fully homomorphic encryption (FHE), which have thrived in the last few years and paved the way toward efficient Privacy-Preserving Machine Learning (PPML) frameworks. This chapter centers around PPML, explaining the main functionalities and building blocks of MPC and FHE, without delving into mathematical nuance. Furthermore, it discusses the latest PPML frameworks and explains how FHE and MPC can be applied to IoT while complying with the hardware and bandwidth limitations in massive IoT.

Chapter 4 carries out an analysis of current federated learning (FL)-based intrusion detection systems (IDSs) in order to provide a roadmap to support the design and development of effective, efficient, and privacy-preserving IDS for protecting the emerging disruptive massive IoT applications in the 6G era. In particular, it provides the fundamentals of IDSs and ML techniques for IDSs in IoT, as well as the limitations that arise from their centralized nature. In addition, it explores the three primary types of FL, based on the distribution of data, along with their corresponding architectures. Furthermore, it presents use cases, definitive examples, as well as state-of-the-art research works of FL-based IDS for IoT that can be expanded into a more massive network scale, specifically that of massive IoT. Moreover, Chapter 4 explores the concept of model aggregation in FL in detail, including model aggregation approaches and model aggregation algorithms. Finally, it presents challenges and open research directions that require further exploration so that FL-based IDSs can become widely adopted for securing massive IoT environments.

Chapter 5 provides a comprehensive study on securing massive IoT with blockchain and network slicing. It shows an overview of the basic background of massive IoT, IoT-relevant blockchain applications, and its consensus mechanisms. Chapter 5 also presents security and privacy requirements and challenges as well as explores security countermeasures that could be applied to secure massive IoT. Moreover, a review of network slicing security, privacy, and trust threats for 5G and beyond is given. Finally, potential blockchain-based approaches for secure network slicing to protect massive IoT are discussed.

Chapter 6 introduces the reader to the concept of PLS and covers a wide spectrum of PLS methods for achieving node authentication, data confidentiality, and malicious node detection. Furthermore, this chapter discusses two approaches of physical-layer-based key establishment that enable any pair of nodes to establish a shared secret key which is a prerequisite to key-based node authentication and data confidentiality schemes. The covered PLS approaches are aligned with the characteristics of IoT devices which makes them suitable solutions for massive IoT networks.

Chapter 7 introduces the concept of Tactile Internet and the challenges that are posed to future communication networks' technologies to enable it. Next, Chapter 7 explains the main characteristics of quantum technologies and paradigms in order to take advantage of their security benefits and enable the Tactile Internet efficiently and effectively. Some critical considerations are mentioned in order to highlight the main pros and cons of each solution.

Chapter 8 provides a detailed review on physical layer attacks against THz communications and examines existing PLS schemes to identify potential PLS countermeasures for protecting THz communications and the massive IoT applications relying on them in the 6G era. First, the fundamental trade-offs associated with data eavesdropping from directional THz links are presented. Then, potential attacks are detailed and analyzed specifically targeting THz communications. Finally, in Chapter 8, the discussion proceeds with possible countermeasures to further increase the level of security in IoT services relying on directional THz communications.

1

Threat Landscape for 6G-Enabled Massive IoT

Maria Papaioannou[1], Georgios Mantas[1,2], Firooz B. Saghezchi[3], Georgios Kambourakis[4], Felipe Gil-Castiñeira[5], Raúl Santos de la Cámara[6], and Jonathan Rodriguez[7]

[1] *Faculty of Engineering and Science, University of Greenwich, Chatham Maritime, UK*
[2] *Instituto de Telecomunicações, Aveiro, Portugal*
[3] *Chair for Distributed Signal Processing, RWTH Aachen University, Aachen, Germany*
[4] *Department of Information and Communication Systems Engineering, School of Engineering, University of the Aegean, Samos, Greece*
[5] *Telematics Engineering Department, University of Vigo, Vigo, Spain*
[6] *R&D Department, HI Iberia Ingenieria y Proyectos, S.L., Madrid, Spain*
[7] *Faculty of Computing, Engineering and Science, University of South Wales, Pontypridd, UK*

1.1 Introduction

The strict network performance requirements of the emerging disruptive massive Internet of Everything (IoT) applications cannot be satisfied by the currently developing 5th Generation (5G) mobile networks due to the inherent limitations of these networks [1–5]. Therefore, the focus on the next 6th Generation (6G) of mobile networks with significantly improved key performance indicators (e.g. 1 Tbps peak data rate, 0.1 ms air latency, 1 μs delay jitter, and 1 Gb/m^2 area traffic capacity) is inevitable to overcome the limitations of 5G and truly meet the stringent network performance requirements of the emerging disruptive massive IoT applications [2, 5].

6G mobile networks are expected to be commercially available from 2030 onwards and be a key enabler to support a great number of applications with a vast number of interconnected IoT devices [1, 2, 4, 5]. This massive number of 6G-enabled IoT devices will serve for reliable and efficient operation of different verticals such as smart cities, smart homes, smart manufacturing, smart education/training, Connected Autonomous Vehicles (CAVs), and intelligent healthcare. For example, 6G-enabled IoT devices can be used for either healthcare monitoring purposes, e.g. to timely detect any health deterioration of patients suffering from chronic diseases, or collecting mass quantities of data from a wide spectrum of smart city applications (e.g. urban planning, garbage collection), to increase the quality of the services provided to the citizens and reduce the operational costs of the public administrations. However, the high volume of 6G-enabled IoT devices and the increasing number of their interconnections will increase the vulnerabilities of the 6G-enabled

Security and Privacy for 6G Massive IoT, First Edition
Edited by Georgios Mantas, Firooz B. Saghezchi, Jonathan Rodriguez, and Victor Sucasas.

massive IoT applications and pose significant security and privacy risks. Therefore, this chapter gives a comprehensive overview of representative massive IoT applications envisioned to be deployed in the 6G era to improve people's lives and identifies the threat landscape of these applications. The aim is to shed light on their potential key threats and identify security and privacy issues that should be addressed before the 6G-enabled massive IoT applications build trust among all relevant stakeholders and realize their full potential.

Following the Introduction, the rest of the chapter is organized as follows: Section 1.2 presents the vision and values of 6G. Section 1.3 discusses a set of representative massive IoT applications envisioned to be deployed in the 6G era to improve people's lives. Section 1.4 gives an overview of a 6G network architecture to enable massive IoT. In Section 1.5, the major security objectives of 6G-enabled massive IoT are discussed. Section 1.6 provides a categorization of the security threats targeting the 6G networks based on the security objectives that they intend to compromise. Finally, the chapter is concluded in Section 1.7.

1.2 6G Vision and Core Values

The 6G vision, as conceptualized by European 6G Flagship Project Hexa-X [2, 6], aims to strengthen and enhance the connections between three distinct worlds: the human world, encompassing our senses, bodies, intelligence, and values; the digital world, comprising information, communication, and computing; and the physical world of objects and organisms. This integration will establish a seamless cyber–physical continuum, leveraging networks as powerful tools to enhance our quality of life. However, it is imperative that future networks, facilitating interactions between these worlds, are designed with core principles in mind, including sustainability, trustworthiness, and digital inclusion. Figure 1.1 presents the Hexa-X 6G vision, illustrating the interactions between these worlds and the fundamental values that 6G should uphold. This vision of technological and societal transformation holds the potential for unprecedented economic growth and addressing societal challenges in the 2030s and beyond, as explored further in the subsequent sections. Moreover, achieving this vision will necessitate a fundamental paradigm shift in the design of mobile networks. Multiple key requirements must be reconciled, including catering to the exponential growth in traffic and the proliferation of devices, subnetworks, and markets

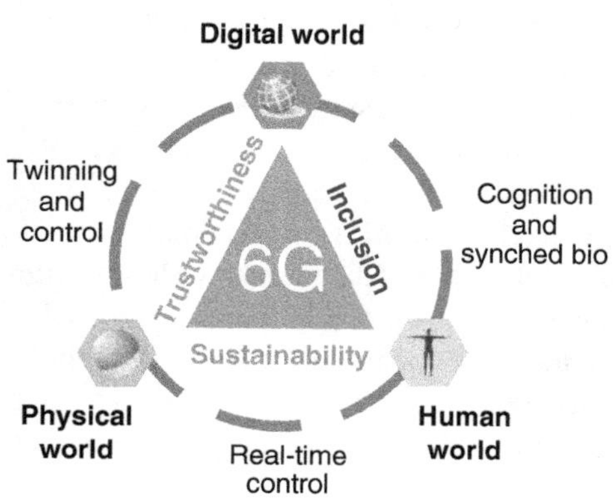

Figure 1.1 The Hexa-X 6G vision of connected worlds and key values. Source: Maria Papaioannou.

while upholding high standards of energy efficiency, security, privacy, trust, and efficient deployment for coverage, capacity, and specialized operations. This will enable sustainable growth and foster innovation across sectors and industries.

Additionally, the trend of enhancing human intelligence will persist through closer integration and seamless intertwining of network and digital technologies. With advancements in artificial intelligence (AI) and machine learning (ML), machines will continue to transform data into reasoning and practical insights, enabling humans to better understand and interact with our world. The interactions between the three worlds will pave the way for advanced sensing capabilities, robust experimentation in the digital world, and high-performance and reliable actuation abilities. As the current dedicated machines in domestic and industrial settings evolve into collaborative multipurpose robots and drones, these future networks will incorporate new interfaces for haptic feedback and human–machine interaction, allowing control and interaction with these machines from anywhere to become an integral part of the network.

1.3 Emerging Massive IoT Applications Enabled by 6G

The new types of upcoming IoT-enabled interactions such as holographic communications, five-sense communications, and Wireless Brain–Computer Interfaces (WBCI) or Wireless Mind–Machine Interfaces (WMMI) are anticipated to lead to new massive IoT applications with far more stringent performance requirements (e.g. sub-ms latency, Tbps data throughput, extreme energy efficiency, and ultralow energy consumption) that will not be matched by 5G capabilities [7]. Therefore, 6G is expected to outperform 5G, meet the new demanding levels of performance requirements of the future massive IoT applications, and become their major key enabler [8, 9]. In this section, we discuss a set of representative 6G-enabled massive IoT applications envisioned to be deployed in the 6G era to improve people's lives.

To organize the discussion of representative 6G-enabled massive IoT applications in a more comprehensive manner, the clustering into use case families (i.e. groups of use cases formed based on the type of usage as well as the research challenges and values addressed), as proposed in [10, 11] in the context of the European 6G Flagship Project Hexa-X [6], has been adopted. In particular, the following six use case families have been considered, as also shown in Figure 1.2, to categorize representative 6G-enabled massive IoT applications: (i) *enabling sustainability*, (ii) *massive twinning*, (iii) *telepresence*, (iv) *robots to cobots*, (v) *trusted embedded networks*, and (vi) *hyperconnected resilient network infrastructures*. These six families of 6G use cases comprise a first baseline to guide the future research directions on 6G, relying on the view of the current European research activities on 6G driven through the 5G PPP and 6G-IA initiatives [3]. Thus, these six use case families are not meant to be exhaustive but are representative of use cases foreseen in the 6G era. In addition, it is worthwhile mentioning that although the use cases described below have been included in the most relevant family, this assignment is not exclusive as there are use cases that may have connections to multiple use case families [11]. Besides, the below-mentioned use cases include several usages characterized as evolutionary or disruptive. The evolutionary ones are those that extend and enhance the 5G usages with new capabilities while the disruptive ones open up new horizons where 6G could benefit and transform society [3].

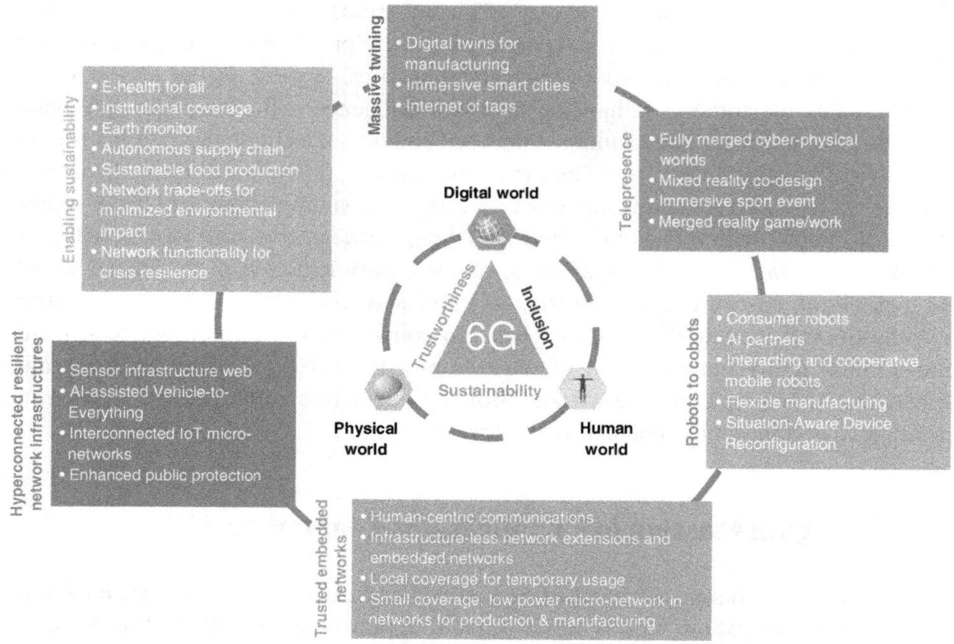

Figure 1.2 Categorization of the representative use case families and respective use cases for 6G-enabled massive IoT applications. Source: Maria Papaioannou.

1.3.1 Enabling Sustainability

The emergence of the 6G system will enable a wide spectrum of applications that go far beyond the current and past generations of mobile networks. Moreover, it is expected that the 6G system will have the ability to make a significant impact on sustainability and societal concerns by contributing to meeting United Nations' Sustainable Development Goals (UN SDGs) and reducing environmental impact for various industries.

Sustainability is explicably the foremost research challenge addressed by this use case family, including environmental sustainability and sustainable development of human societies. Nevertheless, to fulfill the UN SDGs, this use case family will also embrace other research challenges, such as extreme performance and global service coverage, to make sustainable development a reality. In particular, it is anticipated that the extreme performance and global service coverage will empower underserved populations, bridge the digital divide, monitor and counteract environmental challenges, and optimize operations for sustainable performance. On top of that, trustworthiness is essential for any system addressing societal and environmental challenges to gain widespread confidence in its operation and execution.

The enabling sustainability use case family comprises the following use cases, as also depicted in Figure 1.3: (i) e-health for all, (ii) institutional coverage, (iii) earth monitor, (iv) autonomous supply chain, (v) sustainable food production, (vi) network trade-offs for minimized environmental impact, and (vii) network functionality for crisis resilience.

Figure 1.3 Enabling sustainability 6G use case family. Source: Maria Papaioannou.

1.3.1.1 E-health for All

UN SDG #3 – "ensure healthy lives & promote wellbeing for all" focuses on the major health challenges that we are facing nowadays. The most important current health challenges that are being addressed are to save lives and reduce deaths. For instance, the associated targets of UN SDG #3 aim to (i) culminate epidemics such as acquired immunodeficiency syndrome (AIDS); tuberculosis, and malaria; (ii) reduce illness and death provoked by drug abuse, traffic, and pollution; (iii) reduce maternal and child mortality rates; and (iv) promote mental health. Nevertheless, as we progressively move toward a more developed world with substantial demographic, environmental, and economic changes, new challenges arise, such as an increase in elderly populations in rural areas due to urbanization, a rise in "welfare diseases" (e.g. obesity and diabetes), as well as a rise in the inequalities between urban areas and less privileged neighborhoods, and the effects of climate change leading to extreme weather conditions and new pandemics. All these challenges are critical considerations, while technology development increases expectations for reproductive and universal healthcare.

To address these challenges, accessible and remote healthcare becomes increasingly important. Nevertheless, for all remote applications, trusting connected services to be reliable, private, and trustworthy is key to accepting technology solutions. Emerging technologies, such as lightweight devices and sensors for on- and in-body monitoring, augmented reality (AR), and haptic and neuro-based interfaces, can provide new opportunities for remote care. Furthermore, Mobile Broadband (MBB) connections to medical expertise can deliver basic e-health services anywhere, complemented by local analysis of samples with dedicated devices and AI agents for first-line support. Local mobile e-health hubs can also provide last-mile connectivity in areas with infrastructure challenges.

Although providing virtual doctor visits to everyone who needs them worldwide would also be an enormous health benefit, to make this reality, a full expansion of cellular network coverage capable of supporting these services is demanded. The development of 6G networks is expected to enable cost-efficient deployments and operations, even in underserved regions, where it is currently cost-prohibitive. The integration of different topologies, such as Non-terrestrial Networks (NTNs) and satellite communication systems, will ensure service availability with affordable solutions. 3GPP Release 17[1] is already moving in this

1 https://www.3gpp.org/specifications-technologies/releases/release-17

direction, and in 6G, we can expect to see a tighter integration between different topologies to address healthcare needs.

1.3.1.2 Institutional Coverage

The primary objective should be to ensure that all communities have access to high-quality wireless services, which is the essence of digital inclusion for networks. A more practical target might be to guarantee that educational institutions, healthcare facilities, and other essential organizations worldwide have access to full 6G services, even in underdeveloped nations and isolated rural areas in developed countries. This encompasses not only video services but also immersive and precise communication, such as telepresence, remote virtual education, and medicine. In certain regions, installing fiber communication networks may be prohibitively expensive, such as in remote locations, islands with long distances, or areas with political instability. By expanding and enhancing the deployment and performance of 5G Fixed Wireless Access (FWA), for example, by incorporating the 3GPP Release 16 Integrated Access/Backhaul (IAB) capabilities and 3GPP Release 17 NTN features, wireless backhaul based on 6G can be provided to specialized localities that will act as complete 6G digital havens. These critical social institutions connected to 6G services can promote local connectivity, benefit local businesses and infrastructure, and ensure that society is included in the digital revolution. In 6G, we predict that the wireless backhaul capacity will increase exponentially, allowing for cost-effective provision of 6G services, even in presently remote areas.

1.3.1.3 Earth Monitor

Near-real-time monitoring of system-critical environmental aspects such as biodiversity or climate change can be granted by incorporating ubiquitous bio-friendly energy-harvesting sensors. It is expected that this kind of sensor can effortlessly be deployed in any place (e.g. isolated areas and islands) with cost-effective connectivity via, for example, NTN, local mesh, or long-range terrestrial networks. Thus, sampling substantial environmental data will be feasible not only in well-connected areas such as big cities but also in truly inaccessible areas located far from infrastructure. This worldwide telemetry system offers a variety of potential applications, including enhancing climate and weather forecasting models, monitoring and surveillance of environmental conditions, and enabling early warning systems to detect and prevent natural calamities like floods and landslides. It can also aid in protecting endangered ecosystems and species from poaching and illegal logging.

1.3.1.4 Autonomous Supply Chains

A fully integrated autonomous supply chain presupposes an automated demand for scoping, ordering, sourcing, packaging, routing, and delivery. This can be accomplished by integrating local and central AI agents into the supply chain to continuously optimize the overall process, for instance, in relation to unforeseen occurrences such as political or social unrest, or natural disasters. It is expected that 6G will enable a fully automated supply chain at rational complexity and cost. Moreover, decreased energy and material consumption, as well as greater resource efficiency, can be achieved through worldwide end-to-end lifecycle tracking of supplies (from goods production, shipping, distribution, and usage to their end-of-life and recycling). In particular, by integrating 6G-connected micro tags on goods, the processes of goods tracking, customs, safety checks, and bookkeeping can be simplified while at the same time, being entirely automated without requiring any kind of human interference.

The successful realization of autonomous supply chains involves using New Radio Reduced Capability (NR-RC) with Network Time Protocols (NTPs) and Narrowband Internet of Things (NB-IoT) and the integration of AI/ML. This will require a global network of integrated systems, incorporating various technologies for unified and cost-effective access, which can orchestrate an efficient end-to-end supply chain. On top of that, this network should be able to adapt locally to any changes using the most suitable connectivity solution. Additionally, the system will track the current state of businesses and factories to forecast when sourcing is required, determine optimal routes for delivery, and provide last-mile delivery. At every stage, the position and condition of the products should be observed to ensure a flexible and adaptable supply chain.

1.3.1.5 Sustainable Food Production

Ending hunger and achieving food security for all, while ensuring sustainable production, comprise a critical aim of the UN SDGs. However, providing remote, rural, and in-sea areas with higher network capacity and performance is a major challenge toward succeeding in this aim. This is necessary for real-time monitoring of micro-locations (soil conditions, microclimate), developing optimized plant treatments, experimenting with various strategies (e.g. spraying strategies, alternate cultivations, removal of plants), and controlling semi-autonomous ground robots. By closely synchronizing digital representations of the physical world, human experts can inspect and experiment with actions in the digital realm to optimize agricultural production and prevent threats. Despite the need for vast amounts of data transfer from underserved areas, close synchronization is essential to address important sustainability, global coverage, inclusion, and opportunity challenges of our time.

1.3.1.6 Network Trade-Offs for Minimized Environmental Impact

Users desire to have the option to use "green" Information and Communications Technology (ICT) services, which have a decreased environmental impact compared to traditional services, in a comprehensive approach. This approach includes considering not just the applications and materials but also the technology and End-to-End (E2E) network design. As environmentally conscious users, they will be encouraged to weigh the potential trade-offs between performance, cost, and environmental impact, allowing them to monitor the overall environmental impact of their products/services. These services aim to mainly tackle the sustainability and global service coverage research challenges.

This service considers the end-to-end energy consumption, taking into account the environmental impact of all aspects of the service, such as the application, network, terminal, and more. These energy-efficient services will require not only energy-optimized networks but also energy-optimized applications, the appropriate upcycling of materials, and other measures. Creating a comprehensive view will necessitate new metrics for environmental impact and the aggregation of these metrics to form a global view.

For instance, regarding video service delivery, users might have the option to choose from a variety of services that have different environmental impacts and trade-offs in terms of video resolution. They can opt for high-resolution videos that require more energy and new terminals or accept a lower-quality resolution with reduced energy consumption. The user's preferences can determine the video quality they desire and the level of experience they are willing to compromise. The end-to-end environmental impact should be considered, including the user equipment. Standardized indicators for energy consumption and footprint are necessary for each network component to enable the network's tuning toward the targeted trade-off between Quality of Experience (QoE) and environmental footprint, such as selecting paths powered by renewable electricity or avoiding paths that involve non-circular material usage.

1.3.1.7 Network Functionality for Crisis Resilience

Mobile networks are progressively used for more and more fundamental services such as healthcare and education, and thus, their resilience is becoming a major requirement, particularly in crises or extreme situations. In most cases, the resilience of network operations depends on their power supply, which might be easily disrupted in extreme situations such as natural disasters (e.g. floods or storms). In these circumstances, small diesel generators are likely to be used for backup power generation. Nevertheless, these kinds of generators provide a limited power source before they need refueling, while they are considered far more polluting than large-scale power production and much less efficient. Consequently, minimizing energy consumption during network operation on backup power would be highly advantageous. This can be achieved by rerouting traffic through nodes still powered by the power grid, reducing power-consuming functionality, and focusing on essential functions such as communication or prioritizing first responder calls. To reduce baseline energy consumption, it may also involve sacrificing latency performance by increasing the periodicity of always-on signals. For wireless devices, which rely on battery power and require frequent recharging during emergencies, the network can instruct them to conserve power by disabling certain functions, such as high-bandwidth services or processing heavy applications.

1.3.2 Massive Twinning

In 6G, it is expected that the application of the fundamental concept of Digital Twin (DT) in a wide spectrum of use cases, also referred to as massive twinning, will become progressively important. Massive twinning aims to create a complete digital representation of our environment, not only in manufacturing but also in areas such as logistics, transportation, digital health, social interactions, entertainment, public safety, and defense.

DTs are virtual models that provide real-time representations of physical assets, including their structures, roles, and behaviors. The use of DTs is essential in different fields, such as manufacturing, urban quality of life assessment, increasing productivity, improving sustainability, and transforming sectors such as healthcare and public security. These areas require vast amounts of data transfer, low delays, high reliability, and capacity levels that exceed current limitations.

In all cases, the creation of a fully synchronized and accurate digital representation of the physical and human worlds is crucial. However, it is important to highlight that this will also be resource-intensive, requiring precise representations of the physical world, enhanced means for generating insights and predictions, and ways to experiment with different scenarios. Therefore, it is necessary to ensure that the solutions are sustainable,

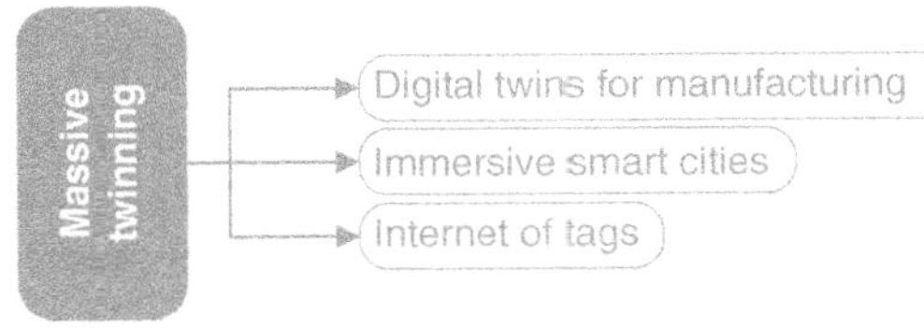

Figure 1.4 Massive twinning 6G use case family. Source: Maria Papaioannou.

reliable, and trustworthy, as well as globally applicable. To achieve this, high-resolution and interactive 4D mapping is necessary, along with means to influence the physical world. High-resolution indoor and outdoor mapping will drive use case scenarios of dynamic DTs and virtual worlds in conjunction with real-time, multisensory mapping and rendering, movement prediction, and real-time analytics. The availability of both the virtual and physical worlds will allow for the mapping and analysis of data and the monitoring of systems to predict performance and operational issues, minimizing downtime and improving contextual awareness.

The Massive Twining use case family comprises the following use cases, as also depicted in Figure 1.4: (i) DTs for manufacturing, (ii) immersive smart cities, and (iii) Internet of tags.

1.3.2.1 Digital Twins for Manufacturing

DTs in production/industrial environments will continue to grow, enabling us to exceed the existing levels of production agility while facilitating more effective interaction of production means. This will allow us to incorporate a larger extent of the respective processes, as well as attain the transfer of massive volumes of data and, often, extreme reliability and performance compared to current levels achieved. In the following, some examples of DTs valuable applications in the realm of logistics and production are given:

(a) To ensure accuracy and efficiency, new products must be designed and automatically twinned to their DT. The process involves creating a virtual version of the product and testing it in a digital environment before moving to physical production. Only when the digital version performs according to specifications, physical manufacturing begins. Once manufactured, the physical product is connected to its DT through sensors and actuators, allowing the DT to contain all the relevant information about its physical counterpart.
(b) Another practical application of DTs includes tracking the history of every part of a system (e.g. production line) through "digital threads." This requires fast and reliable communication at higher capacity levels than what 5G can provide, due to increasing adoption. To increase the chances of commercial success, Ultra-reliable Low-Latency Communications (URLLC) is necessary, with high capacity and efficiency. Trustworthiness is also crucial for widespread adoption, which includes performance, dependability, security, and resource efficiency/cost.
(c) The individual in charge of the production facility is experimenting with different scenarios using the DT's "what-if" feature. These scenarios are aided by real-time communication with the physical world. The operator chooses a specific configuration that is implemented quickly and reliably. In case of an emergency, the DT analyzes the situation using advanced techniques such as reasoning, ML, and prediction methods, as well as conventional polling and alarm functions. The DT then identifies the best

course of action and enforces it at the highest possible performance levels with the utmost security.

(d) To ensure that abnormalities in the physical world are efficiently detected and properly mitigated through dynamic adaptations of the production process and/or system reconfigurations, the successful communication and collaboration of multiple DTs in a flexible production process will also be required. This means that the respective DTs need to ensure access to all pertinent communication, production, and sensing data that are essential to take the necessary mitigation actions in case of system abnormalities or to increase efficiency, while also guaranteeing essential requirements of the production processes: privacy, safety, and real time. This use case is also closely related to the "From robots to cobots" use case family, discussed in Section 1.3.4.

(e) Using AI, it is expected that 6G will attain ultrareliable and ultramassive connectivity enabling fully automated control of processes, devices, and systems. On top of that, with the incorporation of Virtual Reality (VR)/AR/holographic communications in industrial environments, 6G will also enable remote maintenance. Hence, workers in physical industrial environments can work together timely with experts in remote locations to solve problems, resulting in lower costs and higher efficiency. Another application includes remote control of industrial facilities, which ensures remote machine control. This ensures the safety of the workers while reducing costs and increasing efficiency. Nevertheless, this demands rigorous low-latency, broadband, and reliable transmission of 6G.

1.3.2.2 Immersive Smart Cities

The livability of a city is determined by a complex set of factors, which are weighted and correspond to various areas including education and culture (e.g. access to education and culture for all), environment (e.g. air quality), healthcare, infrastructure (e.g. networks and roads), safety/stability, and others. Managing these factors effectively poses technical challenges but also provides societal and business opportunities. For example, the mapping and planning of smart cities is a potential use case for DT technology. In the future, cities will be dynamic systems with various elements that can be modeled and managed using real-time feedback and an interactive 4D map. This can improve the efficiency and sustainability of city operations, from traffic management to utility control and more. By overlaying physical modeling and historical data, the 4D map can be used to predict and manage behaviors and activities, while human and AI operators can modify and manage tasks and schedules in real time. Ultimately, this optimized management can contribute to the transformation of cities into more sustainable models. For this particular use case, there is one more time the need to transmit large quantities of data within specified timeframes, ranging from ultralow latency for health-related purposes to near-real time. The main objective is to enable actions that will have a positive impact on the city's functioning and improve its overall livability. Simultaneously, it is crucial to achieve the utmost levels of sustainability while ensuring trustworthiness, which is a fundamental requirement and will be expected by the public.

In the following, some valuable applications in the realm of immersive smart city use cases are given:

(a) **Smart Education/Training**: The utilization of 6G wireless systems will bring significant advantages to smart education and training [1]. The incorporation of cutting-edge technologies such as holographic communications, five-sense communications, high-quality VR and AR, mobile-edge computing, and AI will contribute to the development of intelligent education and training systems. These advancements will enable students to experience 3D representations of structures and models and even receive instruction from renowned teachers located remotely, resulting in interactive and immersive online education. In terms of training, the application of holography allows learners to observe live processes and interact with objects or trainers. This approach enhances information retention, reduces costs, and eliminates the need for exposure to hazardous environments that are commonly associated with conventional training methods. Additionally, 6G technology enables the creation of intelligent classrooms, where sensors collect data and transmit them to cloud or edge cloud platforms for analysis. The insights obtained can then be utilized to enhance the quality of education and foster improved student interaction.

(b) **Super Smart City/Home**: The advanced functionalities of 6G will bring about significant enhancements in quality of life, intelligent monitoring, and automation, thereby expediting the development of highly intelligent cities and homes [1]. A city attains the status of being "smart" when it can operate intelligently and autonomously by gathering and analyzing vast amounts of data from various sectors, ranging from urban planning to waste management. This enables more efficient utilization of public resources, improves the quality of services (QoSs) provided to residents, and reduces operational costs for public administrations. The incorporation of smart mobile devices, autonomous vehicles, and similar technologies in 6G will contribute to the intelligence of cities. On the other hand, a smart home is not solely characterized by its connectivity to internet-enabled devices that assist in managing and monitoring appliances and systems via mobile phones. It is also an intelligent entity that possesses instantaneous and distributed decision-making capabilities. Additionally, individuals can control lighting, heating, multimedia entertainment, and more through voice commands or even Brain–Computer Interfaces (BCI), or delegate these tasks to AI systems, which can analyze behavioral patterns to enhance safety and convenience in people's lives. 6G will play a crucial role in transforming the concept of smart homes into a reality. However, achieving the vision of smart cities and homes poses significant challenges for 6G, particularly in terms of connectivity and coverage capabilities, given the large number of sensors and intelligent devices involved.

1.3.2.3 Internet of Tags

Tags are expected to appear everywhere, enabling numerous capabilities and operations to facilitate daily life. In particular, they will obtain data through the use of label tags and then monitor and influence the environment by employing more advanced tags with not only communication but also sensing and actuation capabilities. The massive deployment and use of tags, as well as energy harvesting, can be extended to various applications, such as tracking merchandise for logistics, monitoring temperature and light to optimize energy consumption, and activating lights and heating with the push of a button. On top of that, this precise monitoring of goods and merchandise will permit stock tracking of the monitored assets and improvement of the general supply chain efficiency, while also being a tool to improve circularity, diminish waste, and improve reuse. To reduce the impact on

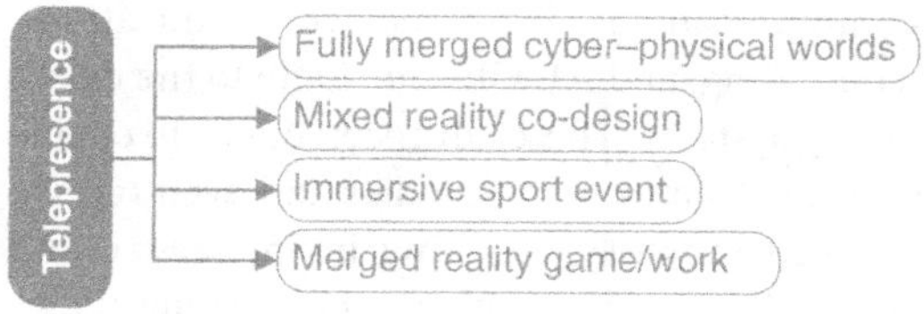

Figure 1.5 Telepresence 6G use case family. Source: Maria Papaioannou.

the environment, the tags will not be powered but will use energy harvested from ambient or renewable sources like Radio Frequency (RF) waves, solar energy, wind, vibration, or mechanical push. Additionally, the use of "zero-environmental-cost" tags, such as those made of bio-degradable materials or printed electronics, can be considered.

1.3.3 Telepresence

Telepresence refers to the "*human experience of being fully present at a live real-world location anytime and anywhere, remote from one's own physical location,*" [2] using all five senses if so desired. It enables enhanced human interactions not only with each other but also with any physical and/or digital object in the physical and digital worlds, respectively. On top of that, telepresence envisions the use of all five human senses to exchange sensory information, as well as extend the current capabilities of these senses. Nevertheless, in order for telepresence to effectively come to realization, there are several research challenges that need to be met. In particular, to provide a seamless telepresence experience, high data rates are required, and low latency and reliability are necessary to avoid negative effects such as incomplete experience or discomfort (e.g. nausea). The basis for this connectivity is a network of networks, while connected intelligence can enhance performance. Again, in this use case family, sustainability and trustworthiness are critical considerations. Telepresence should be available globally and delivered sustainably, which can contribute to meeting the UN SDGs by reducing the need for travel.

The telepresence use case family comprises the following use cases, as also depicted in Figure 1.5: (i) fully merged cyber–physical worlds, (ii) mixed reality co-design, (iii) immersive sport event, and (iv) merged reality game/work.

1.3.3.1 Fully Merged Cyber–Physical Worlds

The use of Mixed Reality (MR) and holographic telepresence will become widespread for both professional and personal communication. With holographic telepresence, one can appear to be in a particular location while in reality being somewhere else, such as appearing to be in the office while actually being in the car. This technology can facilitate collaboration and remote working for white-collar workers, enhance diagnosis during teleconsultations, improve teacher-student interactions in e-learning classes, and enable virtual travel and telepresence meetings with friends and family. The user will experience the world where their hologram is, with a highly sophisticated sensory experience, synchronized to devices on their body. Users desire high-quality communication with distant individuals, with realistic interaction and perception of body language, surrounding sounds, expressions, intonation, and other senses, including touch.

2 https://en.wikipedia.org/wiki/Telepresence

Through MR telepresence, one can interact with both physical and digital objects in physical reality, whether they are near or far. Wearable devices, including earbuds and clothing embedded with devices and other novel interfaces, will enable this experience. People will carry multiple wearables that work seamlessly with one another, providing natural and intuitive interfaces. Typing on touchscreens will likely become outdated, as gesturing and speaking to devices will progressively become the norm. The devices we use will be context-aware, and the network will become increasingly sophisticated at predicting our needs. This context awareness, along with new human–machine interfaces, will make our telepresence interactions intuitive and efficient.

In the following, potential new forms of interactions enabled by 6G in the realm of fully merged cyber–physical world use cases are given:

(a) **Holographic Communications**: The first new form of interaction enabled by 6G that is being broadly discussed is holographic communications, which enables the projection of real-time, full-motion 3D images [1]. This technology captures images of individuals or objects, whether they exist in reality or are located remotely, and transmits these images and accompanying sounds to the receiver. As a result, it allows objects or people, along with their real-time audio, to appear in front of users, creating a more realistic experience compared to VR and AR. Holographic communications are expected to have significant applications in various fields, such as industry, agriculture, education, and entertainment. However, due to the high level of detail involved in transmitting holographic media, the data requirements can be substantial even after compression. This necessitates extremely high throughput, reaching hundreds of gigabits per second or even terabits per second, to support such capabilities. Furthermore, networks must provide reliability and timeliness to avoid any disruptions that could degrade the performance of interactive applications. In essence, holographic communications demand a combination of URLLC and Enhanced Mobile Broadband (eMBB) to ensure both low latency and high data rates.

(b) **Five-Sense Communications**: In addition to the conventional exchange of voice, images, and videos through tactile transmission and traditional human interaction, researchers have identified a growing trend toward new forms of remote human interactions that go beyond holographic communications [1]. These interactions are referred to as immersive five-sense or five-dimension (5D) communications. The concept behind 5D communications involves integrating all human senses, including sight, hearing, touch, smell, and taste, into the media. This technology captures sensations from both the human body and the surrounding environment and combines them through neurological processes. The resulting information is then transmitted to a receiver located remotely, enabling a more immersive experience of a distant environment. When combined with VR/AR or holographic communications, these multisensory applications, such as remote surgery, have the potential to offer truly immersive services in the context of 6G. Achieving such immersive experiences requires a holistic approach that considers not only engineering aspects like wireless communication, computing, and storage but also perceptual factors such as human senses, cognition, and physiology. These factors collectively determine the performance of the service. In terms of latency and data rate, the requirements for five-sense communications are a fusion of traditional URLLC and eMBB, with an emphasis on accurate localization capabilities and incorporation of perceptual elements.

(c) **Wireless Brain–Computer Interfaces (WBCI)**: Another potential type of interaction enabled by 6G is known as WBCI, or WMMI [1]. These interfaces enable machines and/or the environment to be controlled using human thoughts. WBCI technology works by capturing neural signals from the human brain using electrodes and then converting these signals into commands that machines can understand. This allows for various functions, such as turning on a light, to be controlled through the mind. WBCI serves as a communication pathway between the brain and external devices, offering a simpler and more intelligent way to control everyday appliances in smart cities, homes, and medical systems. In addition to healthcare and smart city/home applications, the emergence of WBCI has opened up new possibilities, including brain-controlled movies and even multi-brain-controlled cinemas. When combined with tactile internet or haptic communications, where emotion-driven devices can match the user's mood, WBCI becomes a significant use case for 6G. WBCI demands high data rates, extremely low latency, and high reliability, as well as powerful computational capabilities. It is similar to VR/AR technology but more sensitive to physical perceptions, requiring guarantees of QoS and QoE.

1.3.3.2 Mixed Reality Co-design

MR co-design refers to "experience before prototyping" and remote cooperation for designing pioneering virtual products in a fusion of the real and virtual worlds. For instance, given a factory scenario, two people from different locations might work together on the design of a sophisticated process and/or product using both relevant available physical and virtual objects.

Context awareness will play an important role in MR co-design process, allowing the designers to focus not only on the design of the process and/or product itself but also to take into consideration its relationship with the external environment. Moreover, MR co-design will involve integrating new forms of interaction between humans and machines. This will involve capturing the designer's movements (e.g. eyes or head movement), emotions, facial expressions, as well as bodily parameters, such as heart rate and blood pressure. This falls under the category of "spatial computing," and the co-design context can be captured through spatial mapping and imaging technology. By the 2030s, we anticipate that designers' behavior and vital signs will be included in the co-design process.

1.3.3.3 Immersive Sport Event

Current sports simulators allow the creation of life-like representations of actual players through motion capture technology. As extended reality (XR) gaming becomes more prevalent, it is expected to enable the creation of 3D models of any simulated sporting event. With the advancement of 6G, it will be possible to capture real-time game footage and create a DT of the entire game. This will allow millions of people worldwide to experience the game live from any possible angle. Despite the fact that most viewers will likely be satisfied with the traditional camera angles provided by professional camera operators, 3D rendering will enable end users to experience the game from any angle, including a 360-degree view following a particular player or even watching the game from the ball's perspective.

To ensure extreme fidelity in rendering the interesting field of view, processing will have to be done locally. The use of AI models could also assist in forecasting the movements of players in the near future by combining pre-rendered models with real-time footage. As a result, it would be possible to watch the game from a virtual bleacher while simultaneously interacting with friends who are also watching the game virtually.

1.3.3.4 Merged Reality Game/Work

Gaming in a public place or in dedicated spaces involves a shared merged reality with a large number of people where the boundaries between what is real and what is virtual are blurred. Some players and objects may exist in the physical world, while others are digitally enhanced with visual, haptic, or olfactory sensations. Additionally, there may be fully digital elements that appear to be real. Participants in the same game share a unified experience and exchange synchronized sensory information that may be genuine or artificial. Through digital meetings, individuals can use hologram avatars of themselves to participate and feel fully present. Tactile and sensory feedback can be provided to participants, and visual information can be experienced through smart contact lenses, for instance Digital co-creation is easily managed in virtual environments, simplifying remote work and training.

1.3.4 From Robots to Cobots

The 6G system enables robots to collaborate and form symbiotic relationships, becoming "cobots" to efficiently complete complex tasks and cater to human needs. Trustworthiness and digital inclusion are important values in human–machine and machine–machine interactions. Collaboration and building symbiotic relationships enable sustainable fulfillment of complex tasks without requiring increasingly complex machinery or additional resources. This allows for intelligent and flexible use of existing capabilities for the benefit of society, which is at the core of this use case family. Additionally, this also allows for new business models, enabling machinery to perform highly individualized on-demand tasks, facilitating lot size one production, and utilizing novel production methods like additive manufacturing.

This use case family targets a number of research challenges. *Trustworthiness* is a fundamental research challenge, particularly as use cases that belong in this family depend on *connecting intelligence*, as well as reaching joint decisions. This requires resource allocation and flexible network topologies, targeting the *network of networks* challenge. For instance, use cases, especially in the industrial environment, might demand *extreme performance* in terms of positioning, dependability, and latency. Sustainability is also a core challenge, specifically when dealing with use cases that require extreme performance. Finally, to address the research challenge of inclusion, it is essential to have meaningful human–machine interaction and AI partners.

The From Robots to Cobots use case family consists of the following use cases, as also depicted in Figure 1.6: (i) consumer robots, (ii) AI partners, (iii) interacting and cooperative mobile robots, (iv) flexible manufacturing, and (v) situation-aware device reconfiguration.

1.3.4.1 Consumer Robots

The future will see a wider range of consumer robots that go beyond just lawn mowers and home vacuum cleaners, with smaller robots working together or autonomous

Figure 1.6 From robots to cobots 6G use case family. Source: Maria Papaioannou.

robots providing convenience. These robots will be equipped with advanced sensing and positioning features, as well as video cameras streaming to a local compute server for real-time processing, and will utilize the connected AI capabilities of 6G for situation-aware collaboration and support. This will result in an increase in the number of devices and higher capacity requirements, demanding seamless connectivity across local networks. Domestic robots will also improve the quality of life for the elderly, allowing them to stay in the comfort of their homes for longer. This use case expands on existing 5G functionality to address more challenging scenarios with higher device numbers, latency demands, positioning, and computational requirements. It also requires greater integration of sensing and AI utilization.

1.3.4.2 AI Partners

As AI technology advances and integrates with 6G networks, AI agents will become more widespread in society and capable of handling increasingly complex tasks. Although humans will still be involved in many tasks, they will work alongside AI agents as equal partners, rather than relying solely on the assistance of machines. Rather than being limited to specific tasks or machines, AI agents will be more versatile and able to interact with other agents (both human and machine) autonomously and adaptively, interpreting their surroundings and intentions to perform tasks. For example, an AI agent could be a stationary machine in a factory or software controlling lighting in different locations by communicating with other AI agents nearby. These interactions require dependable cyber–physical systems that can cooperate effectively with groups of humans and intelligent machines, achieving precise action. If an AI agent interacts with other AI agents, communication may shift toward a new form of user-plane communication to enable and benefit from distributed computing functionality.

1.3.4.3 Interacting and Cooperative Mobile Robots

In scenarios involving multiple robots, as described in Section 1.3.4.1, automated communication is necessary for machines to identify and connect with each other, exchange intent, and negotiate action. Examples of coordination between robots include avoiding collisions in consumer-oriented settings, coordinating movements in construction/building scenarios, and swarms of simpler robots performing tasks through emergent action. In industrial environments, production tasks can be conducted through collaboration among mobile machinery, such as robots mounted on Automated Guided Vehicles (AGVs). Coordination in three dimensions is necessary to avoid collisions and enable collaboration among robots evolving in the air, such as drones. In addition to coordination among entities, real-time exchange of process data is required to meet synchronization requirements. High-energy performance, reliability, functional safety, latency, and positioning requirements must be met during all steps, even if trajectories are blocked or need to be altered.

This use case extends 5G functionality to more challenging scenarios with an increased number of devices, increased latency demands, and a high share of local communication. Coordination among interacting entities is moved to the entities themselves, and 3GPP Release 17[3] moves toward this through capability exposure for time-sensitive communication and QoS and a fully distributed configuration model for Time-Sensitive Networking (TSN). Coordination among entities relies on use case enabling services related to distributed AI functionality exposed to applications.

1.3.4.4 Flexible Manufacturing

As personalization, modularization of production, and flexibility of manufacturing systems are progressively increasing, the necessity for powerful wireless communication and localization services and dynamic, flexible configuration of communication services in the network is becoming a critical consideration. The configuration of machinery and associated communication will be done dynamically for each production task, either through a production system or by direct collaboration among mobile production machines. This requires AGV orchestration, as greater flexibility in the production process demands greater flexibility in logistics. Real-time communication services must be configured dynamically, potentially initiated by the end systems themselves, and executed in a distributed manner. A flexible framework must assign corresponding communication resources and capabilities, such as local compute, Device-to-Device (D2D) communication, and frequency ranges. High availability and functional safety requirements must be met, and production process data must remain secure and private.

This use case builds on existing industrial 5G functionality in denser industrial environments with greater flexibility, self-organization capabilities, local processing, and direct communication among entities. As mentioned in Section 1.3.4.3, "Interactive and cooperative mobile robots use case," 3GPP Release 17 is moving in this direction.

1.3.4.5 Situation-Aware Device Reconfiguration

This use case is based on a device's ability to sense and understand its environment and adapt its communication capabilities to meet the needs of the service it is providing by changing its function. This involves adjusting device settings and configurations to meet performance requirements that differ from those of the previous operating environment. With the proliferation of robots that are capable of performing a wide range of tasks, this use case becomes increasingly important and allows a robot to perform various tasks beyond its original intended purpose in a particular environment. For instance, an industrial robot could switch between critical and noncritical tasks, such as changing from welding (requiring high localization accuracy and low latencies) to long-range movement (requiring moderate localization accuracy and latencies).

Moreover, another application of this use case is when either robots or humans enter or leave a specific group of collaborating entities that are working toward a common task. In

3 https://www.3gpp.org/specifications-technologies/releases/release-17

this scenario, the communication entity involved will have to adapt to the communication requirements within the particular group, which may differ from previous device configurations. The purpose of this use case is to allow for flexible and effective changes in device type, such as switching from a consumer device (like a smartphone) to an industrial IoT device or a Vehicle-to-Everything (V2X) device. For instance, when a user with a consumer device (i.e. smartphone) enters an area where V2X communication is being used, the device changes its type and enables safety-related communication, such as warning a pedestrian of potential danger or providing vehicle access to Vehicle-to-Network (V2N) services via the smartphone. The service also benefits the industrial sector, where a robot may have to switch between critical and noncritical actions, such as welding or long-range movement, respectively. The device's ability to change its type and configuration based on the active service requirements is vital, and sensing capabilities can predict the need for such changes and the appropriate timing. The service aligns with the network of networks and trustworthiness research challenges.

The use case being discussed surpasses the current state of 5G standardization, where a device is assigned to a network slice in a static manner; for instance, a consumer device is allocated to a specific network slice(s) once. The challenge for 6G systems lies in the exceedingly high-performance requirements of new applications, which might require an improvement on current 5G solutions. The concept of "device slicing" may emerge as a means of addressing this, where flexible device type change could evolve the current network slicing approach in 5G.

1.3.5 Trusted Embedded Networks

Trusted embedded networks use case family gathers these use cases that involve subnetworks, or networks of networks, and require a high level of trustworthiness. Mobile communications have traditionally relied on cellular networks, but certain use cases require local or private communication capabilities for very sensitive information that is tightly integrated within wide-area networks. These scenarios demand network topologies and security concepts that go beyond classical cellular architectures. To protect sensitive information, trusted embedded networks and independent subnetworks such as body area networks or on-board networks of AGVs need to be seamlessly integrated into wider networks or remain as private networks on-premises.

The trusted embedded network use case family consists of the following use cases, as also depicted in Figure 1.7: (i) human-centric communications, (ii) infrastructure-less network extensions and embedded networks, (iii) local coverage for temporary usage, and (iv) small coverage, low-power micro-network in networks for production and manufacturing.

1.3.5.1 Human-Centric Communications

Currently, medicine typically uses a one-size-fits-all approach to develop disease treatment and prevention strategies for the average person. In contrast, precision medicine considers individual variability in genes, environment, and lifestyle for each person. To understand the environment and lifestyle of individuals, continuous monitoring of vital parameters through numerous wearable devices will be helpful. Those interested in self-tracking and monitoring their personal analytics, or "quantified self," can use in-body devices to perform these tasks.

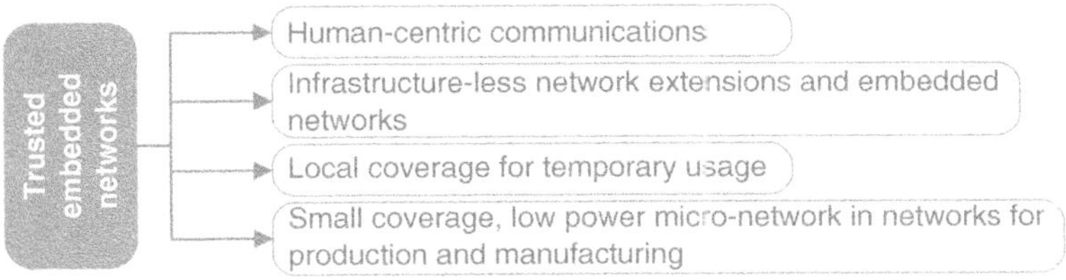

Figure 1.7 Trusted embedded networks 6G use case family. Source: Maria Papaioannou.

Precision medicine will enable personalized diagnosis and treatment through continuous health monitoring, diagnosis, and therapy. Topical, implanted, injected, or ingested sensors and devices can be used for continuous health monitoring, and adjustive measures, such as medicine dispensers, can be implemented if needed. Bodily and sensory reactions like organ malfunction and pain can be represented and analyzed in the digital domain. A 6G telemedical paradigm will be enabled by in-body sensing and analytics, along with a wide-area connectivity option. However, high privacy requirements, including local anonymization and pseudonymization, will need to be met, as well as considering application domain-specific regulations.

This use case focuses on precision healthcare, where the use of in-body and on-body sensors is generalized beyond health purposes. This requires the interaction of a trusted embedded network with the wide-area network and access to information available in other networks under predefined filtering rules. A split of network transport, control, and security is also needed. Regulatory compliance, especially with regard to the medical sector, will be challenging. The integration of medical devices into wide-area networks is currently disruptive and requires careful consideration.

1.3.5.2 Infrastructure-less Network Extensions and Embedded Networks

In situations where network coverage is limited, such as during agriculture harvesting campaigns, temporary network coverage extensions may be necessary to ensure reliable connectivity between the multiple agriculture vehicles. This connectivity should remain intact even if the vehicles move out of the network coverage area but are still operating within the harvesting campaign.

Similarly, an industrial vehicle manufacturer may require a reliable networking solution for their shop-floor vehicles, even when they are not connected to the wide-area network. To achieve this, a local private infrastructure-less network can be established, with authorized access to the spectrum of a local nonpublic or public network. External network control should also be enabled to ensure seamless integration of networking islands and temporary, ad-hoc security solution deployments.

While Long Term Evolution (LTE) and 5G offer D2D solutions, which are often used in Public Protection and Disaster Relief (PPDR) scenarios, some industries such as construction, agriculture, and tactical services may require coverage extending concepts beyond D2D, such as mesh networks or multi-D2D. Direct Mode Operation (DMO) solutions, which are already present in several standards, could be seen as a first step toward achieving this goal. The main aim is to ensure seamless reintegration of networking islands of several devices rejoining cellular networks.

Figure 1.8 Hyperconnected resilient network infrastructures 6G use case family. Source: Maria Papaioannou.

1.3.5.3 Local Coverage for Temporary Usage

Industries such as PPDR and Program Making and Special Events (PMSE), roadwork, and harvesting campaigns often require high-quality local networking coverage to support applications such as massive video transmissions. In situations where cellular coverage is limited or unavailable, temporary or semi-permanent network nodes can be deployed on vehicles, drones, High-Altitude Platforms (HAPs), or other means to ensure connectivity. Access to required spectrum resources can be guaranteed through automated licensing processes.

Although temporary network deployments for PPDR and PMSE are currently in use, reducing deployment costs and administrative burdens, for instance, through automated licensing, could encourage their wider adoption.

1.3.5.4 Small Coverage, Low-Power Micro-Network in Networks for Production and Manufacturing

A manufacturer of machines aims to connect a significant number of sensors within their machine using non-Industrial, Scientific, and Medical (ISM) spectrum for greater reliability. Low-power devices with limited coverage can be used to establish an underlay network, with one of the sensors authorized to access the public or nonpublic network providing the spectrum. This approach can be considered a shared spectrum access concept, with the incumbent retaining full control over the usage of the spectrum. The incumbent can choose to disable spectrum usage by signaling.

1.3.6 Hyperconnected Resilient Network Infrastructures

Hyperconnected resilient network infrastructures use case family gathers the use cases that involve subnetworks, or networks of networks, and require a high level of resilience. Cellular networks have been the conventional means of communication, but there are specific situations where localized or private communication is necessary to safeguard highly sensitive information. These particular circumstances call for network structures and security principles that surpass the traditional cellular frameworks. To ensure the protection of sensitive data, reliable embedded networks and separate subnetworks like body area networks or on-board networks of AGVs must be smoothly incorporated into larger networks or maintained as private networks on-site. The hyperconnected resilient network infrastructures use case family consists of the following use cases, as also depicted in Figure 1.8: (i) sensor infrastructure web, (ii) AI-assisted V2X, (iii) interconnected IoT micro-networks, and (iv) enhanced public protection.

1.3.6.1 Sensor Infrastructure Web

An autonomous vehicle lacking or having limited sensor capabilities navigates the environment while relying on external third-party sensors as if they were installed on-board. The vehicle collects data from external sensors and receives navigation instructions through the network with full confidence in the accuracy, timeliness, and security of the information. In addition, it can share its own sensor data, enabling the aggregation of sensor data across different systems, even on devices that do not have their own sensor capabilities. The network can advertise locally relevant and reliable sensor information accessible to all connected devices, such as vehicles.

However, the current 3GPP framework prohibits the sharing or dissemination of sensor data within predefined local environments and networks or network segments that are under external security management. Depending on the implementation, this use case may require the division of network ownership, control, transport, and security. Furthermore, the current framework does not allow networks to distribute sensor data provided by third-party sources in well-defined local areas.

1.3.6.2 AI-Assisted Vehicle-to-Everything (V2X)

Safety and security are critical concerns for all modes of transportation, particularly road transport, which is prone to accidents. Several efforts have been undertaken to promote rules, technical standards, and awareness campaigns aimed at reducing fatalities resulting from road accidents, as evidenced by the European Commission's Road Safety website[4] [5]. Previous studies and trials have demonstrated the potential of AI in making roads safer. Therefore, there is a need to explore the potential of AI algorithms further in improving automotive services offered by future 6G networks. These algorithms will be utilized to process big data from various sensors, including those installed in and outside vehicles and radio stations in the operators' networks, to monitor, shape, and recommend actions to drivers of connected vehicles and potentially control automated vehicles to minimize traffic. This will have a significant impact on society by improving safety for drivers and passengers and reducing traffic congestion. Compared to existing C-V2X technologies based on LTE and New Radio (NR), the processing of the vast amount of data gathered by automotive services in communication networks requires AI-based algorithms to dynamically control and shape traffic, creating a digital replica of the actual traffic scenario. However, this poses a significant challenge as it requires network capabilities that are currently not available. The integration of digital and physical worlds as envisioned in [10, 11] in the context of the European 6G Flagship Project Hexa-X [6] will improve both safety and sustainability in a human-centric manner.

The aim of future 6G networks is to enhance automotive services by utilizing AI algorithms. These algorithms can process large amounts of data collected not only by in-car and external sensors but also by radio stations in operator networks. The processed data can be used to monitor traffic flow, make suggestions to drivers, and even directly control automated vehicles to improve safety and reduce traffic. This use case is similar to the "sensor infrastructure web" use case in how data are collected, but the difference is

4 https://road-safety.transport.ec.europa.eu/index_en
5 https://transport.ec.europa.eu/transport-themes/security-safety_en

that the AI-processed data are used to provide recommendations to drivers in real time. For example, a vehicle could provide a Mobile Network Operator (MNO) with information about its location in an urban area, which is used to create a digital replica of the traffic scenario. The most appropriate network node would then collect and process information from other vehicles, pedestrians, and infrastructure monitoring entities located within the same area to provide recommendations to the vehicle and dynamically control and shape traffic to reduce congestion. The coordination between network nodes in the urban area is done by sharing AI-processed data.

In the following, the valuable application of fully autonomous driving in the realm of AI-assisted V2X use case is given. By leveraging multiple high-definition cameras and high-precision radar sensors, as well as the support of 6G technology, it becomes possible to achieve fully autonomous driving [1]. This advancement opens up new possibilities for mobility, particularly benefiting individuals with disabilities [1]. The core functions of autonomous driving encompass perception, planning, and control [1]. Various sensors, such as image sensors or cameras, millimeter-wave/terahertz radar, and Light Detection and Ranging (LIDAR), generate essential information for the perception layer. This information includes precise vehicle location and target recognition, serving as a prerequisite for enabling autonomous driving. The planning layer provides instructions for actions like following, overtaking, and accelerating, which rely on input information from the perception layer and feedback from the control layer. The control layer is responsible for executing specific control commands for the vehicle, such as throttle, brake, and gear control, based on the instructions issued by the planning layer. The major challenge in achieving full autonomy lies in meeting stringent safety requirements across diverse driving conditions. However, with the aid of AI, Multi-access Edge Computing (MEC), and the reliable and low-latency transmission capabilities of 6G, complex information can be effectively processed in a timely manner. Furthermore, the use of blockchain technology can ensure secure information sharing to enhance the performance of autonomous driving systems.

1.3.6.3 Interconnected IoT Micro-Networks

The growth of smart cities, which involve managing aspects like energy, traffic, and safety, will require a large number of communicating objects. Smart city administrators aim to provide extensive network coverage while reducing energy consumption and minimizing the number of base stations. To achieve this, they require self-adaptive networks that use objects as relays and manage information flows from local devices like robots. In 5G, network slices and private networks with their own nodes already exist. These networks could include micro-networks with varying ownership and security management, which share parts of the infrastructure with wide-area networks. For instance, a private network could have partially owned infrastructure and a private trust policy integrated into a public network.

Furthermore, this use case goes beyond smart cities and can also benefit other fields such as agriculture. As agriculture looks toward more sustainable and efficient practices, the use of interconnected massive IoT micro-networks is gaining attention. These networks, comprising sensors, drones, and other connected devices, can provide valuable data on soil quality, plant health, and other important factors. However, deploying these micro-networks in rural areas presents unique challenges, such as limited connectivity and power sources. To address these challenges, new self-adaptive network solutions need to be developed that can leverage existing infrastructure and rely on devices within the micro-networks to act as relays. With these solutions, farmers and other agricultural stakeholders can gain valuable insights and make more informed decisions about their operations, ultimately leading to more sustainable and profitable practices.

1.3.6.4 Enhanced Public Protection

Insufficient handling of security events often results from a lack of information, which leads to a lack of situational awareness. However, 6G technology can help address this issue through various building blocks, as discussed in [10, 11] in the context of the European 6G Flagship Project Hexa-X [6]. For example, a "network of networks" approach could enable the creation of a network federation to connect responders and other stakeholders who need information. Additionally, advanced radio and management solutions can provide the necessary capacity, even in the event of faults. The most critical factor is the coupling of protection with aspects such as monitoring and tracking. Therefore, the development of AI-governance, privacy-preserving techniques, secure identity management, and leak-free networks, all of which are being envisioned for 6G infrastructures, plays a vital role in ensuring that these solutions are properly, ethically, and trustworthily applied, limited to the time, place, and context in which they are necessary.

Moreover, wireless cameras are expected to be extensively deployed as sensors in the near future. With the help of AI and machine vision technologies, these cameras will be able to recognize objects and people, allowing them to serve as a universal sensor that can be placed almost anywhere. To address privacy concerns, access to data will be restricted, and information will be anonymized. Additionally, other sensing modalities, such as acoustics and radio waves, will be utilized to collect environmental data. To enhance security screening procedures and minimize wait times, a variety of sensing modalities will be combined to screen individuals as they move through crowded areas, rather than only at entrance points. Radio sensing will be particularly vital in achieving this objective, as it will enable the network to detect potential threats and identify metallic objects carried by people or robots in a crowded area. The communication systems of the future will support this approach, allowing for the automatic detection of potential threats in a given environment.

1.4 Overview of a 6G Network Architecture to Enable Massive IoT

The network architecture for the upcoming 6G massive IoT, as proposed in [12] and depicted in Figure 1.9, encompasses space, air, ground, and sea-underwater networks [1]. This integration expands upon traditional cellular networks to cover various natural spaces, including space, air, land, and ocean [1, 12]. Its goal is to establish pervasive connectivity, ensuring that any user can access any subnetwork and enjoy information assurance. By utilizing terahertz spectrum, visible light spectrum, and unlicensed spectrum for both communication and sensing purposes, this network enables ultrahigh data rates and low latency for data-intensive and mission-critical massive IoT applications. Edge computing plays a crucial role by equipping small IoT devices with additional computational and caching capabilities through offloading computation tasks [1]. Unlike cloud computing,

Figure 1.9 Space–air–ground–sea/underwater network architecture for 6G-enabled massive IoT. Source: Maria Papaioannou.

which involves outsourcing storage and processing functions to third-party servers, edge computing brings storage and processing capabilities closer to the IoT devices themselves. This approach bridges the gap between centralized clouds and distributed IoT, addressing issues such as single points of failure, reachability, lack of location awareness, and latencies associated with core networks. Consequently, the incorporation of edge computing enables applications with requirements such as mobility management, location awareness, geo-distribution, scalability, and ultralow latency [1].

Figure 1.9 illustrates the Space-Air-Ground-Sea/Underwater Integrated Network (SAGSIN), which comprises four primary segments: space, air, ground, and sea/underwater networks. These segments operate autonomously or collaborate with each other to facilitate efficient data communication and deliver versatile end-to-end services to users. SAGSIN exhibits characteristics such as a large-scale, three-dimensional multilayer topology, high heterogeneity, and support for a wide range of businesses [12]. SAGSIN primarily relies on the ground network as its foundation and extends into the space network [12]. In the following section, we will provide the reader with more details on each different segment.

1.4.1 Space Network

The space network encompasses various satellite types, constellations, and associated ground infrastructure, including ground stations and control centers [1, 12]. Satellites are categorized based on their altitudes, including Geostationary Earth Orbit (GEO), Medium Earth Orbit (MEO), Low Earth Orbit (LEO), and Very Low Earth Orbit (VLEO) satellites [12]. GEO satellites are utilized for applications such as television broadcasts, international telegrams, and microwave communication. MEO and LEO satellites operate closer to the earth's surface and are commonly used for high-resolution imaging, topographic reconnaissance, resource detection, and meteorological monitoring. VLEO satellites, being the lowest-orbiting satellites, hold promise to provide high-speed data services and precise positioning.

While satellites offer advantageous positions and long transmission links, they face challenges in providing global mobile communication to users [12]. To address this, the American company Iridium developed a system that employs low-orbit satellites to enable global voice and data connections for mobile users, including dynamic networks for ground vehicles [12]. However, the significant signal propagation delay in satellite-ground links increases vulnerability to attacks and compromises the QoS for real-time interactive applications. As a key technology for future 6G, the SAGSIN aims to deploy a substantial number of VLEO satellites in space. This deployment will offer extensive geographical coverage, ample capacity, and a wide range of broadband communication services for residential and commercial institutions, governments, and professional customers worldwide.

1.4.2 Air Network

The air network consists of aircraft, Unmanned Aerial Vehicles (UAVs), airships, and air balloons [1, 12]. This network enables the automated routing of data between aircraft and ground stations, aircraft and the sea network, and facilitates data exchange with the satellite layer. Among them, the HAP is a common type of air network situated approximately 20 km above the ground in the stratosphere. HAPs offer advantages such as mobility, quick communication response times, and cost-effectiveness compared to satellite communication platforms [12]. They are commonly employed for services like broadcast/multicast, emergency communications, disaster relief operations, and large-scale temporary events [1, 12].

The UAV communication platform, being closer to the ground, possesses characteristics such as rapid response times, high throughput, reliable line-of-sight transmission, and enhanced mobility and flexibility [1, 12]. Multiple UAVs can be employed to establish an aerial subnet that assists ground vehicles in transmitting road information. Additionally, the cameras and GPS information on UAVs can be utilized for obstacle detection and navigation purposes [12]. However, due to the dynamic nature of its network topology and link channels, an effective coordination mechanism must be designed to ensure network security while providing high throughput and low-latency network access for ground users [12].

1.4.3 Ground Network

The ground network encompasses various subnetworks, including cellular networks, ad-hoc networks, Wireless Local Area Networks (WLAN), Worldwide Interoperability for Microwave Access (WiMAX), coastal base stations, and more [1, 12]. Over time, ground network technologies have become highly developed and mature, thanks to the rapid advancement of global mobile communication systems. Compared to other segments within SAGSIN, the ground network offers users high data transmission rates and significant throughput [1, 12]. However, it has limitations in terms of network coverage in remote areas [12]. Moreover, the ground network is susceptible to natural disasters and infrastructure damage caused by human activity [1, 12]. Consequently, relying solely on the ground network is insufficient to meet the growing demands for high-quality services in the future [12].

1.4.4 Sea/Underwater Network

The sea network comprises two main components: the surface network and the underwater network.

1.4.4.1 Surface Network

The surface network includes ships, buoys, and Unmanned Surface Vessels (USVs) [1, 12]. Among these, USVs exhibit greater mobility compared to buoys [12]. Ships, buoys, and USVs are equipped with communication capabilities and can provide edge services [12]. Ships, in particular, offer higher computing power than buoys and USVs. However, it is important to note that the computing performance of these nodes is limited compared to remote cloud servers due to communication constraints in maritime environments, limited bandwidth, and unstable channel quality [12]. Thus, leveraging the computing resources of these nodes becomes crucial for completing time-sensitive tasks at sea [12].

1.4.4.2 Underwater Network

The underwater network typically consists of an underwater observation network and an underwater self-organizing network [12]. The underwater observation network relies on optical and submarine cables for communication [12]. However, due to the challenging marine conditions and the high costs associated with construction and maintenance, deploying a large-scale underwater observation network is difficult [12]. On the other hand, the underwater self-organizing network expands the underwater network through wireless communication. It comprises fixed sensor nodes and mobile nodes such as Autonomous Underwater Vehicles (AUVs), Remotely Operated Vehicles (ROVs), and Unmanned Underwater Vehicles (UUVs). Acoustic waves are commonly used for communication in this network, but it suffers from limitations such as limited bandwidth and long delays.

According to [12], currently, maritime communication methods have low efficiency and narrow bandwidth, which do not meet the increasing demands of economic activities and life at sea. Therefore, the use of coastal base drones and the establishment of new communication systems are necessary and significant [12]. Through the implementation of SAGSIN, global sea coverage can potentially be achieved by adopting the Global Maritime Distress and Safety System (GMDSS) and Inmarsat [12].

1.5 Security Objectives in Massive IoT in 6G

In the context of the 6G networks (i.e. SAGSIN), the following five major security objectives have been identified [4, 12, 13], as also depicted in Figure 1.10.

1.5.1 Confidentiality

Assures that confidential information is not made available or disclosed to unauthorized entities [13]. In the context of the 6G networks (i.e. SAGSIN), confidentiality refers to the protection of data transmission that might contain sensitive information, for instance, related to user behaviors [12]. Thus, malicious actors might infer such sensitive information incidentally by using accessible information that is accidentally disclosed [12]. For instance, if the confidentiality of transmitted data is compromised, it means that an adversary could exploit vulnerabilities between an IoT device (i.e. sender) and a gateway

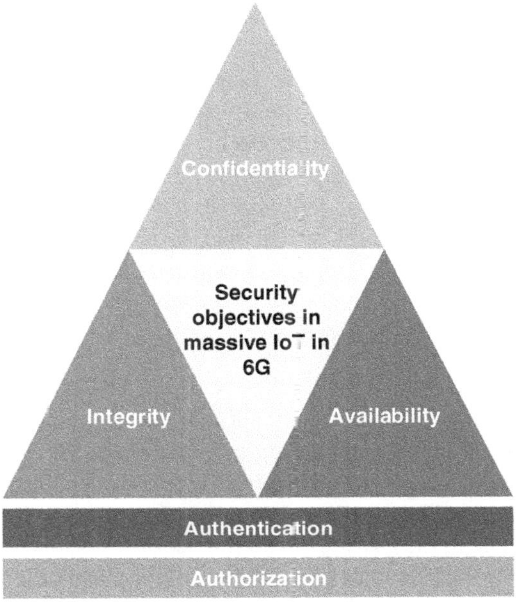

Figure 1.10 Fundamental Security Objectives in Massive IoT in 6G. Source: Maria Papaioannou.

(i.e. receiver) to intercept the data being transmitted and gain unauthorized access to sensitive information. To safeguard confidentiality in 6G networks, various measures can be implemented, including physical protection mechanisms and the utilization of cryptographic algorithms. These algorithms work by transforming the data into an unintelligible form, making it inaccessible to unauthorized individuals.

1.5.2 Integrity

Assures that data have not been altered and/or destroyed in an unauthorized manner [13]. Applied to 6G networks (i.e. SAGSIN), integrity preserves the accuracy of data transmitted over wireless networks, for example, guaranteeing that the data received are exactly the same as the data sent in the message transmission process. In use cases such as "AI-assisted Vehicle-to-Everything (V2X)" or "Precision healthcare," human safety is at stake, depending on the integrity of the network [10]. It is expected that many other future use cases will likewise handle a wealth of highly sensitive data whose compromise of integrity might pose a significant risk [10]. The attack on data integrity is distinct from attacks aimed at confidentiality, as it is characterized by a higher level of sophistication rather than direct aggression [12]. Essentially, this type of attack seeks to manipulate genuine information within a system and disrupt the smooth flow of information across a network [12]. To elaborate further, if an adversary successfully alters or destroys a legitimate node, it can lead to detrimental consequences for data integrity [12].

1.5.3 Availability

Ensures that systems work properly and services are not denied to authorized users based on the requests at anytime and anywhere [13]. In the context of the 6G networks (i.e.

SAGSIN), the network's ability to provide reliable service, even in the face of attacks, is crucial [12]. For instance, attacks that aim to disrupt availability, such as Denial of Service (DoS) attacks, seek to hinder or completely interrupt the transmission of information, rendering network resources inaccessible to communication nodes. On the other hand, jamming attacks are a specific type of DoS attack that occurs at the physical layer, where an unauthorized node maliciously interferes with communication between legitimate users [12]. To combat jamming attacks, existing wireless systems often employ spread spectrum technology, which extends the signal's bandwidth using unrelated code sequences, surpassing the minimum required bandwidth [12].

1.5.4 Authentication

In the context of 6G networks (i.e. SAGSIN), authentication plays a crucial role in verifying both entities (i.e. entity authentication) and transmitted information (i.e. message authentication) [13]. Entity authentication ensures that one communicating entity can trust the claimed identity of another entity involved in the interaction [13]. On the other hand, message authentication verifies that an entity is indeed the original source of specific data generated in the past [13]. With the rise of IoT devices in massive IoT in 6G networks, which often have limited memory and processing power, there is a growing need for lightweight authentication protocols. These protocols aim to provide authentication functionalities while minimizing the computational and memory requirements typically associated with traditional authentication protocols [13].

1.5.5 Authorization

In the context of 6G networks (i.e. SAGSIN), authorization refers to the process of granting official permission to an entity to access specific network services or resources [13]. It ensures that only entitled entities, such as trusted IoT devices or authorized personnel, can gain access to services like accessing collected data or issuing commands to IoT devices [13]. Access control is a widely used security technique that enforces authorization, making sure that only authorized entities are granted access to the desired functionalities or resources within the network [13].

1.6 Security Threats in Massive IoT in 6G

6G networks are expected to handle a wide range of applications and services, including those that involve sensitive information, such as: (i) healthcare applications where 6G networks may transmit sensitive patient data (e.g. medical records, diagnostic images, or real-time health monitoring data); (ii) financial applications where data related to financial transactions, personal identification information, and account details may be transmitted and processed within the network; or (iii) applications in other sectors (e.g. government, defense, and industrial applications), which may also involve sensitive data that need to be securely transmitted, processed, and stored in 6G networks [4, 12, 13]. Therefore, where the transmitted, processed, and stored data are sensitive, the concerns about security and privacy are increased [4, 12, 13]. In this section, we provide a categorization of the security threats, targeting the 6G networks (i.e. SAGSIN) based on the security objectives that they intend to compromise.

1.6.1 Security Threats to Data Confidentiality

It is expected that 6G networks will comprise hundreds of billions of interconnected IoT devices with limited resources [4, 12, 13]. These constraints make it challenging to implement resource-intensive cryptographic solutions, such as traditional data encryption and decryption, to ensure high levels of data confidentiality. Consequently, the network becomes susceptible to threats that target the confidentiality of exchanged or stored data within 6G networks (i.e. SAGSIN) [4, 12, 13]. For instance, an adversary can exploit eavesdropping techniques to intercept information during communication [4, 12, 13]. By monitoring and accessing the contents of transmitted packages, the adversary can gain unauthorized access to sensitive data. This can occur when, for example, a wearable sensor wirelessly transmits a patient's vital signs to an IoT gateway, such as a smartphone. Through techniques like traffic analysis, the adversary can passively intercept the communication and extract confidential data for malicious purposes.

Moreover, interrogation attacks, which involve impersonation, pose a threat to data confidentiality. In this scenario, a malicious actor pretends to be a legitimate entity and sends requests to other entities, aiming to extract private information about users. These attacks compromise the confidentiality of data by exploiting vulnerabilities in the network's authentication mechanisms.

1.6.2 Security Threats to Data Integrity

Man-in-the-Middle (MitM) attacks pose a significant threat to the integrity of 6G networks (i.e. SAGSIN) [4, 12, 13]. These attacks occur when an attacker interferes with the communication between two parties and can modify the exchanged data without detection. This type of attack jeopardizes the integrity of the network. For example, in an Internet of Medical Things (IoMT) network, the collected medical data can be transmitted to a remote server or stored locally in wearable devices' internal memory [13]. During transmission, a MitM attacker can intercept and modify medical data, compromising its integrity. This alteration of data can have serious consequences in terms of accuracy and trustworthiness [13].

Furthermore, physical attacks on the IoT devices themselves can also target the integrity of the applications supported by 6G networks (i.e. SAGSIN) [4, 12, 13]. If an adversary gains physical access to an IoT device, they can manipulate its structure to alter its behavior. This type of attack not only disrupts the provided services but also modifies the stored data, posing a significant risk to the integrity of the system [4, 12, 13].

Additionally, the lack of lightweight malware detection mechanisms for the hundreds of billions of interconnected IoT devices leaves them vulnerable to integrity attacks [4, 12, 13]. Attackers can compromise the integrity of these devices by executing malicious code and exploiting security vulnerabilities in the networking software and hardware. This type of attack can cause harm to the IoT devices and potentially compromise the overall integrity of the application supported by 6G networks (i.e. SAGSIN) [4, 12, 13].

Last but not least, unlike typical networks, SAGSIN exposes more channels to potential adversaries, making it easier for them to establish links and pose threats, ultimately violating its integrity. For instance, in the satellite-ground link, legitimate nodes can be illicitly altered into malicious nodes, leading to tampering with data information in the link. Consequently, these compromised nodes can launch network attacks, leading to the destruction of data integrity during satellite-ground transmission. Detecting such attacks on network nodes proves challenging since the attacked nodes maintain valid identities while running malicious codes, presenting an urgent and complex problem to address. Thus, it becomes crucial to enhance intrusion detection technology [12].

1.6.3 Security Threats to Availability

It is expected that IoT technology will be increasingly utilized in healthcare applications to overcome the limitations of centralized cloud-based healthcare systems [13]. However, healthcare systems that heavily rely on numerous interconnected IoT devices face resource and computational constraints, presenting challenges in maintaining the availability of services. Various types of attacks can target different network layers and have distinct impacts on the availability of the services supported by 6G networks. These types of attacks may include jamming attacks, battery drainage attacks, collision attacks, congestion, and IoT-botnet attacks [4, 12, 13].

Jamming attacks aim to overload communication channels or computing resources by transmitting large volumes of messages, hindering the normal usage of services [4, 12, 13]. When the availability of an IoT network is compromised, it becomes ineffective in providing real-time healthcare services and can pose risks to patients' health. For example, in an IoT-enabled healthcare alert system, if communication channels within the IoT network are jammed, a critically ill patient may not receive the necessary care, putting their life at risk [13]. In addition, battery drainage attacks target the limited battery resources of resource-constrained IoT devices by maliciously sending false messages to drain the device's battery [13].

Collision attacks occur when two nodes simultaneously transmit data on the same frequency channel, leading to identification mismatches at the receiving end [13]. This results in the discarding of corrupted received data packets and their subsequent retransmission, wasting network resources within the IoT network. On the other hand, channel congestion attacks involve the massive transmission of useless messages, causing high traffic in channels and rendering time-related IoT services and data unavailable.

Lastly, Distributed Denial of Service (DDoS) attacks can also target the availability of 6G networks. In these attacks, an attacker employs an IoT botnet (meaning potentially large group of computers owned by unsuspecting users but compromised and controlled by an attacker [10]) to flood the target device, such as a gateway, with multiple requests, overloading it and disrupting the provided services [4, 12, 13]. Attack tools became more and more diverse, trying out multiple exploits against a victim, and more automated, suitable to be applied to huge numbers of targets [10]. This way, botnets were created to perform hard-to-counter DDoS attacks. It is important to note that IoT networks are more vulnerable to DoS attacks compared to cloud platforms in IoT-based systems due to the constrained resources of the devices [13]. In [12], the authors highlight that in recent times, there has been a significant increase in the utilization of numerous bots for executing distributed attacks. These botnet-style attacks deliberately restrict the access of legitimate users and cause damage to their ability to use resources effectively.

1.6.4 Security Threats to Authentication

Authentication is a critical security requirement in massive IoT networks within 6G. Traditional authentication solutions based on Public Key Infrastructure (PKI) are not suitable for the ubiquitous nature of massive IoT devices, as they are inefficient and lack scalability. Adversaries can exploit the weaknesses in authentication systems to gain unauthorized access to resources by impersonating legitimate users without genuine credentials [4, 12, 13].

Various attacks target the authentication process in massive IoT networks. For example, a Sybil attack may involve an IoT device claiming multiple fake identities, allowing rogue devices to impersonate legitimate ones. This enables the rogue node to connect with multiple IoT devices, maximizing its influence and potentially misleading the system to draw incorrect conclusions. Sybil attacks can be categorized based on communication of nodes, identity origin, and simultaneity, considering factors such as direct or indirect communication, fabricated or stolen identities, and synchronous or asynchronous connection of Sybil identities.

Furthermore, device cloning/replication attacks focus on taking over a sensor device, extracting encrypted information, and using it to create multiple clones within the network. These clones can be used to compromise authentication and security objectives. Device cloning attacks are successful when the authentication process lacks location-based schemes that would prevent devices located in the exact same location.

Masquerading attacks can also target massive IoT edge networks. In this type of attack, an adversary either pretends to be a legitimate user by inserting rogue devices to gain unauthorized access to services provided by IoT devices or presents themselves as an IoT device to offer fake services to users. Masquerading attacks pose a significant risk, especially in the healthcare sector, where the services provided by IoMT devices are critical and life-dependent for many patients [13].

To ensure the security of massive IoT networks in 6G, robust authentication mechanisms need to be employed to protect against attacks such as Sybil attacks, device cloning attacks, and masquerading attacks. Additionally, incorporating location-based schemes and implementing stringent security measures are vital to enhance the authentication process and mitigate the risks associated with these attacks.

1.6.5 Security Threats to Authorization

Adversaries may exploit weak authorization mechanisms within 6G networks to gain unauthorized access to network resources. Due to users' lack of security training and awareness, IoT devices can be vulnerable to social engineering attacks, allowing malicious actors to impersonate legitimate entities and gain access to IoT devices within 6G networks [4, 12, 13]. This poses a significant risk, especially for critical applications such as those in healthcare domain where there are IoT devices that monitor vital signs, whose compromise might endanger human lives.

Furthermore, malware attacks can compromise connected IoT devices within 6G networks by exploiting vulnerabilities, including weaknesses in authorization mechanisms [4, 12, 13]. Once infected, these compromised IoT devices can be utilized as bots to launch additional attacks on other devices within the IoT network [13]. This enables the attacker to gain access to network services, such as controlling multiple IoT devices or accessing valuable resources like sensitive data transmitted, processed, and stored in 6G networks.

It is crucial to address these authorization vulnerabilities in the design and implementation of 6G-enabled massive IoT applications to prevent unauthorized access and ensure security of IoT devices and the sensitive data they handle. Educating users about security practices and employing robust authorization mechanisms are essential steps in safeguarding 6G-enabled massive IoT applications and ensuring the security and safety of all entities involved.

1.7 Conclusion

The strict network performance requirements of the emerging disruptive massive IoT applications cannot be satisfied by the currently developing 5G mobile networks due to the inherent limitations of these networks [1–5]. Therefore, the focus on the next 6G of mobile networks with significantly improved key performance indicators (e.g. sub-ms latency, Tbps data throughput, extreme energy efficiency, and ultralow energy consumption) is inevitable to overcome the limitations of 5G and truly meet the new demanding levels of performance requirements of the future massive IoT applications and become their major key enabler [2, 5].

In this chapter, we first provided the vision and core values of 6G, as outlined by the European 6G Flagship Project Hexa-X in [10, 11]. The 6G vision focuses on enhancing the integration of three key worlds: the human world (including senses, bodies, intelligence, and values), the digital world (including information, communication, and computing), and the physical world (including objects and organisms). This integration aims to create a seamless cyber–physical continuum, using networks to significantly improve our quality of life. On the other hand, the core values guiding the development of these future 6G mobile networks include sustainability, trustworthiness, and digital inclusion. 6G mobile networks are expected to handle the challenges of rapid traffic growth, an increase in connected devices, and diverse network demands, all while maintaining high standards of energy efficiency, security, privacy, and trust. The overarching goal is to drive sustainable economic growth and spur innovation across various sectors.

Furthermore, in this chapter, we discussed a set of representative massive IoT applications envisioned to be deployed in the 6G era to improve people's lives. To organize the discussion of representative massive IoT applications in a more comprehensive manner, the clustering into use case families, as proposed in [10, 11] in the context of the European 6G Flagship Project Hexa-X [6], has been adopted. In particular, the following six use case families have been considered, to categorize representative massive IoT applications in 6G era: (i) enabling sustainability, (ii) massive twinning, (iii) telepresence, (iv) robots to cobots, (v) trusted embedded networks, and (vi) hyperconnected resilient network infrastructures. These six families of 6G use cases comprise a first baseline to guide the future research directions on 6G, relying on the view of the current European research activities on 6G driven through the 5G PPP and 6G-IA initiatives [3].

In addition, in this chapter, we presented a proposed 6G network architecture in [12], which encompasses space, air, ground, and sea-underwater integrated networks (i.e. SAGSIN) to enable massive IoT applications [1]. This integration expands upon traditional cellular networks to cover various natural spaces, including space, air, land, and ocean [1, 12]. Its goal is to establish pervasive connectivity, ensuring that any user can access any subnetwork and enjoy information assurance.

Finally, after presenting the five major security objectives, namely confidentiality, integrity, availability, authentication, and authorization [4, 12, 13], in the context of the 6G networks (i.e. SAGSIN), this chapter identified the threat landscape for 6G-enabled massive IoT applications. The aim of this was to shed light on their potential key threats and identify security and privacy issues that should be addressed before the 6G-enabled massive IoT applications build trust among all relevant stakeholders and realize their full potential.

References

1. Guo, F., Yu, F.R., Zhang, H. et al. (2021). Enabling massive IoT toward 6G: a comprehensive survey. *IEEE Internet of Things Journal* 8 (15): 11891–11915.
2. Uusitalo, M.A., Rugeland, P., Boldi, M.R. et al. (2021). 6G vision, value, use cases and technologies from European 6G Flagship Project Hexa-X. *IEEE Access* 9: 160004–160020.
3. The 5G PPP Architecture Working Group (2022). The 6G Architecture Landscape European perspective Version 1.0, December 2022. White Paper.
4. Porambage, P., Gur, G., Osorio, D.P.M. et al. (2021). The roadmap to 6G security and privacy. *IEEE Open Journal of the Communications Society* 2: 1094–1122.
5. Saad, W., Bennis, M., and Chen, M. (2020). A vision of 6G wireless systems: applications, trends, technologies, and open research problems. *IEEE Network* 34 (3): 134–142.
6. Hexa-X Partners, "Hexa-X. – Connecting human, physical, and digital worlds," 2023. Project webpage: https://hexa-x.eu/about/, Project leaflet: https://hexa-x.eu/wp-content/uploads/2022/02/Hexa-X_leaflet_A4_printable.pdf.
7. Nguyen, D.C., Ding, M., Pathinara, P.N. et al. (2022). 6G Internet of Things: a comprehensive survey. *IEEE Internet of Things Journal* 9 (1): 359–383.
8. The 5G Infrastructure Association (2021). *European Vision for the 6G Network Ecosystem*. 5GIA.
9. Malik, U.M., Javed, M.A., Zeadally, S., and ul Islam, S. (2022). Energy-efficient fog computing for 6G-enabled massive IoT: recent trends and future opportunities. *IEEE Internet of Things Journal* 9 (16): 14572–14594.
10. Hexa-X Partners, "Deliverable D1.2 - Expanded 6G vision, use cases and societal values – including aspects of sustainability, security and spectrum," 2021. Online: https://hexa-x.eu/wp-content/uploads/2022/04/Hexa-X_D1.2_Edited.pdf.
11. Hexa-X Partners, "Deliverable D1.3 - Targets and requirements for 6G – initial E2E architecture," 2022. Online: https://hexa-x.eu/wp-content/uploads/2022/03/Hexa-X_D1.3.pdf.

12 Guo, H., Li, J., Liu, J. et al. (2022). A survey on space-air-ground-sea integrated network security in 6G. *IEEE Communication Surveys and Tutorials* 24 (1): 53–87.

13 Papaioannou, M., Karageorgou, M., Mantas, G. et al. A survey on security threats and countermeasures in Internet of Medical Things (IoMT). *Transactions on Emerging Telecommunications Technologies* 33 (1): e4049.

2

Secure Edge Intelligence in the 6G Era

Tanesh Kumar[1], Juha Partala[2], Tri Nguyen[3], Lalita Agrawal[4], Ayan Mondal[4], Abhishek Kumar[3], Ijaz Ahmad[5], Ella Peltonen[6], Susanna Pirttikangas[3], and Erkki Harjula[7]

[1] *Department of Information and Communications Engineering, Aalto University, Espoo, Finland*
[2] *Center for Machine Vision and Signal Analysis, University of Oulu, Oulu, Finland*
[3] *Center for Ubiquitous Computing, University of Oulu, Oulu, Oulun Yliopisto, Finland*
[4] *Department of Computer Science and Engineering, Indian Institute of Technology Indore, Indore, India*
[5] *VTT Technical Research Centre of Finland, Espoo, Finland*
[6] *M3S Research Unit, University of Oulu, Oulu, Finland*
[7] *Center for Wireless Communications, University of Oulu, Oulu, Finland*

2.1 Introduction

Despite the ongoing commercial deployment of the 5G technology, the research community has already started to explore the potential of future mobile and wireless communication networks beyond 5G systems. Future 6G networks are expected to enable a smart, connected, and intelligent digital ecosystem where numerous stakeholders and network entities will require secure and trusted communication-computing continuum to ensure the ubiquitous availability of novel services to consumers in an efficient way. In order to ensure successful deployment and acceptance of 6G networks, it is highly important to consider and address various strict requirements, such as ultralow latency, security, and privacy, embedded trust, automated operations, network management, big data management, as well as intelligent data analysis and decision-making, among many others [1].

To ensure the smooth development of future 6G networks, strong support is required from various advanced and disruptive communication-related technologies, such as virtualization, artificial intelligence (AI), softwarized/programmable networks, edge and fog computing, quantum computing, and distributed ledger technologies (DLTs)/Blockchain [2]. Among these, edge intelligence (EI) is a centric enabler for various novel 6G-based applications, which combines the capabilities of edge computing and AI to provide crucial features, such as ultralow latency, high reliability, high resource and energy efficiency, and high level of security and privacy [3]. The integration of EI with 6G networks will enable the efficient use of local computing, storage, and processing capabilities along with cloud data centers for optimal placement of data processing, analysis

Security and Privacy for 6G Massive IoT, First Edition.
Edited by Georgios Mantas, Firooz B. Saghezchi, Jonathan Rodriguez, and Victor Sucasas.

and decision-making. This is vital for various future 6G-enabled critical Internet of Things (IoT) applications in different verticals, such as real-time process control in Industry 5.0, virtual reality use cases in computer-aided surgery or remote vehicle operation, self-driving vehicles, etc., all requiring highly reliable, efficient, scalable, and secure communication and computing infrastructure for their operations and processes [4, 5].

Along with all the above benefits, the integration of edge, AI, and 6G networks brings wider than-before threat landscape for adversaries to launch various security and privacy attacks. For example, during the dynamic computational task offloading in a multitier edge-cloud architecture, a malicious entity can fetch or alter the user's sensitive information that is transmitted among various computational nodes [6]. Moreover, when AI algorithms are either fully or partially running on various distributed edge nodes, several security and privacy threats will rise. For example, poisoning attack is one of the most common attacks against machine learning (ML) algorithms to alter or change their training models [7]. Furthermore, since 6G networks combine various other enabling technologies, such as DLT/Blockchain and quantum computing, there will be a high probability of existing and new security and privacy attacks [8]. Finally, since the 6G-EI ecosystem will comprise several entities (users, devices, service providers, applications, etc.), dynamic and embedded trust management mechanisms will be required.

To address these challenges and secure the EI for future 6G networks, the literature has recently pointed out some potential technologies that will enhance security and privacy [9]. For example, ML-based security and privacy algorithms will play a significant role in providing intelligent and real-time intrusion detection, prevention, and mitigation mechanisms [10]. Moreover, the addition of various security-enabling technologies, such as DLTs/Blockchain and quantum-resilient cryptography, will enhance the overall security and trust of future 6G network architecture.

The key objective of this research work is to analyze what is still missing in the current mobile generation in terms of security, privacy, and trust challenges and what kind of novel and advanced technological developments are required to cope with the new and strict requirements of future massive-scale IoT applications and critical infrastructures. This chapter mainly explores the potential security vulnerabilities in the context of the future EI and 6G networks. Section 2.2 provides background details about the fundamentals of edge computing, the evolution of EI, and the need for secure EI for 6G. Sections 2.3–2.5 discuss various potential security, privacy, and trust threats for 6G EI-enabled systems, respectively. We briefly present the ongoing and future vision for the security standardization activities in 6G and Edge AI in Section 2.6 and conclude the chapter in Section 2.7.

2.2 Background

2.2.1 Edge Computing and Its Importance

Cloud computing allows various users to access services or resources from a remote location via the Internet. Based on the level of software and hardware resources provided, the cloud services are categorized into three categories – Software-as-a-Service (SaaS), Platform-as-a-Service (PaaS), and Infrastructure-as-a-Service (IaaS) [11, 12]. These cloud services allow users to create applications and use hardware, software, and operating systems in a virtualized environment at a large scale.

IoT is one of the fastest-growing domains of Information and Communications Technology (ICT), and IoT devices generate rapidly growing amounts of data to be handled by the cloud. However, due to the long distance from IoT data sources and end-users to the cloud, there are many challenges related to transferring, processing, and storing massive data in the cloud, such as high latency, high burden on networks, and high energy consumption [11, 12]. Edge computing is envisioned to overcome the aforementioned challenges of cloud computing for IoT applications by bringing cloud computation capacity closer to data sources and end-users, as shown in Figure 2.1.

In edge computing, the processing of data generated from different IoT devices occurs at the edge, i.e. between the cloud and the IoT devices. Edge computing platform is distributed in nature and provides networking, storage, and computation services for IoT applications and services, [13–16]. The edge computing platform addresses the following key issues faced by the cloud computing platform:

1. **Performance (Latency, Throughput):** Edge nodes are available in the access or local network. Therefore, the distance between edge nodes and IoT devices is less than between the cloud servers and the IoT devices. Hence, edge computing serves the user requests with less delay, and the bandwidth is not limited by as many potential bottlenecks.
2. **Efficiency (Resource, Cost, and Energy Efficiency):** Edge computing decentralizes the processing of data. Therefore, data coming from IoT devices can be processed closer to their sources, and consequently, not as much data need to be sent over. This reduces energy consumption while also core network congestion is relieved.
3. **Dependability (Reliability, Security, and Privacy):** In addition to decentralized data processing, edge computing also decentralizes decision-making. This brings resilience against server/data center failures, network failures, and congestion, as well as denial of service or other types of security attacks. Furthermore, edge computing helps limit the propagation of private data, reducing its probability of being exposed to potentially hostile parties.

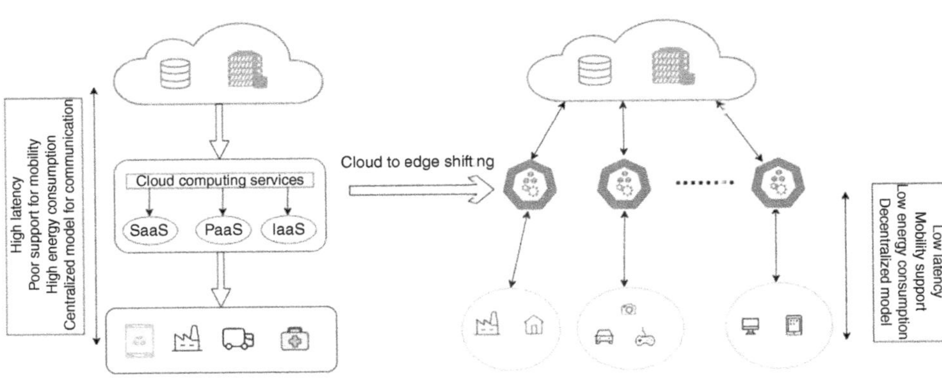

Figure 2.1 Cloud computing to edge computing shifting for IoT applications. Source: Lalita Agrawal.

2.2.2 Emergence of Edge Intelligence

To support the decentralized edge computing platform, AI technologies are envisioned to serve as a support system for fast data processing, i.e. analyzing and extracting meaningful information/patterns from a large amount of data near the source. AI-based optimization methods, such as federated learning (FL) and reinforcement learning (RL), can help in enhancing the overall system efficiency at large. Additionally, the integration of AI into edge supports multiple heterogeneous applications in terms of data types, data handling, and offloading while ensuring a minimum latency. Hence, EI comes into existence by incorporating AI techniques and models to support the edge computing platform, [17, 18]. The edge nodes enabled with AI techniques and platforms are called intelligent edge nodes. Figure 2.2 represents a schematic diagram for EI. It is to be highlighted that EI is not only a simple integration of edge computing and AI. To understand EI, we need to find answers to the following fundamental questions:

1. How can AI techniques provide a solution for edge computing challenges?
2. How can edge computing optimize the operation of AI techniques?

To answer these questions, researchers have presented multiple architectures, systems, and enabling technologies for EI [17–19]. For example, [17] presented two aspects of EI, i.e. the first one is "AI for edge," which mainly concentrates on efficiently utilizing AI-based solutions and technologies to enhance the edge computing paradigms, e.g. a better, intelligent, and more secure solution to data collection and computation offloading for mobile

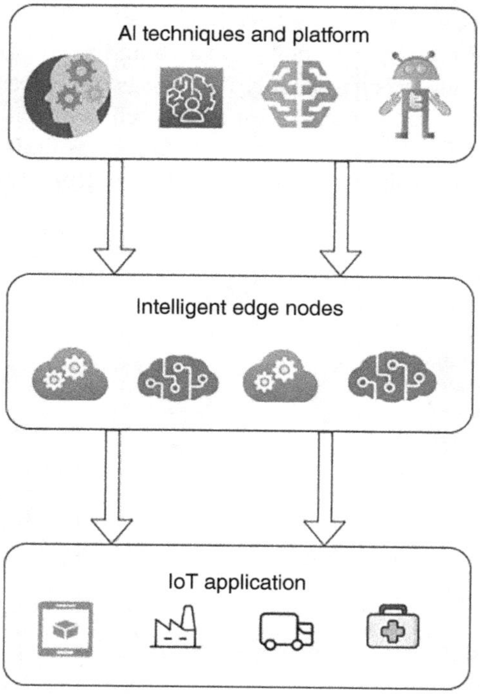

Figure 2.2 Schematic diagram of edge intelligence. Source: Lalita Agrawal.

environments. The other aspect is "AI on edge," which deals with constructing AI models for integrating and executing AI techniques in the edge computing environment.

The edge computing platform enhances AI applications to work in heterogeneous and distributed networks, i.e. shift from cloud to edge network. Due to the physical limitations of the edge nodes, EI suffers from the following challenges:

- **Constrained Resources:** Due to having limited computational and storage capacity, executing AI algorithms on edge nodes is challenging in a distributed edge computing architecture.
- **Inconsistency:** Cloud computing is capable of supporting multiple platforms for executing AI techniques. Edge nodes have limitations in provisioning these AI-compatible platforms.

To overcome the aforementioned issues, [18] proposed a framework, named OpenEI. The authors constructed a model using lightweight packages of deep RL algorithms. The OpenEI framework aims to find appropriate AI techniques corresponding to edge environments to remove the inconsistency. In another work [20], it is identified that the distributed learning has significant importance, i.e. FL, for edge nodes. Using FL, the authors addressed several issues using the amalgamation of AI and edge computing.

2.2.3 Fusion of Edge Intelligence and 6G

The future 6G network is expected to be AI-native, which will potentially usher in a new era of transformation within the telecommunications industry. Recent efforts toward making interfaces of radio access network (RAN) open, thanks to O-RAN alliance, indicate that network operators would have an opportunity to deploy third-party AI/ML-empowered applications (in the form of xapps and rapps). This advanced AI-native open network ecosystem will integrate a diverse array of established and innovative technologies, alongside sophisticated sensing and communication advancements. These developments will fuel the emergence of future large-scale IoT applications, including Industry 5.0, virtual reality scenarios, and autonomous driving. To accomplish the 6G vision, the research community has already started to drive deep into the various requirements, e.g. ultralow latency, reliability, massive connectivity, autonomous and self-adapting networks, peak data rate, optimal spectrum usage, and maturity in the relevant enabling technologies, among others [21, 22]. In addition, various open challenges require suitable solutions such as robust security and privacy solutions, energy-efficient and sustainable solutions, optimized resource utilization, and implementing AI/ML solutions for future communication networks [23].

The fusion of EI and 6G is poised to facilitate pervasive intelligence on a large scale [24]. This entails the convergence of communication, sensing, and localization, coupled with decentralized learning and inference at the edge across a comprehensive environment. This integration is expected to give rise to a cognitive network architecture capable of meeting the diverse needs of 6G verticals. Additionally, it will streamline the development and implementation of verticals capable of leveraging resources for real-time learning and inference in a programmable environment. EI will enable the incorporation of autonomous computing functionalities, such as self-healing, self-configuration, and self-optimization, within 6G verticals, thus enhancing the resilience of 6G against unexpected incidents and disruptions.

2.2.4 The Need for Secure Edge Intelligence

The 6G ecosystem will be a collective, collaborative platform ecosystem with a plethora of advanced enabling technologies and multiple stakeholders, such as network operators and service providers, that may utilize integrated computation and communication platforms for various services.

The integration of EI with the future 6G will enable massive IoT applications by offering distributed intelligence, reliable connectivity, and faster data analysis/decision-making. To make this vision a reality, novel security solutions are required to protect entire EI-enabled 6G ecosystems for various applications. Some of the security solutions for such smart ecosystems may also utilize different existing security technologies and approaches proposed for 5G systems. For example, blockchain-based/integrated/empowered security mechanisms and frameworks have been widely proposed and developed for the 5G networks.

2.3 Security Challenges in 6G EI

2.3.1 Computational Offloading

In an untrusted IoT computing environment where several sensor nodes/devices are connected to a cloud server through edge nodes/devices, the outsourcing of resources/computations, or sensitive data from end devices to potentially untrusted edge nodes or devices can create challenges regarding the confidentiality and integrity of the computation. Dishonest serving nodes can modify both the input data and the actual computation and return plausible but false results. The untrustworthiness of the computing party is also a problem if the input data contain sensitive information. This type of environment, where we cannot assume inherent trust within a network and its entities, should be dealt with a zero-trust approach (ZTA), [25], requiring strict identity verification and continuous authentication for all users, devices, and applications.

To ensure confidentiality, homomorphic encryption (HE) can be applied to compute encrypted data [26, 27]. Fully HE enables computations to be performed on encrypted information, and those computations translate to the unencrypted domain. However, fully HE is computationally demanding and may be too costly for edge devices, even if advances are made in the development of these schemes. Flexible mechanisms are needed to determine whether certain data can be transmitted for outsourced computation.

Verifiable computing is seen as one of the potential solution candidates to the integrity problems. In verifiable computing, the computing party generates proof that the required computation was performed correctly [28, 29]. The proof is significantly easier to check than to repeat the actual computation. Thus, computation can be outsourced even to an untrusted party with its integrity guaranteed. The generation of such proofs is, however, computationally expensive and can be prohibitively expensive for edge devices.

2.3.2 Security of Machine Learning

It is evident that ML plays an integral part in 6G EI. Therefore, the trustworthiness of the applied models is essential in the correct operation of the network. However, there are multiple challenges regarding the security of ML. In particular, we have to ensure that (i) the training data are correct and not disclosed to unauthorized parties; (ii) the model training is executed correctly and no false data are inserted into the training set; and (iii) the applied models do not enable malicious parties to infer information about the training data from them or to exploit the model, for example, through evasion attacks [30]. These challenges are discussed in detail below and have been depicted in Figure 2.3.

EI can function correctly only if the applied models are correct. Regarding AI and ML, a correct model requires correct training data. Poisoning attacks insert malicious data into the training set in order to skew the trained model. As a result, the model may exhibit poor generalization and behave incorrectly. In order to have trustworthy models, poisoning attacks need to be detected, and the false data need to be removed from the training set. In general, protecting against poisoning is hard [31], and there is no generic method that works for any use case. Additional research is needed to have satisfactory protection mechanisms for the EI use case. When using the FL proposed by [20], there are risks of potential poisoning attacks to occur [32].

While the integrity of training data is important, the same applies to the integrity of the trained model. For EI, the training is performed on the edge, potentially by an untrusted entity. Such delegation of computing can raise serious questions of trust. The training party has complete authority over the training process and can potentially insert its own data into the training set or create a backdoor into the model. Such a backdoor can have incorrect and unexpected behavior on certain inputs specified by the malicious party. Furthermore, such a backdoor can be *provably undetectable* even if the user tries to validate the model's accuracy and robustness [33].

As discussed in Section 2.3.1, one of the potential solutions to the integrity issues is verifiable computing, which generates computational proofs and verifies the correctness of the outsourced information. However, the generation of proof for the correctness of training can be prohibitively expensive.

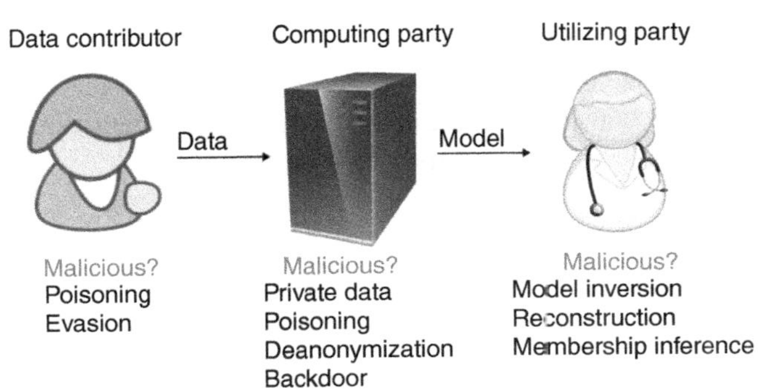

Figure 2.3 Challenges related to the security of machine learning. Source: Juha Partala.

2.3.3 Post-quantum Cryptography

Quantum computing has taken large steps toward practicality during the last decade. It is widely known that certain problems are easier to solve on a quantum computer than on a traditional one. Some of those problems underlie the contemporary public-key encryption algorithms. Therefore, a lot of research has recently concentrated on the security of cryptographic algorithms in the quantum computing model.

Since quantum computing will have a significant role in the context of 6G and EI, the employed security mechanisms should be designed in such a way that they should guarantee and fulfill the desired security requirements for the quantum computing models. Cryptographic primitives that are secure both in the standard computational model and the quantum model are called *post-quantum secure*. There have been recent efforts to standardize post-quantum primitives by various institutions, such as the National Institute of Standards and Technology (NIST) in the United States. However, post-quantum security requires a trade-off in terms of performance and efficiency.

In general, cryptography comes in two flavors: symmetric and asymmetric (or public-key cryptography). Quantum computing affects both, but mostly public-key primitives. Starting from 5G, public-key protocols have become an integral part of the core of wireless networks. Contemporary asymmetric cryptography is based on mathematical problems such as integer factorization and the discrete logarithm problem that can be solved quickly on a quantum computer. These methods need to be replaced with post-quantum-secure versions.

Symmetric cryptography involves primitives, such as encryption and message authentication. Fortunately, contemporary primitives are only lightly affected by quantum algorithms. However, some modifications are needed in order to preserve the current security level. Due to Grover's algorithm [34], the key length needs to be doubled in the quantum model (i.e. 128-bit encryption to 256-bit encryption). In practice, such an increase incurs a modest performance penalty. Otherwise, the same symmetric cryptographic primitives can be applied.

Asymmetric cryptography involves primitives such as key exchange, public-key encryption, and digital signatures. Contemporary primitives are typically based on elliptic curves and the elliptic curve Diffie–Hellman (ECDH) problem, which is relatively efficient and has compact public and private keys. These methods are not secure in the quantum computing model and need to be replaced with more efficient ones. Currently, there are different alternatives with their own pros and cons.

With regard to public-key cryptography in 6G EI, choices need to be made between methods that provide efficient key generation, relatively compact keys or signatures, or efficient operations (encryption, signing, etc.), but not all at the same time. Their efficient use in post-quantum-secure wireless networks remains a future challenge.

2.4 Privacy Challenges in 6G EI

The protection of privacy on the edge can be challenging. Devices collect huge amounts of personally identifiable information (PII), which can potentially be used to monitor every action of an individual. Thus, end-to-end encryption should be applied to protect the users from those who seek to enact harm, and computation should be performed only by trusted entities. Privacy laws, such as General Data Protection Regulation (GDPR), also require that PII needs to be protected. The GDPR also mandates that the system has to adopt a *privacy-by-design* approach; the whole system has to be designed from the ground up to protect the privacy of its users.

For ML models that are trained exclusively on sensitive information, the protection should not refer only to the samples used for training or to the input to a classifier. In many situations, the ML model itself needs to be protected since PII can be inferred directly from the model using, e.g. model inversion or extraction, where information about the training data is deduced from the model [35, 36] or membership inference, where the owner of a specific training data point is deduced [37]. In the contemporary setting, the training data are typically seen by the training party.

Privacy-preserving ML attempts to enable model training without the disclosure of the private training set. In general, there are two approaches to the protection of PII: its removal using anonymization techniques and protection using cryptographic protocols. Anonymization techniques [38–40] remove any PII from the training data by grouping individuals so that it is impossible to identify a single individual from the group. However, for certain use cases, such a grouping may render classification tasks impossible or limit their accuracy.

The most widely used anonymization approach to prevent the leakage of PII is differential privacy [41]. In general, differential privacy inserts noise into the training data in order to prevent the extraction of PII. It is an efficient method of preventing PII leakage but can lead to performance issues regarding training [42], as well as poor performance of the final model. In addition, differential privacy cannot be applied in all use cases; it has strict limitations for certain types of data [43] and can be totally inapplicable if the model is trained on data from a single individual.

Another widely adopted method for limiting access to the training data is to apply FL. In FL, training data are not transferred to a centralized server. Instead, a local model is trained close to the data collection. Then, multiple local models are transferred to the server, and a final global model is computed using the local models. Since the training data do not leave the collection site, FL offers better privacy guarantees than the centralized training approach. However, it should be noted that FL does not offer perfect privacy guarantees; the local models can be abused to infer information about the training data using, e.g. model inversion and membership inference attacks [37]. FL is also highly susceptible to poisoning attacks.

On the other hand, cryptographic schemes can also be applied to protect sensitive training data. For certain models, HE [26, 27] can be applied to perform the training on encrypted data. Thus, the confidentiality of both the training data and the final model is preserved. The deployment of the fully HE requires significant computational capabilities, but, in some use cases, its limited or lightweight version may be efficient and effective [44–46].

Functional encryption (FE) [47] is similar to HE, but the output of the computation is not hidden. It is also significantly more efficient than HE, but there are limitations: currently, it is not known how to enable efficient FE for other than linear or quadratic functions.

Secure multiparty computing (MPC) is another approach that can protect the confidentiality of the training data. In MPC, multiple parties jointly compute a function on hidden

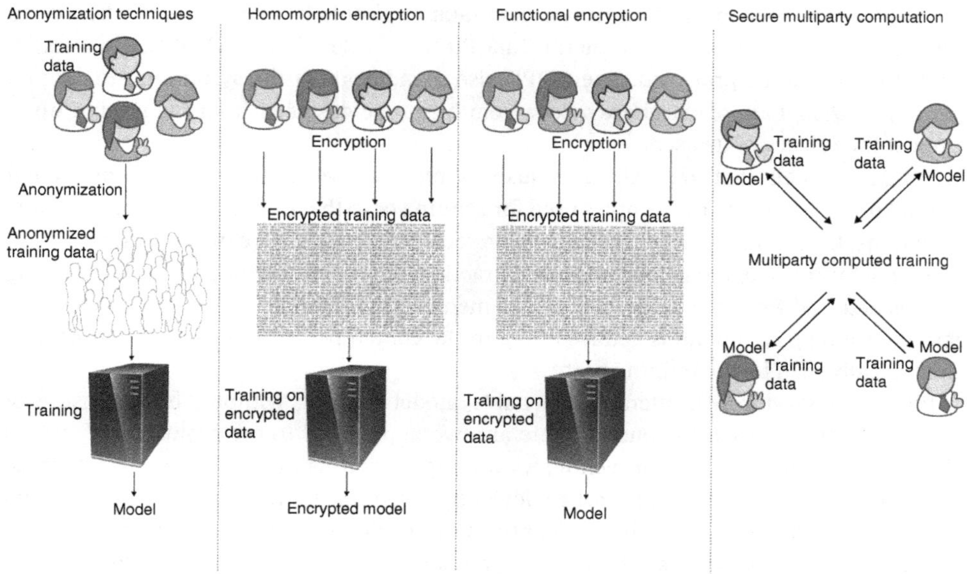

Figure 2.4 Methods for privacy-preserving machine learning. Source: Juha Partala.

data and learn only the output of the computation. Thus, the training data can remain confidential while the model training is performed by running MPC. The drawback of MPC is the large amount of communication required while running the protocol. If the number of parties is limited, the computation is relatively efficient – especially compared to HE – but does not scale well with the number of participants due to the communication costs. Contrary to HE and FE, secure MPC is an interactive protocol; it requires the computing parties to be online and is a feasible choice only if interactivity between the data contributors is possible.

The benefit of the cryptographic approach compared to anonymization is that no noise is added to the training data, resulting in better accuracy. However, it should be noted that the discussed methods are costly in terms of computation and communication, and further studies are needed in order to assess their feasibility in EI. Privacy-protection methods presented in this section have been collectively depicted in Figure 2.4.

2.5 Trust Challenges in 6G EI

Trust raises challenges for integration via the usage of EI in the 6G era. For example, 6G is expected to provide an interconnected computing platform where several network elements and service providers will require frequent collaboration and communication; it is highly vital to establish an embedded trust management mechanism among them. Also, EI systems demand communication and exchange of data and local models, which raise trust challenges in collecting, storing, and processing. Notably, trust formation is done via advanced technologies, ensuring security.

Trust is critical in the 6G era, especially when EI is in use. Applying EI cannot guarantee trust in a cloud-based system when the system needs communication from a set of different devices and resources. Also, with the next generation of IoT systems, the requirements for collaboration [48] raise the trust in communication, data generation, and computation. Hence, if a suitable trust management mechanism is placed, it will minimize the need for intermediaries and eventually help in the overall cost reduction of the heterogeneous massive-scale IoT application. Notably, the swift growth and the numerous demands of FL in EI raise trust challenges to deploy the technology in the 6G era. Therefore, the 6G white paper [49] mentions the importance of trust in increasing security in 6G via defining trust modeling, policies, and mechanisms. Also, [50, 51] point out the need for trust network topologies for forming EI services, including authentication, access control, data integrity, and mutual platform verification. The 6G architecture proposed by [52] follows the heterogeneous cloud (het-cloud) computing environment, which will comprise various subnetworks, edge clouds, private clouds, central clouds, and public clouds. In such an environment, trustworthiness challenges arise due to the heterogeneous and distributed het-cloud architecture components connected from different locations and operated/managed by diverse stakeholders. Besides, the huge scale of millions of subnetworks and billions of data sources (sensors) raises further trust challenges. ZTA [25] emphasizes trust management in every component of a communication system by requiring continuous verification and validation of the identity and security posture of users, devices, and applications at every interaction, ensuring that trust is never assumed and security is maintained throughout the communication process. Therefore, the creation of multivendor trust domains via the cloud stack and topologies that connect untrusted domains as subnetworks of data generation is required. In particular, the data generation side, including sensors, devices, and human–machine interfaces, is the origin of threats in trust data collection. From the network side, trust is a question for subnetworks that require secure definitions, such as authentication and identity.

The trust related to EI comes from data and models via procedures, including collection, storage, and processing. Due to the vulnerabilities of the sensor networks to attacks, the guarantee of trust in collected data is a huge question, especially when applying EI [53]. For example, due to the diversity of sensor networks, it may become hard to protect all the devices, and thus, the possibility of adversary attacks also increases. Besides that, bad environmental conditions can also cause hurdles in ensuring the correctness of collected data. Therefore, to maintain the required trust in collected data, it is highly critical to detect and recognize the sensor failures, which will help eliminate the noise data and improve the overall system performance. In addition, maintaining data storage raises security issues concerning the integrity and therefore requires suitable solutions. For instance, since the storage of collected data in centralized locations leads to security concerns due to data integrity or single-point failure of storage, a distrusted storage mechanism can be useful to increase the availability of data/resources and can improve the performance by performing dynamic load balancing. Interestingly, leveraging EI in heterogeneous networks leads to issues related to trust in exchange for training models. For example, when deploying FL, the most significant challenges affecting the trust are single-point failure of aggregators, malicious clients or false data, and lack of incentives [54].

The formation of trust is frequently based on reputation; however, with the growth of advanced technologies, trust can be shaped by novel technologies. In particular, the trust in current works is mainly based on logs revealing the history of operations. Blockchain technology [49, 51, 55] can be a promising solution for trust formation in the 6G era, especially when applying EI. Particularly, the collaboration of service providers requires trust, which can be enabled by leveraging blockchain technology [56, 57]. Moreover, due to the immutability and transparency, blockchain supports audits by keeping track of records to build trust in communication and verification. Also, via the integration between blockchain and FL [58–60], the trust of EI in 6G is an example of the promise of this technology to address trust challenges for 6G EI. In the context of trust-based hardware, trusted execution environments (TEE) are another notable trusted solution for confidentiality and integrity in specific network environments [48]. Like this consideration, [49] mention predefined liabilities as the trusted party to handle various liability-related challenges, e.g. safety and work health.

2.6 Security Standardization for EI and 6G

Security-related standardization for EI and 6G is still in its infancy. However, for data privacy and security in general, various standards, such as ISO/IEC TS 27570 [61] and ISO/IEC DIS 27400 [62], have been developed by, e.g. the International Organization for Standardization (IOS) and the International Electrotechnical Commission (IEC). The former provides guidelines and recommendations for the management of privacy and the usage of standards, while the latter provides guidance for principles and controls to provide private and secure IoT systems, services, and solutions [62]. Furthermore, the European Telecommunications Standards Institute (ETSI) has recently unveiled ETSI EN 303 645 [63] to provide cybersecurity standards and a baseline for IoT consumer products and certification schemes. While these guidelines do not specifically refer to Edge AI for 6G, they are applicable for developing private and secure edge AI models and algorithms to provide trustworthy products and services.

ETSI industry specification group (ISG) on multi-access edge computing (MEC) recently published a white paper on MEC security [64] highlighting the status of the ongoing standardization activities, as well as potential future trends for MEC security. Along with the evolution of the next-generation 6G networks, the report presented insight into the security standardization needed for edge computing-related technologies. For example, the privacy-preserving AI/ML algorithms at the edge network are vital for securing the users' critical information. Therefore, standardization efforts are needed to develop trustworthy AI/ML models for edge networks. Moreover, DLTs are an important security enabler for 6G and EI-enabled systems, e.g. by providing decentralized trust. To maximize the benefits of DLTs, efforts are required for standardizing various security-related processes, e.g. for smart contract-based mechanisms.

NIST has been at the forefront of working on various standards for post-quantum cryptography (PQC) solutions, e.g. standards for PQC algorithms that are acceptable for both quantum and classical systems [65]. Related to the 6G security standards, International Telecommunication Union (ITU) recently published a technical paper on the roadmap for 5G security standardization, where the emphasis was also given to analyzing and identifying potential gaps in 5G security standards [66]. 5G Infrastructure Association (5G IA) released a white paper presenting the European vision and recommendation on future 6G ecosystems for policymakers and businesses that cover various dimensions of 6G networks, including the need for network and service security and trustworthiness for next-generation 6G networks [67]. Furthermore, various research projects worldwide, such as 6G Flagship, HEXA-X, and MSIT 6G, among others, are aiming to contribute toward various security standardization activities in the coming future [68–70].

2.7 Conclusion

The incorporation of EI with future 6G networks is vital from the perspective of enabling massive-scale smart and connected IoT applications with the key requirements of ultralow latency, reliability, faster data analytics, and intelligent distributed decision-making. Such smart and hyperconnected digital environments are assumed to be highly complex and, therefore, vulnerable to various threats. For this, it is crucial to design and implement sufficient security mechanisms for securing intelligent 6G-enabled applications and communication architectures. In this chapter, we have provided an overview of the potential security, privacy, and trust issues for 6G EI systems, such as threats related to computational offloading and distributed service provisioning on the edge, vulnerabilities due to the complex integration of various enabling technologies, and trust issues when multiple devices and other stakeholders (service providers and network operators) share information and resources using a common platform. In addition to the identified security, privacy, and trust issues, the work also briefly presents some of the potential solutions to these issues. Moreover, a brief discussion about ongoing security-related standardization activities for EI and 6G systems is provided.

References

1 Akyildiz, I.F., Kak, A., and Nie, S. (2020). 6G and beyond: the future of wireless communications systems. *IEEE Access* 8: 133995–134030. https://doi.org/10.1109/ACCESS.2020.3010896.

2 Akhtar, M.W., Hassan, S.A., Ghaffar, R. et al. (2020). The shift to 6G communications: vision and requirements. *Human-centric Computing and Information Sciences* 10 (1): 1–27.

3 Letaief, K.B., Shi, Y., Lu, J., and Lu, J. (2022). Edge artificial intelligence for 6G: vision, enabling technologies, and applications. *IEEE Journal on Selected Areas in Communications* 40 (1): 5–36. https://doi.org/10.1109/JSAC.2021.3126076.

4 Bhat, J.R. and Alqahtani, S.A. (2021). 6G ecosystem: current status and future perspective. *IEEE Access* 9: 43134–43167. https://doi.org/10.1109/ACCESS.2021.3054833.

5 Khan, L.U., Saad, W., Niyato, D. et al. (2022). Digital-twin-enabled 6G: vision, architectural trends, and future directions. *IEEE Communications Magazine* 60 (1): 74–80. https://doi.org/10.1109/MCOM.001.21143.

6 Zhu, P., Xu, J., Li, J. et al. (2021). Learning-empowered privacy preservation in beyond 5G edge intelligence networks. *IEEE Wireless Communications* 28 (2): 12–18. https://doi.org/10.1109/MWC.001.2000331.

7 Mukherjee, M., Matam, R., Mavromoustakis, C.X. et al. (2020). Intelligent edge computing: security and privacy challenges. *IEEE Communications Magazine* 58 (9): 26–31. https://doi.org/10.1109/MCOM.001.2000297.

8 Siriwardhana, Y., Porambage, P., Liyanage, M., and Ylianttila, M. (2021). AI and 6G security: opportunities and challenges. In: *2021 Joint European Conference on Networks and Communications & 6G Summit (EuCNC/6G Summit)*, 616–621. https://doi.org/10.1109/EuCNC/6GSummit51104.2021.9482503.

9 Adhikari, M., Munusamy, A., Hazra, A. et al. (2022). Security in edge-centric intelligent internet of vehicles: issues and remedies. *IEEE Consumer Electronics Magazine* 11 (6): 24–31. https://doi.org/10.1109/MCE.2021.3116415.

10 Sun, Y., Liu, J., Wang, J. et al. (2020). When machine learning meets privacy in 6G: a survey. *IEEE Communications Surveys & Tutorials* 22 (4): 2694–2724. https://doi.org/10.1109/COMST.2020.3011561.

11 Botta, A., de Donato, W., Persico, V., and Pescapé, A. (2016). Integration of Cloud computing and Internet of Things: a survey. *Future Generation Computer Systems* 56: 684–700.

12 Stergiou, C., Psannis, K.E., Kim, B.-G., and Gupta, B. (2018). Secure integration of IoT and cloud computing. *Future Generation Computer Systems* 78: 964–975.

13 Dizdarevic, J., Carpio, F., Jukan, A., and Masip-Bruin, X. (2019). A survey of communication protocols for Internet of Things and related challenges of fog and cloud computing integration. *ACM Computing Surveys* 51: 1–29. https://doi.org/10.1145/3292674.

14 Laroui, M., Nour, B., Moungla, H. et al. (2021). Edge and fog computing for IoT: a survey on current research activities & future directions. *Computer Communications* 180: 210–231. https://doi.org/10.1016/j.comcom.2021.09.003.

15 Pan, J. and McElhannon, J. (2018). Future edge cloud and edge computing for Internet of Things applications. *IEEE Internet of Things Journal* 5 (1): 439–449.

16 Tefera, G., She, K., Shelke, M., and Ahmed, A. (2021). Decentralized adaptive resource-aware computation offloading & caching for multi-access edge computing networks. *Sustainable Computing Informatics & Systems* 30: 100555.

17 Deng, S., Zhao, H., Fang, W. et al. (2020). Edge intelligence: the confluence of edge computing and artificial intelligence. *IEEE Internet of Things Journal* 7 (8): 7457–7469.

18 Zhang, X., Wang, Y., Lu, S. et al. (2019). OpenEI: An open framework for edge intelligence. *IEEE 39th International Conference on Distributed Computing Systems (ICDCS).*

19 Nain, G., Pattanaik, K.K., and Sharma, G.K. (2022). Towards edge computing in intelligent manufacturing: past, present and future. *Journal of Manufacturing Systems* 62: 588–611.

20 Zhu, G., Liu, D., Du, Y. et al. (2020). Towards an intelligent edge: wireless communication meets machine learning. *IEEE Communications Magazine* 58 (1): 19–444.

21 Qadir, Z., Le, K.N., Saeed, N., and Munawar, H.S. (2022). Towards 6G Internet of Things: recent advances, use cases, and open challenges. *ICT Express* 9 (3): 296–312.

22 Ray, P.P. (2021). A perspective on 6G: requirement, technology, enablers, challenges and future road map. *Journal of Systems Architecture* 118: 102180.

23 Zhang, S. and Zhu, D. (2020). Towards artificial intelligence enabled 6g: State of the art, challenges, and opportunities. *Computer Networks* 183: 107556.

24 Taleb, T., Benzaïd, C., Lopez, M.B. et al. (2022). 6G system architecture: a service of services vision. *ITU Journal on Future and Evolving Technologies* 3 (3): 710–743.

25 Syed, N.F., Shah, S.W., Shaghaghi, A. et al. (2022). Zero trust architecture (ZTA): a comprehensive survey. *IEEE Access* 10: 57143–57179. https://doi.org/10.1109/ACCESS.2022.3174679.

26 Craig Gentry. Fully homomorphic encryption using ideal lattices. In *Proceedings of the 41st Annual ACM Symposium on Theory of Computing*, STOC '09, page 169–178, New York, 2009. *Association for Computing Machinery*. ISBN 9781605585062. https://doi.org/10.1145/1536414.1536440.

27 Rivest, R.L., Adleman, L., and Dertouzos, M.L. (1978). On data banks and privacy homomorphisms. *Foundations of secure computation* 4 (11): 169–180.

28 Babai, L., Fortnow, L., Levin, L.A., and Szegedy, M. (1991). Checking computations in polylogarithmic time. In: *Proceedings of the 23rd Annual ACM Symposium on Theory of Computing*, STOC '91, 21–32. New York: Association for Computing Machinery. ISBN: 0897913973. https://doi.org/10.1145/103418.103428.

29 Gennaro, R. (2017). Verifiable outsourced computation: a survey. In: *Proceedings of the ACM Symposium on Principles of Distributed Computing*, PODC '17, 313. New York: Association for Computing Machinery. ISBN: 9781450349925 https://doi.org/10.1145/3087801.3087872.

30 Biggio, B., Fumera, G., and Roli, F. (2014). Security evaluation of pattern classifiers under attack. *IEEE Transactions on Knowledge and Data Engineering* 26 (4): 984–996. https://doi.org/10.1109/TKDE.2013.57.

31 Pitropakis, N., Panaousis, E., Giannetsos, T. et al. (2019). A taxonomy and survey of attacks against machine learning. *Computer Science Review* 34: 100199. https://doi.org/10.1016/j.cosrev.2019.100199.

32 Tolpegin, V., Truex, S., Gursoy, M.E., and Liu, L. (2020). Data poisoning attacks against federated learning systems. In: *Computer Security – ESORICS 2020* (ed. L. Chen, N. Li, K. Liang, and S. Schneider), 480–501. Cham: Springer International Publishing. ISBN: 978-3-030-58951-6.

33 Goldwasser, S., Kim, M.P., Vaikuntanathan, V., and Zamir, O. (2022). Planting undetectable backdoors in machine learning models. https://arxiv.org/abs/2204.06974.

34 Grover, L.K. (1996). A fast quantum mechanical algorithm for database search. In: *Proceedings of the 28th Annual ACM Symposium on Theory of Computing*, STOC '96, 212–219. New York, NY, USA: Association for Computing Machinery. ISBN: 0897917855 https://doi.org/10.1145/237814.237866.

35 Fredrikson, M., Jha, S., and Ristenpart, T. (2015). Model inversion attacks that exploit confidence information and basic countermeasures. In: *Proceedings of the 22nd ACM SIGSAC Conference on Computer and Communications Security*, CCS '15, 1322–1333. New York: Association for Computing Machinery. ISBN: 9781450338325. https://doi.org/10.1145/2810103.2813677.

36 Fredrikson, M., Lantz, E., Jha, S. et al. (2014). Privacy in pharmacogenetics: an end-to-end case study of personalized warfarin dosing. In: *23rd USENIX Security Symposium (USENIX Security 14)*, 17–32. San Diego, CA: USENIX Association. ISBN: 978-1-931971-15-7.

37 Shokri, R., Stronati, M., Song, C., and Shmatikov, V. (2017). Membership inference attacks against machine learning models. In: *2017 IEEE Symposium on Security and Privacy (SP)*, 3–18. https://doi.org/10.1109/SP.2017.41.

38 Li, N., Li, T., and Venkatasubramanian, S. (2007). t-Closeness: privacy beyond k-anonymity and l-diversity. In: *2007 IEEE 23rd International Conference on Data Engineering*, 106–115.

39 Machanavajjhala, A., Kifer, D., Gehrke, J., and Venkitasubramaniam, M. (2007). L-diversity: privacy beyond k-anonymity. *ACM Transactions on Knowledge Discovery from Data* 1 (1): 3–es. https://doi.org/10.1145/1217299.1217302.

40 Samarati, P. and Sweeney, L. (1998). Protecting privacy when disclosing information: k-anonymity and its enforcement through generalization and suppression.

41 Dwork, C., McSherry, F., Nissim, K., and Smith, A. (2006). Calibrating noise to sensitivity in private data analysis. In: *Theory of Cryptography* (ed. S. Halevi and T. Rabin), 265–284. Berlin, Heidelberg: Springer-Verlag. ISBN: 978-3-540-32732-5.

42 Wei, K., Li, J., Ding, M. et al. (2020). Federated learning with differential privacy: algorithms and performance analysis. *IEEE Transactions on Information Forensics and Security* 15: 3454–3469. https://doi.org/10.1109/TIFS.2020.2988575.

43 Liu, B., Ding, M., Shaham, S. et al. (2021). When machine learning meets privacy: A survey and outlook. *ACM Computing Surveys* 54 (2): https://doi.org/10.1145/3436755.

44 Gilad-Bachrach, R., Dowlin, N., Laine, K. et al. (2016). CryptoNets: Applying neural networks to encrypted data with high throughput and accuracy. In: *Proceedings of the 33rd International Conference on Machine Learning, Proceedings of Machine Learning Research*, vol. 48 (ed. M.F. Balcan and K.Q. Weinberger), 201–210. New York: PMLR. https://proceedings.mlr.press/v48/gilad-bachrach16.html.

45 Liu, X., Lu, R., Ma, J. et al. (2016). Privacy-preserving patient-centric clinical decision support system on naïve Bayesian classification. *IEEE Journal of Biomedical and Health Informatics* 20 (2): 655–668. https://doi.org/10.1109/JBHI.2015.2407157.

46 Phong, L.T., Aono, Y., Hayashi, T. et al. (2018). Privacy-preserving deep learning via additively homomorphic encryption. *IEEE Transactions on Information Forensics and Security* 13 (5): 1333–1345. https://doi.org/10.1109/TIFS.2017.2787987.

47 Boneh, D., Sahai, A., and Waters, B. (2011). Functional encryption: Definitions and challenges. In: *Theory of Cryptography* (ed. Y. Ishai), 253–273. Berlin, Heidelberg: Springer-Verlag. ISBN: 978-3-642-19571-6.

48 Ziegler, V., Schneider, P., Viswanathan, H. et al. (2021). Security and trust in the 6G era. *IEEE Access* 9: 142314–142327. https://doi.org/10.1109/ACCESS.2021.3120143.

49 Ylianttila, M., Kantola, R., Gurtov, A. et al. (2020). 6G white paper: research challenges for trust, security and privacy. https://arxiv.org/abs/2004.11665.

50 Porambage, P., Gür, G., Osorio, D.P.M. et al. (2021). 6G security challenges and potential solutions. In: *2021 Joint European Conference on Networks and Communications & 6G Summit (EuCNC/6G Summit)*, 622–627. https://doi.org/10.1109/EuCNC/6GSummit51104.2021.9482609.

51 Porambage, P., Gür, G., Osorio, D.P.M. et al. (2021). The roadmap to 6G security and privacy. *IEEE Open Journal of the Communications Society* 2: 1094–1122. https://doi.org/10.1109/OJCOMS.2021.3078081.

52 Ziegler, V., Viswanathan, H., Flinck, H. et al. (2020). 6G architecture to connect the worlds. *IEEE Access* 8: 173508–173520. https://doi.org/10.1109/ACCESS.2020.3025032.

53 Zhang, G., Wang, T., Wang, G. et al. (2021). Detection of hidden data attacks combined fog computing and trust evaluation method in sensor-cloud system. *Concurrency and Computation: Practice and Experience* 33 (7): e5109. https://doi.org/10.1002/cpe.5109.e5109cpe.5109.

54 Bagdasaryan, E., Veit, A., Hua, Y. et al. (2020). How to backdoor federated learning. In: *Proceedings of the 23rd International Conference on Artificial Intelligence and Statistics*, Proceedings of Machine Learning Research, vol. 108 (ed. S. Chiappa and R. Calandra), 2938–2948 (26–28 August 2020). PMLR.

55 Nguyen, T., Lovén, L., Partala, J., and Pirttikangas, S. (2021). *The Intersection of Blockchain and 6G Technologies*, 393–417. Cham: Springer International Publishing. ISBN: 978-3-030-72777-2. https://doi.org/10.1007/978-3-030-72777-2.18.

56 Nguyen, H., Nguyen, T., Leppänen, T. et al. (2022). Situation awareness for autonomous vehicles using blockchain-based service cooperation. In: *Advanced Information Systems Engineering* (ed. X. Franch, G. Poels, F. Gailly, and M. Snoeck), 501–516. Cham: Springer International Publishing. ISBN: 978-3-031-07472-1.

57 Nguyen, T.H., Partala, J., and Pirttikangas, S. (2019). Blockchain-based mobility-as-a-service. In: *2019 28th International Conference on Computer Communication and Networks (ICCCN)*, 1–6. https://doi.org/10.1109/ICCCN.2019.8847027.

58 Kim, H., Park, J., Bennis, M., and Kim, S. (2020). Blockchained on-device federated learning. *IEEE Communications Letters* 24 (6): 1279–1283. https://doi.org/10.1109/LCOMM.2019.2921755.

59 Lu, Y., Huang, X., Dai, Y. et al. (2020). Blockchain and federated learning for privacy-preserved data sharing in industrial IoT. *IEEE Transactions on Industrial Informatics* 16 (6): 4177–4186. https://doi.org/10.1109/TII.2019.2942190.

60 Nguyen, D.C., Ding, M., Pham, Q.-V. et al. (2021). Federated learning meets blockchain in edge computing: opportunities and challenges. *IEEE Internet of Things Journal* 8 (16): 12806–12825. https://doi.org/10.1109/JIOT.2021.3072611.

61 ISO/IEC (2022). ISO/IEC TS 27570:2021. Privacy protection – Privacy guidelines for smart cities. *Guidelines.*

62 ISO/IEC (2022). ISO/IEC 27400:2022. Cybersecurity – IoT Security and Privacy – Guidelines. Guidelines.

63 ETSI (2020). CYBER; Cyber Security for Consumer Internet of Things: Baseline Requirements. *ETSI EN 303 645 V2.1.1 (2020-06).*

64 ETSI (2022). MEC security; Status of standards support and future evolutions. *ETSI White Paper No. #46.*

65 Alagic, G., Alperin-Sheriff, J., Apon, D. et al. (2020). Status report on the second round of the NIST post-quantum cryptography standardization process. US Department of Commerce, NIST.

66 ITU (2022). XSTP-5GsecRM 5G security standardization roadmap. ITU-T, Technical Paper.

67 The 5G Infrastructure Association (2021). European Vision for the 6G Network Ecosystem. *White Paper*.

68 Castro, C. (2020). Korea lays out plan to become the first country to launch 6G. https://www.6gworld.com/exclusives/korea-lays-out-plan-to-become-the-first-country-tolaunch-6g/ (accessed 25 March 2023).

69 Katz, M., Matinmikko-Blue, M., and Latva-Aho, M. (2018). 6Genesis flagship program: building the bridges towards 6g-enabled wireless smart society and ecosystem. In: *2018 IEEE 10th Latin-American Conference on Communications (LATINCOM)*, 1–9. https://doi.org/10.1109/LATINCOM.2018.8613209.

70 Uusitalo, M.A., Ericson, M., Richerzhagen, B. et al. (2021). Hexa-X the European 6G flagship project. In: *2021 Joint European Conference on Networks and Communications & 6G Summit (EuCNC/6G Summit)*, 580–585. https://doi.org/10.1109/EuCNC/6GSummit51104.2021.9482430.

3

Privacy-Preserving Machine Learning for Massive IoT Deployments

Najwa Aaraj, Abdelrahaman Aly, Alvaro Garcia-Banda, Chiara Marcolla, Victor Sucasas, and Ajith Suresh

Cryptography Research Center, Technology Innovation Institute, Abu Dhabi, UAE

3.1 Introduction

Internet of Things (IoT) is changing the way data is gathered and processed. In particular, massive IoT deployments of sensing devices will gather vast sources of data in real time, and actuators will automate the response without human interaction. The combination of sensors and actuators can be a cost-effective solution to complex problems, for a diverse set of scenarios such as agriculture, urban management, supply chains, body area networks, etc. This will fuel the market penetration of new paradigms such as Smart Farming, Smart Cities, Smart Manufacturing or eHealth, among others. It is clear that massive IoT will be a cornerstone component in the path toward making old technologies "Smart." However, new technologies also bring new challenges, and in the case of massive IoT, one of these challenges is *privacy*.

The problem of privacy in IoT stems from the limited hardware capabilities of IoT sensor networks. Sensors have low computational power and are battery limited. Hence, data processing is frequently pushed to the cloud or to edge devices connected to the power grid. In terms of security, data in transit from IoT devices to the cloud can be protected with conventional cryptography, i.e. encrypted and authenticated, hence preventing data extraction and tampering. However, for data processing, the data must be decrypted, and this opens a window of opportunity for untrusted entities. The privacy concern is twofold: (i) external attackers can extract data from the cloud service; and (ii) even if not extracted, the cloud service learns the data, demanding a high level of trust in the service provider (see Figure 3.1).

This privacy conundrum can be solved by adopting *private* data processing techniques that enable operations to be performed on data without learning it. Some solutions have already been proposed in the form of synthetic data, data aggregation, randomization, and masking [1–3]. Conceptually, these ideas are based on: (i) transforming original data into new "random-looking" data but with similar statistical properties; (ii) removing identifiers from data sets; or (iii) pushing part of the processing to the edge, into the trusted area. However, the accuracy of the data processing algorithms can be significantly degraded with these approaches and the privacy guarantees are questionable.

Security and Privacy for 6G Massive IoT, First Edition.
Edited by Georgios Mantas, Firooz B. Saghezchi, Jonathan Rodriguez, and Victor Sucasas.

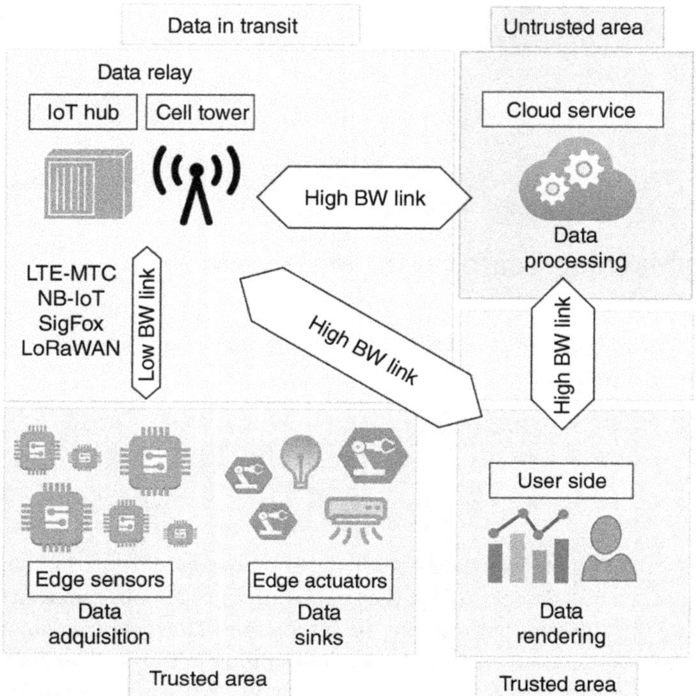

Figure 3.1 IoT data flow and trust. Data at rest in the edge (sensors, actuators, and user's devices) is secure, if located in our trusted area. Data in transit can be protected by means of conventional cryptography. Data processing in the cloud or external edge devices requires trust in external entities.

In this chapter, we explore two novel techniques within the realm of Privacy Enhancing Technologies (PETs) that address privacy concerns in IoT, and that can scale to massive IoT. Specifically, we focus on secure Multiparty Computation (MPC) and Fully Homomorphic Encryption (FHE), which have thrived in the last few years and paved the way toward efficient Privacy-Preserving Machine Learning (PPML) frameworks. This chapter centers around PPML, explaining the main functionalities and building blocks of MPC and FHE, without entering into mathematical nuance. It also discusses the latest PPML frameworks and explains how FHE and MPC can be applied to IoT while complying with the hardware and bandwidth limitations in massive IoT.

3.2 PPML for IoT

Due to limited computational resources, IoT devices often outsource complex machine learning (ML) algorithms to cloud servers, risking privacy breaches through data leakage. PPML aims at solving this problem by enabling oblivious data processing, i.e. without learning the data. Note that privacy is necessary not only during the process of data classification (also called inference), but also during the training phase of ML algorithms.

Specifically, when the ML model requires data of sensitive nature from diverse scattered locations. For instance, an ML model detecting a specific disease may need diverse positive and negative samples from multiple medical institutions worldwide, which may be protected by different privacy regulations. A similar scenario can be considered for IoT networks as well, where sensors from scattered locations and diverse ownership may participate collaboratively in training a single ML model. In addressing this problem, various Privacy Enhancing Technologies (PETs) have emerged, including secure MPC, Fully Homomorphic Encryption (FHE), Zero-Knowledge Proofs, Differential Privacy, and Federated Learning. However, in this chapter, our focus will be on MPC and FHE. These two techniques have received significant research attention in the field of PPML over the past few years.

Regarding the adoption of MPC or FHE to the IoT scenario, both technologies offer privacy but with different costs and guarantees. MPC is an interactive technique that requires multiple communication rounds between the IoT device and the cloud server. Also, majority of the MPC protocols require similar computational load in all parties, i.e. in the IoT device/sensor and the cloud server, which can be problematic for a hardware-limited IoT device when evaluating a complex ML model. The computational power requirements in FHE schemes, on the other hand, are asymmetric. This means that the IoT sensor simply needs to encrypt the data, and the cloud server handles all of the FHE overhead for private data processing. Moreover, FHE requires transmitting the encrypted data only once. On the downside, FHE offers less security and privacy guarantees than MPC and has the same side effect of significantly increasing bandwidth consumption. Although FHE only sends

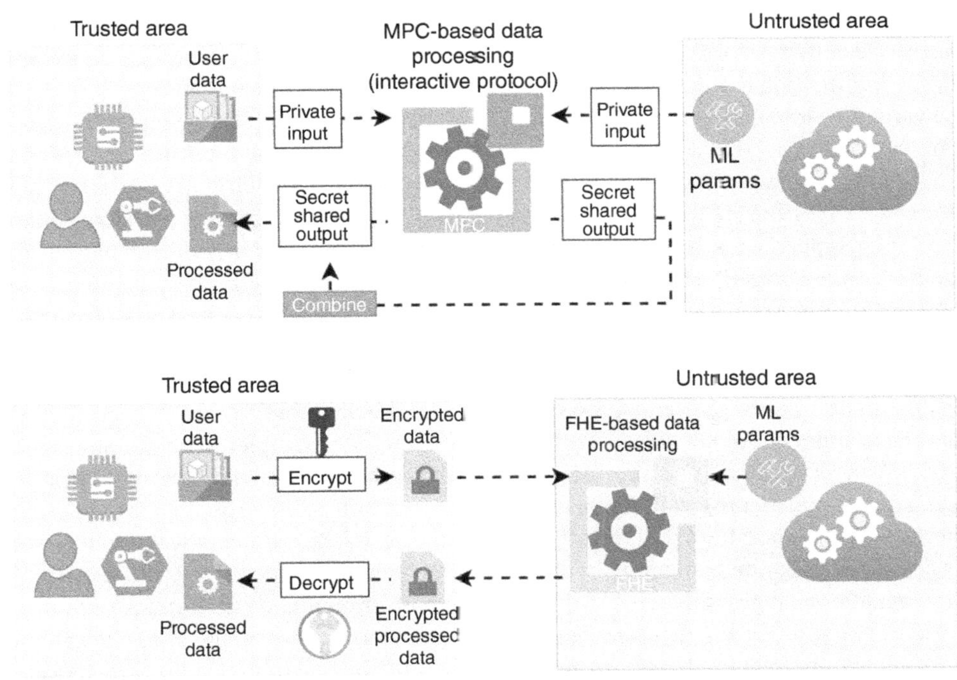

Figure 3.2 Private data processing by means of: (a) MPC and (b) FHE.

data once, for every bit of data encrypted, the ciphertext contains a bit-string several orders of magnitude larger than the data in clear. This is called ciphertext expansion, and it is a serious impairment for bandwidth limited scenarios such as IoT. Section 3.8 details how to bypass these limitations in order to make MPC and FHE compatible with the hardware and bandwidth constraints of IoT. Figure 3.2 details the data flows of both FHE and MPC for the IoT scenario.

3.3 Secure Multiparty Computation

Secure MPC was initially conceived in 1982 by Andrew Yao when exposing the well-known millionaires problem. Yao described a scenario where two millionaires wanted to know who is richer without disclosing any additional information about the actual values in their accounts. There was no feasible solution using conventional cryptography, which led to Yao coming up with the first-ever description of a theoretical solution to secure two-party computation in the form of Garbled Circuits (GC) [4]. Yao devised a bit-wise comparison circuit composed of interconnected Boolean gates that are garbled into encrypted truth tables; these tables are evaluated (decrypted) with keys obtained by evaluating previous garbled gates, except for the initial layer of the circuit that requires a series of Oblivious Transfer (OT) protocols [5, 6] between the two parties. Although Yao's GCs were considered impractical due to its computational complexity, his idea triggered a new research paradigm and fostered novel designs for GCs [7, 8] followed by an extension to scenarios with more than two parties and an entirely new branch of MPC protocols based on secret sharing [9, 10].

MPC is a technique that evaluates a function in a distributive fashion, between a set of parties, such that parties cannot obtain any information about other parties' inputs except for the information that can be inferred directly from the function's output.

Figure 3.3 shows a toy example of garbling a circuit composed of only one AND gate. One party acts as garbled circuit generator (garbler) by encrypting the truth table of the AND gate. Each input wire is associated with two random symmetric keys representing "0" and "1" (in the figure, reflected in subindex of w_0 and w_1). Using these keys, the garbler encrypts the truth table corresponding to the four possible input combinations of the AND gate. The garbled gate is given to the other party for evaluation, together with the key representing the garbler's private input "0" or "1" (note that the key representing the garbler's input is a random value, hence it does not leak information about the garbler's private input). The second party, the evaluator, must obtain the key corresponding to its private input from the garbler to proceed with the local evaluation. Toward this, the parties use OT technique that allows a party (the evaluator in our example) to obtain one value among several values offered by a second party (the garbler). OT guarantees that the party picking one value learns nothing about the unchosen values, whereas the party offering the values learns nothing about the other party's choice. In the last step, the evaluator uses its keys to decrypt one of the rows of the table and obtain the output of the gate. Note that the evaluator must give this value to the garbler if both entities need to learn the output.

Figure 3.4 shows a simple example of how to extend a garbled circuit from one gate to three interconnected AND gates. Instead of encrypting the real output value of each gate, the garbled gates contain encryptions of the wire keys for the subsequent gates (note that tables of first two gates encrypt w_0^g or w_1^g and w_0^h or w_1^h, respectively). This approach can be

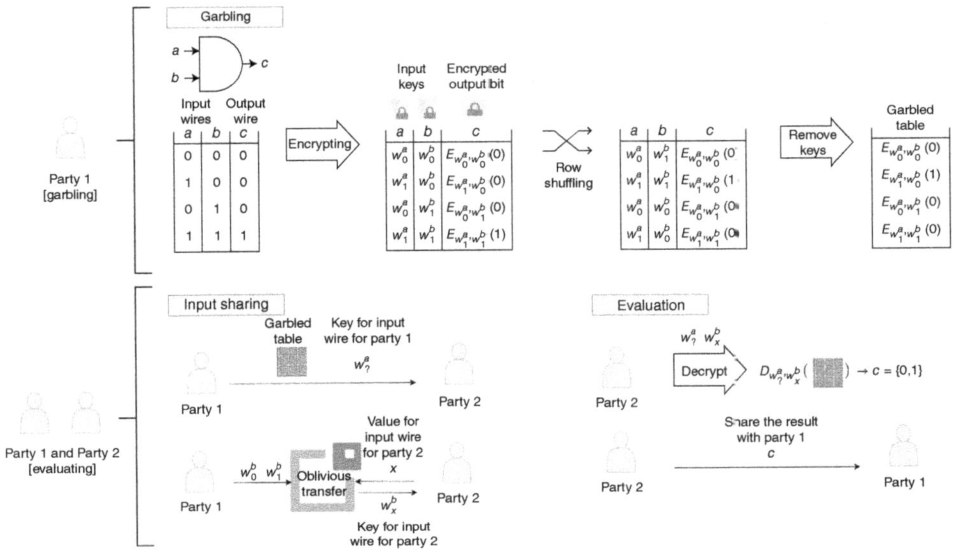

Figure 3.3 Example of Garble Circuits composed of a simple AND gate.

used to garble unlimited number of gates of different nature (not only ANDs), and hence get garbled circuits. It is worth commenting that many optimizations have been proposed to reduce the size of the garbled circuits [11, 12], and also to make them work with more than two parties [13, 14].

Garbled circuits are quite convenient for Boolean operations, or for arithmetic circuits that require bit-wise decomposition (e.g. to perform private comparisons). However, secret sharing-based MPC is more efficient for other arithmetic operations. Values can be secret-shared over either a ring or field, and the private computations are done directly over the secret shares held by different parties. Reconstruction of secret-shared values is only performed at the end of the protocol to obtain the output. Figure 3.5 shows a simple MPC protocol, BGW [9], based on Shamir secret sharing [15]. To share a private input x, a party selects a random polynomial $p()$ of degree "t" with $p(0) = x$. Then it distributes the shares, where each party P_i gets the value $p(i)$ as its share. Note that if any of the $(t + 1)$ parties disclose their shares, the value x can be reconstructed as $p(0) = \sum_{i=1}^{t+1} \Delta_i \cdot p(i)$.[1] Note that

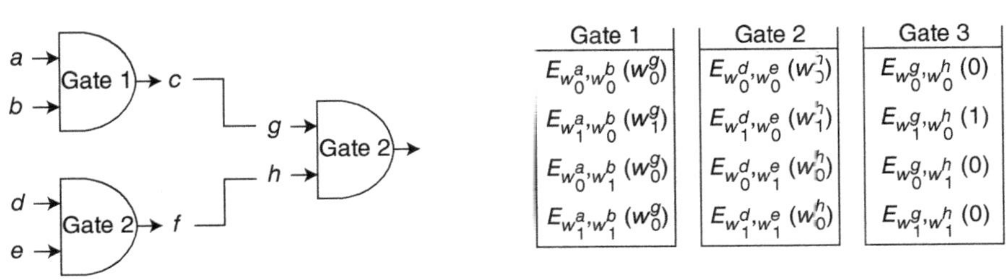

Figure 3.4 Example of a Garbled Circuit (GC) with three AND gates. Input garbled gates (and also intermediate gates in bigger circuits) encrypt keys of the garbled gates in the subsequent level.

1 The value Δ denotes the Lagrange coefficient.

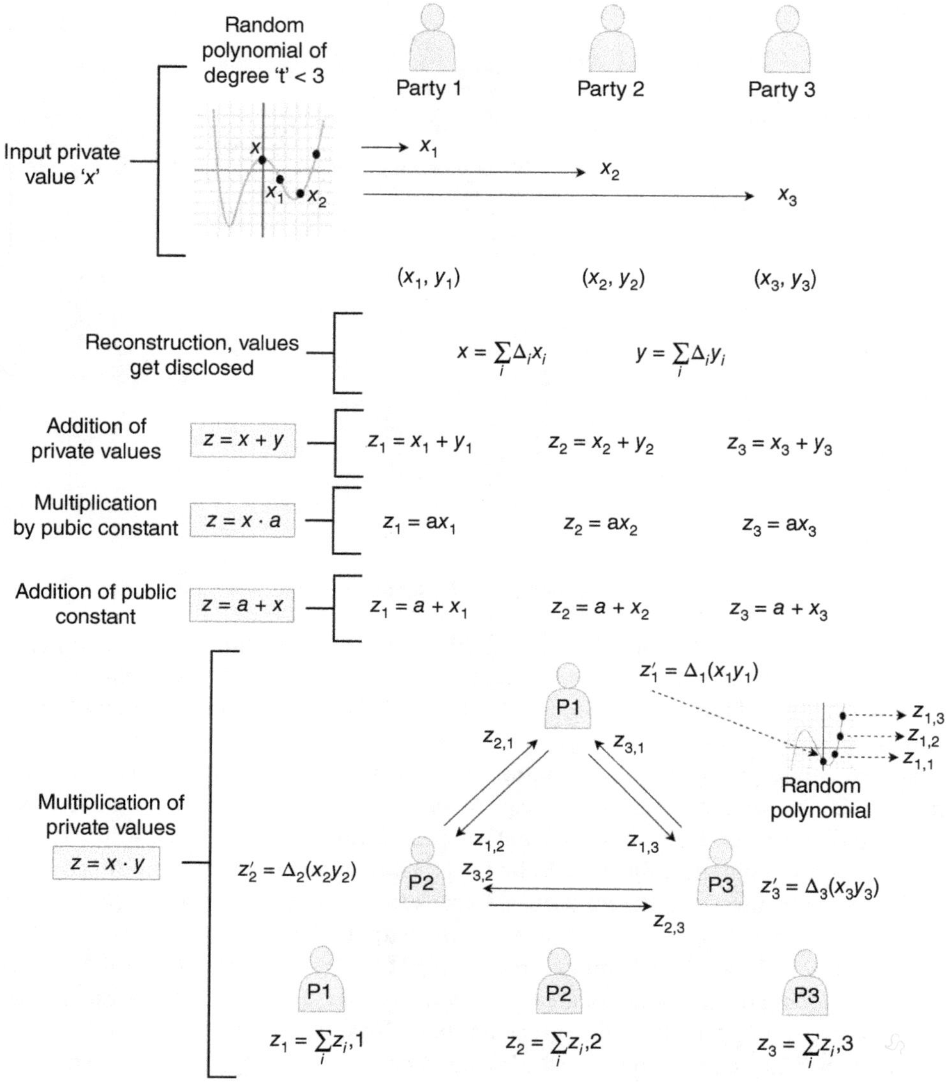

Figure 3.5 Basic operations in a secret-shared MPC scheme with Shamir secret sharing. There is an initial phase of input sharing where each party selects a random polynomial to split its input in shares. Then several operations can be performed with local computations (no communication). However, multiplication of private values requires communication between parties.

the polynomial degree must fulfill $t < n$, where n is the number of parties, otherwise the value cannot be reconstructed. As described in Figure 3.5, addition of private values and multiplication or addition by public constants are straightforward and requires only local computations.

However, for multiplication of private values, the parties must perform several communications. Locally multiplying private shares of two private inputs x and y, i.e. $x_i y_i$, yields secret shares of the multiplication $z' = xy$, but with a polynomial of degree $2t$. This implies that the reconstruction requires $2t + 1$ shares, hence $2t + 1$ parties. Although it is possible to select $t < n/2$, every consecutive multiplication would reduce t even further. To solve this problem, each party P_i re-shares the value $\Delta_i(x_i y_i)$ using a new random polynomial of degree t, as shown in Figure 3.5.

Later, in [16], a new form of multiplication for secret sharing-based MPC protocols was proposed, which has been adopted as the norm in modern protocols. It is based on multiplication triples often referred to as *Beaver triples*, and it pushes the communication rounds to a preprocessing stage, independent of the actual inputs. Parties first obtain secret-shares of random elements a, b, along with their multiplication $c = ab$. The tuple $[a, b, c]$ is called a multiplication triple, and it can be used to multiply any two private values x and y with only two broadcast communications (hence avoiding the pairwise channels). The values a and b act as one-time mask for the inputs x and y, respectively, as shown in Figure 3.6. Once the values $(x - a)$ and $(y - b)$ are publicly reconstructed, the parties can locally compute the secret shares for the output $z = xy$. Multiplication triples can be obtained in several ways: (i) with an MPC protocol over random values; (ii) with an OT protocol [17]; (iii) with Levelled Homomorphic Encryption (LHE) [10]; and (iv) with additive homomorphic encryption [18].

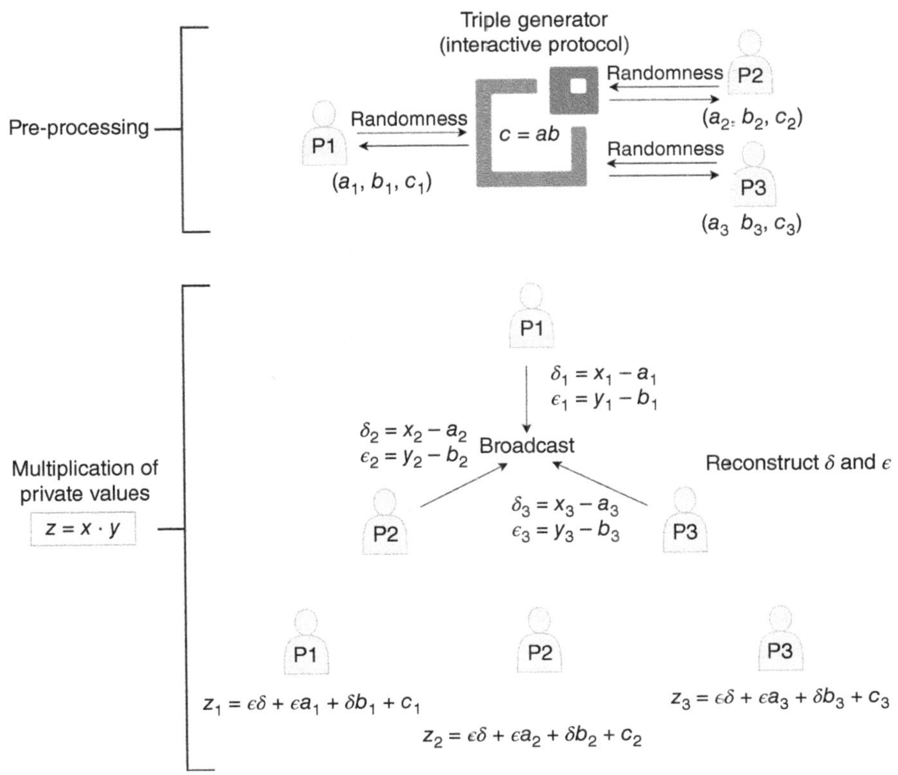

Figure 3.6 Optimization of private multiplication in secret-sharing MPC, based on multiplication triples. There is a preprocessing phase where triples are obtained, and then an online phase where each multiplication consumes one triple.

The protocols described above are toy examples, and present protocols that work with passive security (i.e. they assume that the parties follow the protocol strictly, see Section 3.3.1). Modern protocols, with higher privacy guarantees and efficiency, have been proposed in the last two decades following the main concepts of secret sharing (commonly used for arithmetic circuits) and garbled circuits (used for Boolean circuits). Also, novel research works have provided new insights into the theoretical boundaries of MPC in different adversarial models while efficient optimizations have paved the way for practical implementations. These implementations cover, among others, statistical evaluation of private data, private market auctions, voting systems, privacy-preserving predictive systems and, more recently, virtualization of cryptographic modules in a distributive manner [19].

3.3.1 MPC: Security Models and Setups

MPC protocols can be designed with different security guarantees, depending on various orthogonal traits. One such trait classifies the protocols based on the capability of the adversary, who takes control over a subset of the parties. The adversary is categorized into one of the following three types:

- **Semi-honest**: The adversary correctly follows the protocol specification but it may attempt to learn additional information by analyzing the transcript of messages received during the protocol execution. Protocols that assume semi-honest adversaries are referred to as passively secure.
- **Malicious**: The adversary may behave arbitrarily, it may deviate from the instructions of the MPC protocol. Protocols that assume malicious adversaries are referred to as actively secure.
- **Covert**: The adversary may behave arbitrarily, similar to the malicious case, but is guaranteed to be caught with non-negligible probability. The main difference between active security and the covert model is that in the former, any cheat or protocol deviation by one of the parties is detected by honest parties, except with some negligible probability.

Protocols in the malicious model guarantee the strongest security while the ones in the semi-honest model provide the weakest as they assume correct behavior from all the parties. The middle ground, namely covert model, are designed to detect cheating parties when honest parties identify any attack, effectively encouraging potential malicious entities to behave correctly [20].

In another trait, MPC protocols are classified by the number of parties—two parties (2PC) or multi-party (MPC)—and the number of dishonest parties they support. There are mainly three categories based on this classification: (i) at least one honest party (normally referred to as dishonest majority); (ii) more than half of parties to be honest (i.e. honest majority); or (iii) more than two thirds of the parties to be honest. The complexity of the protocol decreases with an increase in the number of honest parties.

In summary, MPC protocols that support active security and deal with a dishonest majority are the most complex. Conversely, passively secure protocols with an honest majority are the simplest. Now, let us examine the privacy and output guarantees of MPC protocols, considering both of these characteristics in combination.

3.3.2 MPC: Privacy and Output Guarantees

Based on the privacy guarantees, MPC protocols can be classified into three categories: (i) perfect, (ii) statistical, and (iii) computational. With perfect security, no adversary can break privacy guarantees even with an unbounded computational power and the security is ensured from an information theoretical perspective. Statistical security relaxes the previous definition, allowing attacks (i.e. leak information, provide wrong output, or whatever the MPC protocol guarantees) with a probability that can be made arbitrarily small by adjusting certain protocol parameters. The term "statistical" refers to the attacker's extracted information being statistically close to that of a hypothetical execution of an ideal protocol with perfect security. On the other hand, computational security is another relaxation, allowing leaks when the attacker exerts computational effort. The amount of computational effort can be adjusted with a protocol parameter. Computational security is weaker than statistical, as it depends on the adversary's computational power. However, in both cases, the protocol parameters can be increased arbitrarily to achieve the required privacy guarantees, at the expense of increasing protocol complexity.

While a secure MPC protocol ensures privacy of the inputs and correctness of the output, an adversary may still cause damage by denying the outputs to a subset of the parties. This is modeled as "output guarantees" and the MPC protocols can then be classified into three types: (i) guaranteed output delivery (GOD) a.k.a. robust, (ii) fairness, and (iii) security with abort. Protocols with GOD ensure that honest parties always receive the output. Fair protocols, on the other hand, ensure that if dishonest parties receive the output, then honest parties do as well. Finally, delivery with abort does not guarantee fairness. It allows a subset of parties to obtain the output, but those who are deprived of the output can detect this situation and label the protocol execution as faulty (i.e. abort). However, depending on the protocol's design, it may or may not be possible to fully identify which malicious party is responsible for this abort; this is known as "identifiable abort." The choice to have identifiable abort is a design decision that adds complexity to the protocol.

While the security model and the number of dishonest parties can be subject to design choices, the privacy and output guarantees remain unaffected by such decisions. In the literature, certain impossibility results have been established, which impose restrictions on the attainable levels of privacy and output guarantees, as illustrated in Table 3.1. In the context of the 2PC (two party computation) scenario involving an IoT device and a cloud server, it falls under the category of the dishonest majority setting ($t < n$). As a result, the actively secure protocols in this scenario can achieve only computational security and provide an output guarantee of delivery with abort.

3.3.3 MPC: Arithmetic

MPC protocols are primarily designed to function using modular arithmetic, which can be applied to Boolean rings ($\mathbb{Z}_2$), ℓ-bit integer rings ($\mathbb{Z}_{2^\ell}$), or finite fields ($\mathbb{F}_p$). The commonly used garbled circuits approach involves evaluating the function as a Boolean circuit, but it may become inefficient when dealing with large-scale linear transformations. On the other hand, secret sharing-based approaches offer efficient solutions for performing transformation operations directly over finite fields or rings. To benefit from both these approaches,

Table 3.1 Privacy and output guarantees for MPC protocols with n parties over a synchronous network.

		Privacy			Output guarantees		
Security	**# dishonest parties**	**Perfect**	**Statist.**	**Comput.**	**GOD**	**Fairness**	**Abort**
Passive	$<n$	×	×	✓	✓	✓	✓
	$<n/2$	✓	✓	✓	✓	✓	✓
	$<n/3$	✓	✓	✓	✓	✓	✓
Active	$<n$	×	×	✓	×	×	✓
	$<n/2$	×	✓	✓	✓	✓	✓
	$<n/3$	✓	✓	✓	✓	✓	✓

"Statist." and "comput." denotes statistical and computational, respectively. Some of these properties are not achievable over an asynchronous network.

recent protocols have adopted a hybrid strategy which entails evaluating specific parts of the protocol in the domain where they are most efficient.

In practical applications of MPC involving real numbers, a crucial step is to encode the real numbers into integers within a field or a ring before performing the computation. This encoding process allows the MPC protocols to handle real numbers effectively and ensures accurate and secure calculations involving them. There are two common approaches to adopt real numbers in MPC protocols: (i) fixed-point; and (ii) floating-point arithmetic [21, 22].

For fixed-point, a fixed number of least significant bits in each secret-shared element is reserved to represent the fractional part of the encoded real number. A real number is represented with one only secret-shared field or ring element containing both the fractional and integer parts, plus a public value defining the size of the fractional part. Additions with this encoding are straightforward, but multiplications increase the number of bits representing the fractional part by a factor of two. This implies that several consecutive multiplications may produce an overflow in the encoding and yield unexpected results. A potential solution, but rather inefficient, would be to scale up the dimension of the ring or field to a higher value after one (or several) multiplications. More efficient solutions are adopted in practice, which rely on truncating the fractional part to its original size after one or several operations. Probabilistic truncation [23] is one such efficient technique, which corrects the fractional part at the cost of a probabilistic error in the last bit (which is assumed as acceptable for a wide range of applications).

Floating-point representation eliminates the need for truncation by adjusting the fractional part in every operation. However, floating-point arithmetic is considerably more intricate compared to fixed-point. Along with the components found in fixed-point representation, floating-point stores the size of the fractional part in a secret-shared element, which is updated after each operation. This introduces complexity in multiplication and addition operations, as they must consider the size of the fractional part. Furthermore, floating-point requires three additional secret-shared elements per private value to account for the sign, indicate a zero value, and express errors resulting from division by zero.

3.3.4 MPC: Preprocessing

In MPC, preprocessing plays a crucial role in achieving a faster protocol execution. Its main objective is to advance certain required operations before the parties' inputs are provided. At this stage, the time complexity is not a major concern because parties are not anticipating any useful output. Therefore, a widely accepted approach is to shift as many operations as possible to the preprocessing phase. Once the parties' inputs are established, the protocol transitions into the "online phase," where the information generated during preprocessing is utilized for computation.

For MPC protocols based on garbled circuits, preprocessing is intrinsic in the protocol design. Garbling, as described earlier, consists of encrypting the truth tables of the circuit's Boolean gates. This can be performed by one party [4] or several [7, 13], before the inputs are defined. However, this approach is not so intuitive for secret sharing-based MPC protocols. Additions and multiplications are performed after the inputs are defined. However, preprocessing can still speed up multiplications, with a technique known as Beaver's multiplication triples [16]. This technique consists of multiplying two random and private elements during the preprocessing, which are later consumed during the online phase. Each multiplication in the online phase requires one such triple, and the triples cannot be reused.

In the context of ML, linear transformation layers, which encompass vector–matrix multiplications and convolutions, are commonly utilized. The distinction between a convolutional layer and a conventional fully connected layer lies in how their inputs are processed. Unlike fully connected layers that involve matrix multiplication with weights as matrix elements, convolutional layers perform convolutions of small-sized filters on input features, generating output mappings for subsequent layers. However, performing matrix multiplications in both convolutions and fully connected layers can be computationally expensive, with the number of multiplications increasing cubically with matrix size. To address this, the authors in [24] propose to use auxiliary dot-product triples that follows a similar approach as that of multiplication triples. Additionally, [17] and [25] introduced the concept of matrix triples and convolutional triples, where two random matrices (or tensors) are pre-multiplied (or convolved) during preprocessing and then utilized during the online phase. This preprocessing technique helps reduce the complexity of matrix multiplications, contributing to more efficient MPC protocols for ML applications.

3.4 Fully Homomorphic Encryption

In 1978, Rivest et al. introduced the concept of FHE, originally called "privacy homomorphism" [26]. FHE is a revolutionary form of encryption that allows computations to be performed on encrypted data without decryption, guaranteeing the confidentiality of information. Several partial homomorphic encryption schemes have been proposed for over three decades, allowing one type or a limited number of operations. Some examples are the RSA cryptosystem by Rivest et al. [27], ElGamal [28], and Paillier [29], and the encryption scheme by Boneh et al. [30]. However, only in 2009, Gentry made a breakthrough by proposing the first FHE scheme [31]. In his PhD thesis, Gentry not only introduced the first fully homomorphic scheme but also presented a general method, called *boostrapping*, for constructing FHE schemes from a *somewhat* homomorphic scheme with limited homomorphic evaluation capabilities.

This achievement has provoked significant interest in the field of cryptography, leading to several novel FHE constructions based on Gentry's initial work. Some examples include the BGV scheme by Brakerski et al. [32], the BFV scheme by Brakerski [33], and Fan and Vercauteren [34], the FHEW scheme by Ducas and Micciancio [35], the TFHE by Chillotti et al. [36], and the CKKS scheme by Cheon et al. [37].

The security of most of the FHE schemes is based on the presumed intractability of the Learning with Errors (LWE) problem and its ring variant (RLWE) [38]. The latter version is often preferred for efficiency reasons. Informally, the (decision version of) RLWE problem consists of distinguishing equations perturbed by a small *error* from random tuples. This random error, also called *noise*, has to be small to guarantee correct decryption. Unfortunately, the intrinsic feature of FHE is that the error increases any time a homomorphic operation is applied. So, due to the error growth, the ciphertext cannot be decrypted correctly after a certain number of multiplications (or additions). To overcome this limitation, Gentry [31, 39] introduced the bootstrapping technique. Essentially, it consists of a re-encryption procedure of a ciphertext c to *refresh* it, i.e. encrypt it again under another key, obtaining a new ciphertext (for the same plaintext) but with a smaller error. This procedure can convert a scheme that is not fully homomorphic (in the sense that it only allows a limited number of operations) into one that is.

It is worth highlighting that bootstrapping is the only known way to obtain FHE schemes. Unfortunately, this technique is computationally complex and requires a large memory space. To overcome these problems, Brakerski et al. [32] proposed a new method to reduce the error by obtaining a so-called *leveled fully homomorphic scheme* (LHE). Roughly speaking, this scheme supports the evaluation of all algebraic functions of a specific degree, meaning, of precise multiplicative depth circuits. Namely, instead of providing an unlimited number of multiplications, LHE is bounded to a limit. It is worth noting that the LHE scheme's parameters must be adjusted to achieve the limit required by the application.

3.4.1 FHE: Bottlenecks and Advantages

Despite significant progress in developing FHE, its complexity remains one of the biggest challenges. Current FHE schemes are still considerably slower compared to traditional encryption schemes, making them impractical for many real-world applications. Additionally, FHE presents a memory challenge: first, we have a considerable *ciphertext expansion*, namely, the ciphertext sizes are substantial compared to the corresponding plaintexts. Second, FHE schemes require the management of large public and private keys.

Moreover, as mentioned before, another (big) issue is the noise management. The ciphertext's error has to be small to be decryptable. However, it grows exponentially with each sequential multiplication. We can enable more multiplications before the noise grows out of bounds by adjusting the scheme parameters. For example, increasing the ciphertext modulus supports more operations but also decreases the security level of the underlying scheme. On the other hand, to increase the security level, we can adopt a higher ciphertext and plaintext length size. However, this comes at the cost of reducing efficiency. There is, therefore, a trade-off between security, performances (small ciphertext modulus), and error margin (big ciphertext modulus), which illustrates the difficulty of finding an optimal parameter set for a specific FHE scheme.

A primary goal of the FHE community is to standardize FHE and establish a common framework for its implementation. To address this challenge, several efforts have been made. In particular, to make FHE accessible to everyone, researchers proposed tools for efficiently selecting parameters [40–42] and the Homomorphic Encryption Standard [43], using the LWE Estimator[2] [38] provides look-up tables that indicate the upper limits on the size of the ciphertext modulus for specific security levels and ciphertext length size. Also, open source FHE libraries, e.g. [44–47], have been proposed, and their optimization has significantly increased the efficiency of FHE applications over the past years. Finally, a number of FHE compilers [48–51] have become available with the objective of converting high-level programs to FHE-based implementations. See Viand et al. [52] for more details about the FHE libraries and compilers.

From a hardware point of view, considerable efforts leveraging specific hardware architectures (e.g. FPGA and ASIC-based) have been made. Such designs are referred to as FHE accelerators [53–57] and provide substantial improvements in FHE schemes' performance in software.

3.4.2 Comparison with MPC

Unlike MPC, FHE offers the incredible advantage of being *non-interactive*. The computations can be performed directly on encrypted data without the need for decryption or user interaction at each step. In the cloud computing paradigm, the client is only required to encrypt and send. All the operations are performed confidentially on the cloud server side without leaking the user's data. In the IoT scenario, this is highly convenient since IoT devices only need to send encrypted data. However, that encryption is still complex for hardware-limited IoT devices. Also, the ciphertext expansion introduces a bottleneck in terms of bandwidth consumption that is significantly deleterious in IoT due to the bandwidth limitations of IoT communication standards. These problems could be addressed through hybrid-FHE, as described in Section 3.8.2. Nevertheless, it is worth mentioning that to the best of the authors' knowledge, there are no sufficiently efficient implementations of hybrid-FHE to enable its application to IoT.

A problem that is present in FHE, and not in MPC, is the lack of *verifiability*. In MPC, all the parties can be certain that a specific algorithm is executed correctly. But in FHE, the cloud service can choose the operations arbitrarily. Namely, it cannot decrypt or learn from the data but can manipulate it at will. This cannot be detected by the user, which can only decrypt the result and accept it (or not) without any proof or guarantee of correct algorithm execution.

Thus, to achieve both privacy and verifiability like in MPC, a novel approach called Verifiable Computation (VC) should be applied on top of FHE [58–60]. However, this comes at the cost of added complexity.

2 The LWE Estimator, which the Lattice Estimator has succeeded, is a software tool to determine the security level of LWE instances.

3.4.3 Generations of FHE Schemes

FHE schemes can be divided into three families (we omit the first generation of schemes since they are impractical) with different features and limitations [61]:

- *Second Generation* scheme, as BGV [32] and BFV [33, 34], are suitable to work with finite fields in the modular exact arithmetic. They are equipped with efficient packing, which enables the use of single instruction multiple data (SIMD) instructions to perform computations over vectors of integers (i.e. batching). Thus, these schemes are excellent candidates when large arrays of numbers are to be processed simultaneously, such as matrix multiplication. However, these schemes are not well-suited for implementing nonlinear functions or for circuits requiring bootstrapping (e.g. circuits with large multiplicative depth). The latter limits their application to only their leveled version (LHE).
- *Third Generation* schemes, namely TFHE-like schemes, can outperform other schemes for bit-wise operations, i.e. when computations are expressed as Boolean circuits [43]. Moreover, the biggest advantage of these schemes is the fast bootstrapping and the capability to evaluate nonlinear functions. This last feature is very useful for ML since it permits to implement activation functions efficiently. In contrast, the main limitation of TFHE is the lack of support for CRT packing (i.e. batching). Hence, TFHE schemes can be outperformed by previous approaches when processing large amounts of data simultaneously.
- *Fourth Generation* schemes, such as CKKS [37], share similar properties with second-generation schemes. The main difference is that these schemes are designed for computations on real or complex numbers. CKKS supports approximate arithmetic operations, which are particularly useful for applications involving ML or data analytics.

A comparison among the families of the FHE schemes is presented in Table 3.2. It is worth clarifying that although TFHE provides the fastest bootstrapping procedure, the batching feature of second and fourth-generation schemes allows for the parallel bootstrapping of several plaintexts. For the specific case of CKKS, it is possible to obtain a more efficient amortized bootstrapping than for TFHE. This is, however, not true for second-generation schemes because the number of slots (plaintexts that can be batched together) is significantly lower than for CKKS, and this renders CKKS bootstrapping more than one order of magnitude (often two) faster than BGV/BFV bootstrapping (this is denoted by the gray color in Table 3.2).

We recommend interested readers to refer to [61], which offers an extensive explanation of FHE, covering its origins, mathematical foundation, and emphasizing advancements and practical applications in the field.

Table 3.2 Usage of FHE schemes according to functionality.

	Fast operations			
Schemes	**Linear functions**	**Nonlinear functions**	**SIMD**	**Bootstrapping**
Second generation	✓	—	✓	—
Third generation	✓	✓	—	✓
Fourth generation	✓	—	✓	✓

3.5 Oblivious Neural Networks (ONN)

Recently, several PPML frameworks have been proposed adopting FHE, MPC, or a combination of both. They address the challenge of implementing private versions of NN blocks from different angles. FHE-based frameworks require polynomial approximations for the activation layers but are efficient for linear transformations (note that works adopting secret sharing-based MPC face a similar problem). On the other hand, frameworks entirely based on GCs, which are suitable for activation functions (i.e. nonlinear functions like comparisons), incur inefficiencies when dealing with linear transformations Some frameworks focus on providing efficient mechanisms to switch from FHE to GCs or from secret-sharing MPC to GCs. However, such switching techniques get considerably complex when active security is required.

This section aims at describing such works conceptually, removing the mathematical nuance and focusing on the main innovations, privacy guarantees (active/passive security), and setting (number of parties and trust level). Most of the studies concentrate on scenarios involving a limited number of parties, specifically 2, 3, and 4. Table 3.3 shows the main PPML frameworks, categorized according to the number of parties (2, 3, 4, or any) and technologies in which they are based: HE – Homomorphic Encryption; GC – Garbled Circuits; SS – Secret Sharing; OT – Oblivious Transfer; FSS – Function Secret Sharing. PPML frameworks are also categorized according to their security model and delivery guarantees (as explained in Sections 3.3.1 and 3.3.2): ○– semi-honest; ◐– malicious with abort; ◕– malicious with fair delivery; and ●– malicious with guarantee of delivery.

3.5.1 Two-Party (2PC)

One of the first works on Oblivious Neural Networks (ONNs), based on LHE is CryptoNets [63]. Soon after, SecureML [18] utilized techniques such as GC and SS to create efficient protocols for PPML inference, including neural networks. Subsequently, several works such as MiniONN [24], DeepSecure [64], Gazelle [65], and XONN [66] focused on improving efficiency by leveraging advancements in underlying primitives. Recent works such as CryptFlow2 [67], Delphi [68], ABY2.0 [69], and Cheetah [70] employ state-of-the-art optimizations in HE and OT extension domains. However, with the exception of XONN,[3] these works only offer security against semi-honest adversaries and are unable to handle malicious corruptions. Some of these works are more thoroughly detailed in Section 3.5.5.

3.5.2 Two-Party with Helper (2PC$^+$)

Two-Party solutions are commonly utilized for PPML, since they match the client-service provider (i.e. model owner). However, there exists a significant computational and communication burden during the preprocessing phase, even for frameworks only catering semi-honest security. To tackle this efficiency issue, some works have explored the use of an external helper during preprocessing [85]. This approach has proven beneficial in enhancing the efficiency as demonstrated in works like Chameleon [71] and Crypten [72].

3 XONN's method can be generalized to the malicious setting, but its underlying cut-and-choose method is expensive for practical use.

Table 3.3 Summary of PPML frameworks (only a representative subset).

Setting	Framework	Security	Techniques
2-party (dishonest majority)	CryptoNets [63]	○	HE
	SecureML [18]	○	HE + GC + SS
	MiniONN [24]	○	HE + GC + SS
	DeepSecure [64]	○	GC
	Gazelle [65]	○	HE + GC + SS
	XONN [66]	◐	GC
	CryptFlow2 [67]	○	HE + OT + SS
	Delphi [68]	○	HE + GC + SS
	ABY2.0 [69]	○	OT + GC + SS
	Cheetah [70]	○	HE + OT + SS
2-party (+helper)	Chameleon [71]	○	OT + GC + SS
	Crypten [72]	○	SS
	LLAMA [73]	○	FSS
3-party (honest majority)	ABY^3 [74]	◐	GC + SS
	ASTRA [75]	◕	SS
	SecureNN [76]	○	SS
	SWIFT [77]	●	SS
	Falcon [78]	◐	SS
4-party (honest majority)	Trident [79]	◕	GC + SS
	FLASH [80]	●	SS
	SWIFT [77]	●	SS
	Fantastic Four [81]	●	SS
	Tetrad [82]	●	GC + SS
n-party	Cerebro [83]	◐	HE + GC + SS
	MPClan [84]	◕	SS

Figure taken from the FANNG–MPC project.
Source: Adapted from Riazi et al. [66].

More recently, protocols like LLAMA [73] have also embraced this setting in their 2PC designs, employing the Function Secret Sharing (FSS) paradigm and demonstrating practicality. However, this approach also implies that both the client and model owner must trust this external party or employ expensive verification mechanisms to ensure the correctness of the computation performed by this party.

3.5.3 Three-Party (3PC) and Four-Party (4PC)

2PC frameworks not only incur considerable communication and computation overhead, but they are also limited to security with abort in a dishonest majority setting (as presented in Section 3.3.2). However, this level of security might not be sufficient for handling sensitive applications, such as healthcare. Some works address this issue by adding an extra party, i.e. focused on a 3-party honest majority setting, where only one party can be maliciously corrupt ($t < n/2$). Works like ABY3 [74] (described in Section 3.5.5.7) and ASTRA [75] demonstrated improvements over the 2PC protocols. Besides, the 3PC protocol in SWIFT [77] utilized function-dependent preprocessing and distributed zero-knowledge proofs to enhance communication and achieve guarantee of delivery. In parallel, works like SecureNN [76] and Falcon [78] provided efficient primitives for NN blocks like Maxpool, normalization, and division.

To further improve efficiency, some works focus in super honest majority setting involving four parties where only one can be corrupt ($t < n/3$). Works like Trident [79], FLASH [80], Fantastic Four [81] (further detailed in Section 3.5.5.8), and Tetrad [82] follow this paradigm.

3.5.4 *n*-Party (MPC)

While the landscape of PPML frameworks is abundant for scenarios involving small parties ($n < 5$), there are only a few that specifically cater to supporting more than four parties. One such work is Cerebro [83], which introduced an end-to-end collaborative learning platform by developing a compiler based on SCALE–MAMBA (SM) and EMP–Toolkit (EMP). Essentially, this platform converts ML-friendly APIs into either SM or EMP code, enabling the execution of various protocols supported by these frameworks. However, Cerebro lacks ML-specific optimizations for SM and inherits the limitations of SM when compiling large programs. In a different approach, MPClan [84] focused on PPML inference in scenarios involving n parties by utilizing function-dependent preprocessing. Nevertheless, their technique is constrained to $n \leq 11$ parties due to the exponential growth of computation and storage with the number of parties.

3.5.5 A Closer Look at PPML Frameworks

While the previous discussion focused on providing an overview of the works in each category, we now discuss some of the prominent works described above in detail to provide a more in-depth insight.

3.5.5.1 CryptoNets

CryptoNets [63], from Microsoft, was the first approach enabling an oblivious NN model for both inference and training. It is entirely based on LHE, specifically, the scheme in [86]. It implements a NN consisting of two convolutional layers, two fully connected layers, two mean pooling layers, and three activation layers. Out of the three activation layers, two are squared activation layers and hence can be implemented using homomorphic encryption with multiplicative properties. An additional activation layer is placed in the output layer to improve accuracy in the error calculation and computation of the gradient descent during the training phase. The activation functions were chosen LHE-friendly, i.e. easy to implement with an LHE scheme. However, from the ML perspective, this type of activation function does not achieve high accuracy. This is one of the main drawbacks of HE for NNs. Activation functions must be approximated with polynomial functions, which diminishes the inference and training accuracy. In fact, training may not even converge when activation functions get closer to a linear function.

The main limitation of the adopted scheme, which is intrinsic in any FHE/LHE scheme, is that the noise grows significantly with each multiplication. This forces the LHE parameters to be large enough to support the noise growth for the required circuit's multiplicative depth. Also, CryptoNets uses an LHE scheme that accepts batching (i.e. many plaintexts are encrypted into the same ciphertext). Although this is very convenient for processing large data in parallel, it also presents a challenge. The batch size depends on the polynomial degree of the LHE scheme, thus it cannot be adjusted according to the size of the input data. This implies a constant computational overhead per batch. Thus, to optimize its performance, CryptoNets requires data processing in batches close to the maximum batch size, which is undesirable when the predictive model is offered in a "Neural Network as a Service" fashion (NNaaS) to different distributed clients sending isolated data samples.

3.5.5.2 SecureML

SecureML [18] provides MPC-based privacy-preserving NN that can be trained with a vertically partitioned database. It is based on secret sharing for the linear transformation layers and GCs for the activation layers. It avoids adopting polynomial approximations for the activation layer, hence solving the limitations previously mentioned in CryptoNets. However, SecureML can only implement Rectifier Linear Units (ReLUs) efficiently. Other complex activation functions are approximated with a combination of ReLUs, hence decreasing performance.

SecureML represents shared decimal numbers in fixed-point arithmetic with shared integers in a large finite field, and performs private multiplications using the conventional approach assisted by multiplication triplets. After each multiplication, the integer shares are truncated, and the occasional offset produced when mapping the integers back to real numbers is treated as negligible error. The adopted MPC protocols work in the semi-honest model exclusively, hence truncation does not require communication.

3.5.5.3 MiniONN

MiniONN [24] was the first MPC solution to be applied to any NN model. It defines an oblivious procedure, relying on a combination of secret sharing-based 2PC and garbled circuits, to perform pooling, linear transformations, and nonlinear activation functions. For arithmetic operations, it uses secret sharing, and for activations and pooling layers requiring comparisons (e.g. max pooling), it adopts GCs.

Activation functions that rely on comparisons, such as ReLUs, are performed obliviously using GCs. But other nonlinear activation functions such as sigmoid $f(y) = \frac{e^y}{e^y+1}$, cannot be implemented efficiently in a GC. Exponentiations and divisions are inefficient operations in GCs. Although sigmoid functions can be approximated with high-order polynomials, GCs can only perform efficiently in low-order polynomials. MiniONN proposes the approximation of smooth functions using sets of piecewise continuous polynomials called splines [87]. The main idea consists of splitting the smooth function into several ranges and using curve fitting with different low degree polynomials in each range. Then a 2PC solution is used to perform several comparisons and evaluate in which range the value falls, and thereafter compute the corresponding low degree polynomial. MiniONN also proposed an innovative dot-product triplet, to perform dot-product vector multiplications, that reduces the complexity of convolutions and fully connected layers.

To pass from secret sharing-based 2PC to GC, the GC has an initial reconstruction circuit where both shares are added up. More explicitly, each party bit-decomposes its share and uses it in as input in the GC, then the GC first adds up the bit-decomposition of both shares and performs the circuit of the activation function, as shown in Figure 3.7.

It is worth clarifying that the reconstruction is a simple addition if the secret sharing-based 2PC protocol works with the same underlying structure as the GC, i.e. binary representation of values. This implies that the 2PC protocol must work over a ring with values modulo 2^K, i.e. all secret-shared values are words of K bits. However, the majority of 2PC/MPC protocols do not work over modulo 2^K. Instead, they work over finite fields, i.e. modulo a prime p. The reason to adopt a field and not a ring is the absence of zero divisors in the former, which makes it easier to design protocols with active security. However, MiniONN adopts a ring structure because it only aims at providing passive security.

It is worth commenting that it is possible to pass from a secret-sharing 2PC protocol based on fields to GCs, but it requires an extra operation on the GCs, the modulo "p" reduction. It is also possible to do this conversion in protocols with active security, but to be efficient it requires preprocessing to obtain randomness represented in both rings (modulo 2^k) and fields (modulo p) [88].

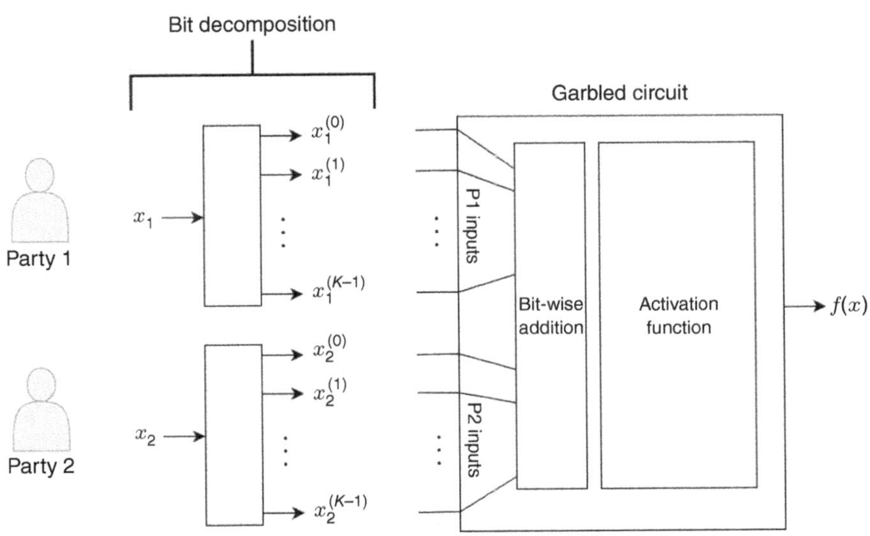

Figure 3.7 Conversion from secret sharing-based 2PC to GC. The output of the function can also be secret-shared.

3.5.5.4 DeepSecure

DeepSecure [64] is entirely based on GCs, and it integrates several SoTA optimizations for GC implementation and proposes a new preprocessing phase based on high-ranked matrix decomposition that reduces the communication costs in the online phase. DeepSecure self-claims to be the first GCs-based privacy-preserving ML approach, and compares itself only to CryptoNets (fully based on HE). This is because the authors were not aware of the parallel works, MiniONN and SecureML, which were published almost at the same time. The advantage of DeepSecure is that it provides a framework for implementation of privacy-preserving NN based on [89], supporting fixed-point and floating-point accuracy.

Contrary to the most common deployment strategy where the model owner garbles the circuit and the client evaluates the GC, DeepSecure implements the garbling step on the client side. Although this seems counter-intuitive since garbling is more computationally demanding than evaluating, it is designed in this way to implement outsourcing of the garbling operation to a third noncolluding honest-but-curious party. The outsourcing of the garbling step in a 2PC setting is straightforward: (i) the client generates a random string s with the same size of its input x (the data sample to evaluate privately); (ii) sends the string s to the third party and $x \oplus s$ the cloud server. The GC has to include an initial layer to reconstruct x during the circuit evaluation, as shown in Figure 3.8. Note that the client must trust the garbler.

For the activation layer, DeepSecure implements a wide set of NN blocks using GCs. The simplest function is the ReLU, which takes 15 XOR-like gates (i.e. XOR, XNOR, and

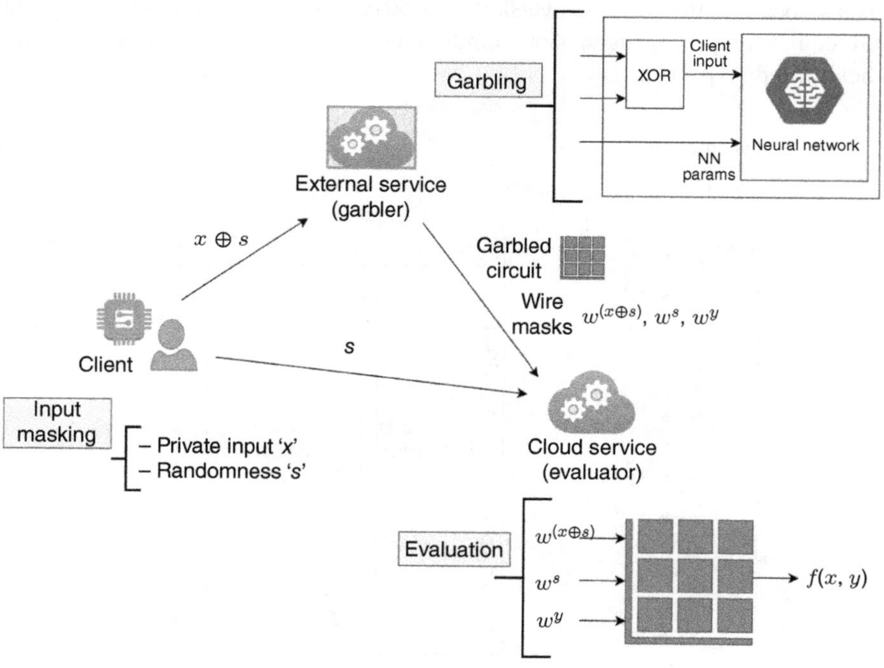

Figure 3.8 GC outsourcing. The client secret shares its input and sends both shares to the cloud server (evaluator) and the external trusted service (garbler). The garbler includes a reconstruction step in the circuit to combine (XOR) both shares.

NOT) and 30 non-XOR. Additions are also simple with 16 XOR and 16 non-XOR. However, a vector–matrix multiplication of $A_{1\times m}B_{m\times n}$, required in a fully connected layer, requires $(397mn - 16n)$ XOR gates and $(228mn - 16n)$ non-XOR gates. Similarly, complex activations like Sigmoids may reach almost 150 K non-XOR gates depending on the version. Note that all these values are adjusted for the specific word bit-length in DeepSecure implementation. It is worth highlighting that we distinguish between XOR gates and non-XOR gates because the former has a convenient optimization method called Free-XOR [90]. This method enables gate evaluation without encryption/decryption operations for XOR, XNOR, and NOT gates.

DeepSecure also performs preprocessing in the client's data, which can be executed without the client's input. First it includes a data projection phase, usually performed to decompose a high-dimensional input matrix into a sequence of low-dimensional subspaces [91]. DeepSecure uses this data projection phase to shape the cloud server's inputs in order to optimize the communication between the client and server in the online phase. For that, DeepSecure requires a retraining of the model with a customized objective function, and it also yields a projection matrix that must be applied to the client's input before using the predictive model. The projection matrix is public and does not disclose information about the model or the training data. It is also independent from the client's input.

3.5.5.5 Gazelle

Gazelle [65] focuses on Convolutional Neural Networks (CNN), which are specially designed for image classification. A convolution can be expressed as a matrix multiplication with some mathematical manipulation. However, Gazelle uses HE and performs the convolution applying filters without transforming them into a single big matrix multiplication. It leverages the fact that homomorphic encryption (HE) can outperform MPC when the computation has a small multiplicative depth. Specifically, it adopts Packet Additive Homomorphic Encryption (PAHE). For other operations in the CNN, Gazelle adopts MPC.

For the HE-based vector–matrix multiplication, required in fully connected layers, the paper proposes a slight modification of the Diagonal Method by Halevi and Shoup [92]. In the Diagonal Method, the input vector is encrypted using PAHE, hence the vector elements are encrypted as concatenated elements into the same ciphertext. The sever encrypts the vector of the diagonal $(w_{0,0}, \dots, w_{n-1,n-1}, w_{n,n})$ and multiplies both encrypted vectors using SIMDScMult (Single Input Multiple Data Scalar Multiplication). Then the sever shifts the diagonal $(w_{0,1}, \dots, w_{n-1,n}, w_{n,1})$ and multiplies with a rotation by one of the input vectors. After $n - 1$ rotations, where n is the input size, the server has n SIMD encrypted vectors that are added. The result is a packet encryption of the output vector. This method improves the Naive Method of multiplying the input vector by the different matrix rows in the sense that it does not need addition of the elements packed in the same ciphertext, which would require $log_2(n)$ SIMD rotations of on n different ciphertexts. The Diagonal Method only requires rotations on the input vector, and therefore admits the optimization proposed in [93] called *Hoisting*.

In a convolutional layer, a bank of filters of small size is iteratively convolved with the input images of some size in terms of pixels and dimension. The process is repeated in each pixel, centered in such pixels and covering adjacent pixels up to the filter size. The output of the convolution is an image of the input size in terms of pixels (depending on the stride and padding), but the number of filters defines the dimensionality. Gazelle caters to an alternative way of performing this operation which is slightly more efficient than previous approaches. Since the whole input image is in raster disposition in the same ciphertext, and the PAHE scheme allows multiplication of the different encrypted elements by different values, it is possible to rotate the encrypted image and multiply some of the elements by the corresponding filter weight and nullify the rest of the elements, as shown in Figure 3.9. This new approach still requires the same number of SIMDScMult and SIMDAdd operations than previous state-of-the-art solutions, but it does not require randomization of the edge elements.

3.5.5.6 Delphi

Delphi [94] follows the same approach as Gazelle, since it relies on LHE for the linear transformations and MPC for the activation layers. The innovation comes in the preprocessing phase, since the LHE-based convolutions and matrix multiplications are moved to the preprocessing stage. The idea is to process some client's random values with the model owner's real parameters using LHE. Then, when the client's inputs are defined, those random values are used to obtain the linear transformations without using encryption (see Figure 3.10). The advantage is twofold: (i) the computational complexity of LHE only affects preprocessing time, hence it reduces delay during the online phase (when the inputs are defined); (ii) the ciphertext expansion of LHE is avoided during the online phase, which reduces communication complexity and hence delay.

Additionally, Delphi also adopts a planner tool to replace some ReLU activations by quadratic approximations, which are much faster when implemented using MPC. However, replacing ReLUs by quadratic approximations degrades the NN performance. In fact, the training phase may not converge. Delphi's planner is optimized to select which ReLUs can be replaced without degrading performance significantly. It is worth highlighting that, since it is based on homomorphic encryption, both Gazelle and Delphi can only achieve semi-honest security.

3.5.5.7 ABY^3

ABY^3 [74] is a framework for PPML with 3-party MPC. It achieves active security and relies on honest majority, i.e. only one of the three parties is assumed to behave dishonestly. ABY^3 leverages the honest majority setting to propose novel efficient protocols to switch from secret sharing MPC (based on replicated secret sharing), binary sharing, and Yao-based GCs. Replicated secret sharing is used for any arithmetic operation, whereas binary sharing or GCs are used for Boolean operations. Binary sharing refers to a form of secret sharing-based MPC but bit-wise, i.e. private elements are shared as single bits. The main difference of binary sharing with respect to GCs is that the former reduces the quantity of bits transmitted, but it requires a number of communication rounds proportional to the circuit depth. On the other hand, GCs requires higher data transmission but it performs in only two rounds regardless the circuit depth. Thus, for wide circuits, where many AND gates can be done in parallel, binary sharing is preferable whereas for deep circuits, or when latency is high, GC is more efficient. It is also worth commenting that the secret-shared part works on rings 2^k, and although it does not relay on triples for multiplications, it still needs

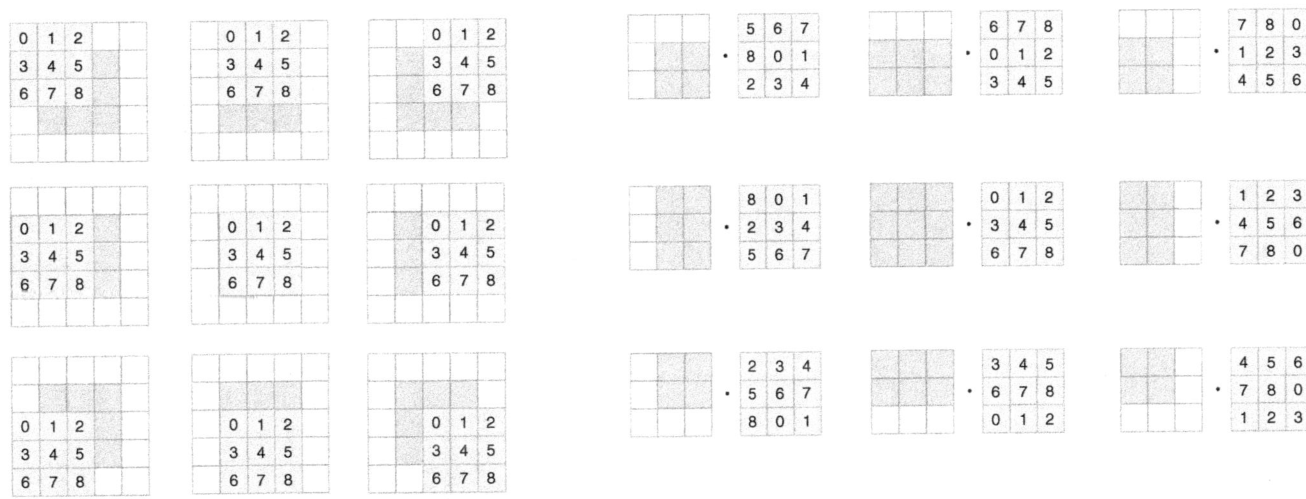

Figure 3.9 The left side shows how a convolution would normally be between a 3×3 filter and a padded picture (with padding 2). The right side shows how the convolution is actually performed. The convolution over a padded 3×3 image with a 3×3 kernel is performed by packing and rotating the filter and applying it to punctured ciphertext (nullifying the not needed elements).

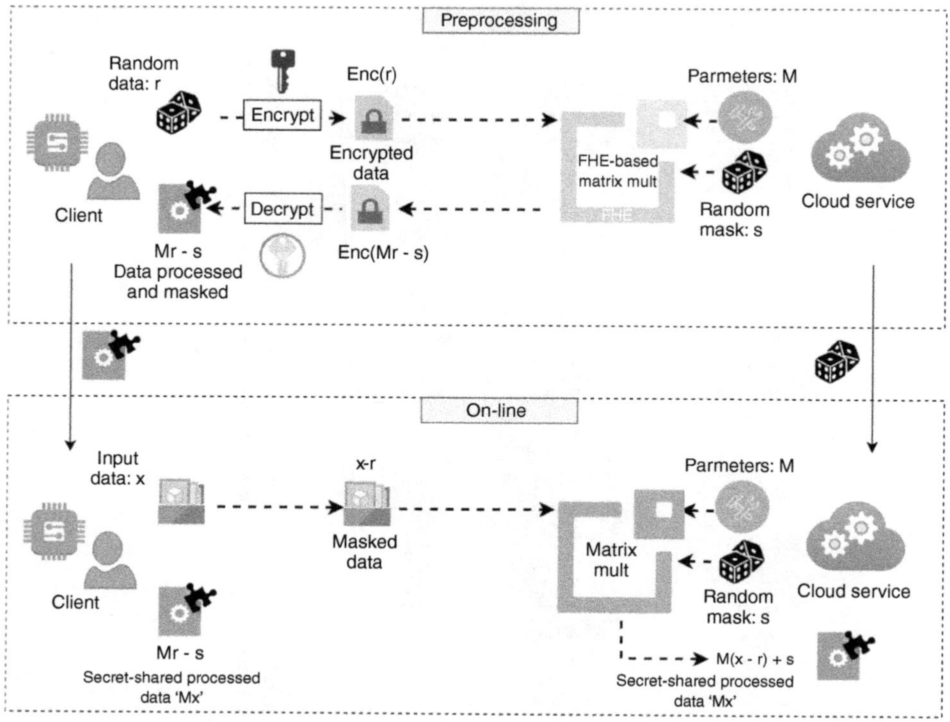

Figure 3.10 On top, Delphi's preprocessing phase for the convolutional layer. During preprocessing, an LHE-based operation is performed over random data and the real parameters of the cloud service. Below, the online phase where the client's inputs are defined and the convolution is performed.

preprocessing to obtain secret shares of zero, i.e. random shares that add up to zero. These are consumed in each multiplication.

The main novelty of ABY^3 is the conversion mechanisms between the three different types of MPC protocols. Also, they have an efficient truncation operation that also leverages the 3-party with honest majority setting. However, this setting cannot be directly applied to the client and model owner (i.e. cloud service) scenario previously described. In ABY^3, both the client and model owner send shares of their inputs to three servers, which are in charge of performing the confidential evaluation of the NN, as shown in Figure 3.11. Note that both the client and the model owner must trust at least two of the servers.

In the same 3-party setting as ABY^3, and simultaneously published, the SecureNN [76] framework also adopts replicated secret sharing (namely the MPC protocol of [95]) to implement oblivious NNs. The main innovation is a set of sub-protocols enabling private comparisons by adopting rings modulo $2^k - 1$, i.e. rings of odd order. They propose a technique to switch between 2^K and $2^K - 1$; the latter is only used for comparisons whereas the rest of the operations are carried out in the former. Note that 2^K is the native representation for most processors (namely for $K = 64$), hence the modulo reduction is

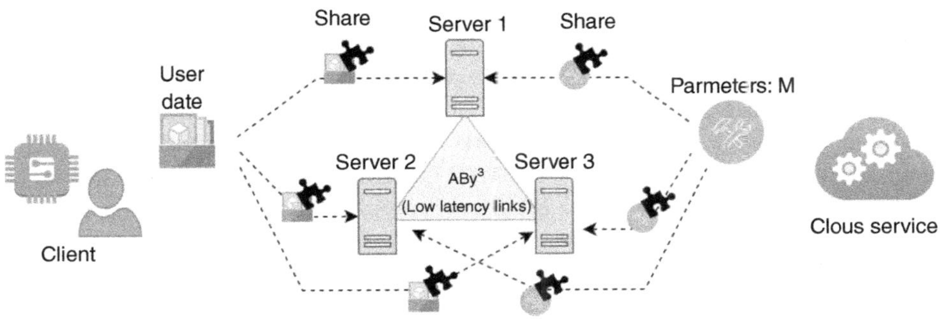

Figure 3.11 ABY^3 set up with three servers. The output is given to the relevant entity, i.e. to the client for private inference and the model owner for private learning.

inherently implemented. Also, the work in [96] proposes replicated secret sharing as ABY^3 and SecureNN, but caters for novel sub-protocols for exponentiation, used in the training phase of NNs (e.g. softmax activations).

3.5.5.8 Fantastic Four

Fantastic Four [81] extends the setting of ABY^3 from three to four parties while maintaining the allowed dishonest parties in only one. Their protocols also rely on replicated secret sharing, and it also splits the ML algorithm in secret sharing and binary sharing for arithmetic and Boolean operations, respectively. The additional party, although it weakens the trust model, enables an optimized preprocessing phase whose size does not depend on the circuit depth.

Also, the paper caters to a new definition of GOD (as described in Section 3.3.2), namely a private version of GOD. In the paper, they coin the term *private robustness*. Some previous schemes providing GOD assume that in the course of an MPC execution, if a malicious behavior is detected, and as a result one of the parties is identified as honest, then that party can be selected to reconstruct all secrets and terminate the protocol successfully. This guarantees delivery, and it does not violate the privacy definition of MPC, since its main purpose is to prevent information leaks to dishonest parties. However, in a real scenario leaking all secrets to an honest party may not be desirable. Fantastic Four not only provides a new notion of GOD, i.e. private robustness, it also caters to an efficient implementation in a 4-party setting while providing active security. Frequently, protocols providing active security require adding extra information to the secret shares in the form of message authentication codes or redundancy in the operations (which is indeed the case of the 3-party protocol they also define in their paper). But in the 4-party setting, there is sufficient information for the honest parties to perform error correction and reconstruct the correct outputs.

3.5.5.9 Crypten

Crypten [72], developed at Facebook Artificial Intelligence (AI) research, leverages well established MPC protocols to cater for a novel PPML framework. The protocols are

SPDZ, GMW, and ABY, hence they adopt secret sharing-based MPC (SPDZ) for linear transformations and garbled circuits (GMW) for activation layers. ABY is a 2-party protocol to switch between both worlds, i.e. SPDZ and GMW. It is worth commenting that, although SPDZ provides active security and there are GMW versions with active security, ABY is semi-honest. Hence the whole framework relies on semi-honest model. The main innovation of Crypten stems not from the MPC implementation, but from the API, since it was the first framework providing a ML-friendly set of functions that abstracts away the MPC layer. This simplifies the adoption of the framework by the AI community.

Following up on the idea to ease development of private versions of NN algorithms, the authors in [83] propose a new framework called Cerebro. It converts ML friendly python code into MPC code to be compiled in Scale–Mamba, a well-known general purpose MPC framework [97]. Depending on the ML block, the code is converted into GCs of secret sharing-based MPC, which follows the same strategy as previous approaches.

3.5.5.10 FANNG-MPC

FANNG [62] is a general purpose MPC framework, customized for private NNs, that stems from Scale–Mamba [97]. It contains all MPC protocols provided by Scale–Mamba, thus it can run in multiple settings. It provides new bytecodes and a new set of libraries with ML blocks including convolutional and fully connected layers (leveraging the work of [25]), folding layers, and activation functions (it also includes a new protocol for private comparisons [98]).

The framework is adapted to provide more efficient preprocessing for NNs than Scale–Mamba. Namely, GCs can be generated in the preprocessing time, which is not possible in Scale–Mamba. Second, the input/output functionality is prepared to load GCs and matrix/convolutional triples generated like [25]. The engine for matrix triple generation is also provided in two favors: software-based and hardware-based. With the aim of improving preprocessing, FANNG also counts with a conversion MPC–2PC to transform matrix/convolutional triples from multiparty to 2-party. The idea is to run the engine for matrix triple generation in separated set of dealers, and transform the triples to 2PC version when required by the client and the model owner.

3.6 Decision Trees

In this section, we explore some of the state-of-the-art and basic design principles of Privacy-Preserving Decision Trees (PDT). Decision Trees (DT's) are type of supervised learning that allow us to perform classification and regression. Training focuses on creating a collection of binary trees with certain properties, whereas inference is achieved by exploring them. Its appeal on the privacy-preserving space stems from the simplicity of the principle on which inference is based. More specifically, the evaluation of a tree could be seen as a deterministic exercise of low complexity e.g. constant, logarithmic. Additionally, the binary nature of the question *what node to explore next* can be synthesized in a simple way by encoding an inequality test. Keeping in line with the concepts revised in this chapter, training has its own unique challenges.

3.6.1 Structure of a Decision Tree

According to the data structure theory, a tree is an acyclic connected graph that in the case of DTs is composed of:

- A *root* node that is at the *root*. It does not have parent node, a characteristic that makes it unique. In DTs, the root is also our starting point when evaluating.
- A set of *internal nodes*. Each one having at most two child nodes and one minimum. Each internal node has a split point that matches a specific feature for a data point traversing the tree. DT models are trained to define what feature and what split point has to be used on each internal node.
- A set of *leaf* nodes. These are nodes that have no child nodes. On classification, they are the ones that contain the labels of the classes.

Figure 3.12 illustrates a simple example of the architecture of a DT. Note that many privacy-preserving techniques use padding to hide the architecture of tree and only expose complete DTs.

3.6.2 Training

When talking about PPML, there might not be a more glowing example of the security/scalability/efficiency trade-off than in training PDTs. In fact, when we discuss training for DTs specifically, more than 70 of the around 80+ different works, surveyed by Chatel et al. [99], are only secure in the presence of semi-honest parties. Security, in this case, seems to be getting the short side of the stick. Intuition would dictate that real-world deployments require real world solutions, thus active security. However, for some scenarios where there is already some trust in the system e.g. intergovernmental entities, and that are limited by legal constraints, semi-honest security may be sufficient.

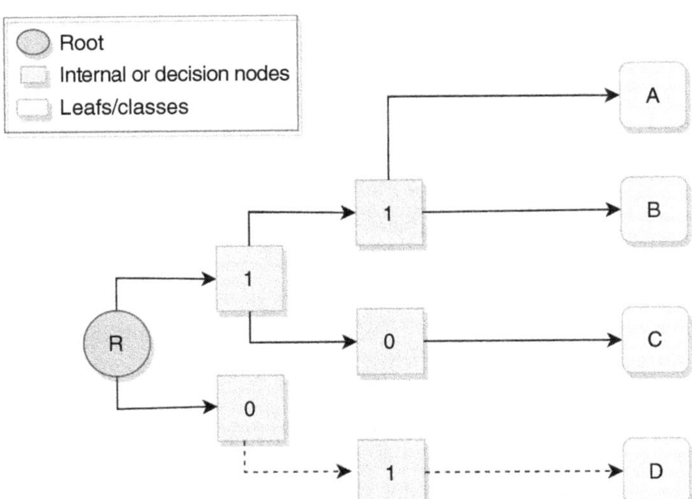

Figure 3.12 Typical decision tree architecture without padding.

Let us explore the different types of training, with some context of what it means in the privacy-preserving space. Basically, training for DTs can be grouped as follows:

- **Greedy Algorithms**: The approaches that focus on trying to build the tree graph from the top of the tree (the root) toward the leaves. At each passing, new weights are assigned until some target approximation is met. There are three notable *greedy* algorithms that are relevant on the privacy-preserving space. These are the following: (i) ID3; (ii) C.45; and (iii) CART. Their main distinction (but of course not their only one) is the splitting technique they employ. Researchers have proposed several privacy-preserving mechanisms for these algorithms.
- **Random Algorithms**: Consider a fixed tree structure. Here, weights and features are randomly selected at each node and tuned at each passing. This process represents a challenge on the privacy-preserving space, hence it makes fewer common appearances in the literature than the other mechanisms described in this Section [100].
- **Random Forests**: This is not properly a training strategy for a DT. Rather than that, it is a way to group together DTs trained via some of the algorithms we mentioned before. In this context, predictions are the product of the aggregation of the results obtained by each tree. We can see them quite present in the privacy-preserving literature, but mainly described on evaluation centered works, such as the Systematization of Knowledge (SoK) on Evaluation of DTs and Random Forests by Kiss et al. [100].
- **Boosting**: These kinds of algorithms rely on a variety of training techniques of a more mathematical foundation. There are several forms of *boosting* in the literature, like Adapting Boosting or the most commonly found in privacy-preserving literature, Gradient-Boosting Decision Trees (GBDT). Authors have also proposed improvements on top of GDBT, the most common derivations of the technique are called XGBoost and LigthGBM. However, compared with their greedy counterparts, there are fewer examples of boosting in the privacy multiparty literature. Nevertheless, a relevant example of this is the XGBoost based HEP-XGB protocol by Fang et al. [101]. On the more federated aggregation/hardware side of things and secure enclaves (namely SGX), we have the results introduced by Leung et al. [102].

3.6.3 Computational Setups

It is also important to consider how both *data and parties* are distributed when training PDTs. Let us start by talking a bit about data distribution. That is to say, where is the data. Works in this area contemplate typically three different scenarios:

- There is only one data holder, that would like to offload the computation to a third party (TP). In this setup, the TP possesses sufficient computational resources to produce the trained model. In our context, the TP should not learn any information derived from the dataset.
- Different actors hold data samples, albeit all share the same feature space. Coming back again to the work by Chatel et al. [99], it is clear that this is one of the more popular setups for training, as having some homogeneity on the data makes things simpler.
- Let us consider a rather similar scenario, as before where different actors hold the data samples, but in this case not all share the same feature space. That is to say, data is not

partitioned by subsets of data samples holding all the same features, but rather parties holding data points on different feature spaces for the same data samples. For instance, public institution A might have collected features of Charlie, while private institution B might also hold a different set of Charlie features. Training in this case happens on the global dataset, meaning, all information that has been collected about Charlie.
- Finally, consider any other way you could mix the cases above. The literature typically calls this arbitrary or mixed data partition. Data distribution adds complexity to the training and algorithmic solutions tend to be proposed on a per case basis. A strong example of this, given that it is based on Arithmetic Circuits, is the work by de Hoogh et al. [103]. Nevertheless, authors only guarantee semi-honest security, although extending it to active security does not seem an uphill battle.

Now let us move to the *when*, and in this case the *when* means *what happens while we train the model*. As you might expect, federated learning is also an important strategy researchers investigate for DTs.

Consider, for instance, a more centralized model where either a single party or a collection of parties transfer their database to an offloading entity, as described in [99]. This centralized entity might be a single semi-trusted party or a collection of computational parties with various levels of trust. A common occurrence is to have 2, 3, or 4 noncolluding parties executing the computations.

A frequent topic found in the literature is the study of a fully distributed model. In this case, parties that hold the data are also the ones performing the computation. Initially, case studies considered the use case when there are only two parties holding data and performing the computation. That is the case of the seminal work presented by Lindell and Pinkas [104]. Nowadays many of these approaches consider different scenarios where the number of parties involved in the computations varies. When we think on an unbounded number of parties, scalability is still an issue that limits the expansion of the state-of-the-art.

Figure 3.13 describes, in general terms, how the setup for the evaluation and training of privacy PDTs would look like in the centralized model. It can be then appreciated how the multiparty model and the fully distributed model can be derived from it. It is worth mentioning that some works on the field might use models outside the ones mentioned here, specially relying on trusted parties to do some partial work or relying on setup phases.

3.6.4 Inference

After dwelling on training, let us discuss a bit about how inference typically works on multiparty setups. Literature is quite uniform on how to tackle the challenge of inference with three well-defined phases. We will explore this later on in this section. As with training, the literature is vast, and success is measured in the number of communication rounds. The preferred techniques are Garbling Circuits, Homomorphic Encryption and Secret Sharing. The prevalence of semi-honest security is not as widespread as in training.

3.6.4.1 The Three Stages Model

Obliviously, exploring a DT can be cumbersome. Through the years, the literature has identified three main challenges. One may find that the difference between different works in

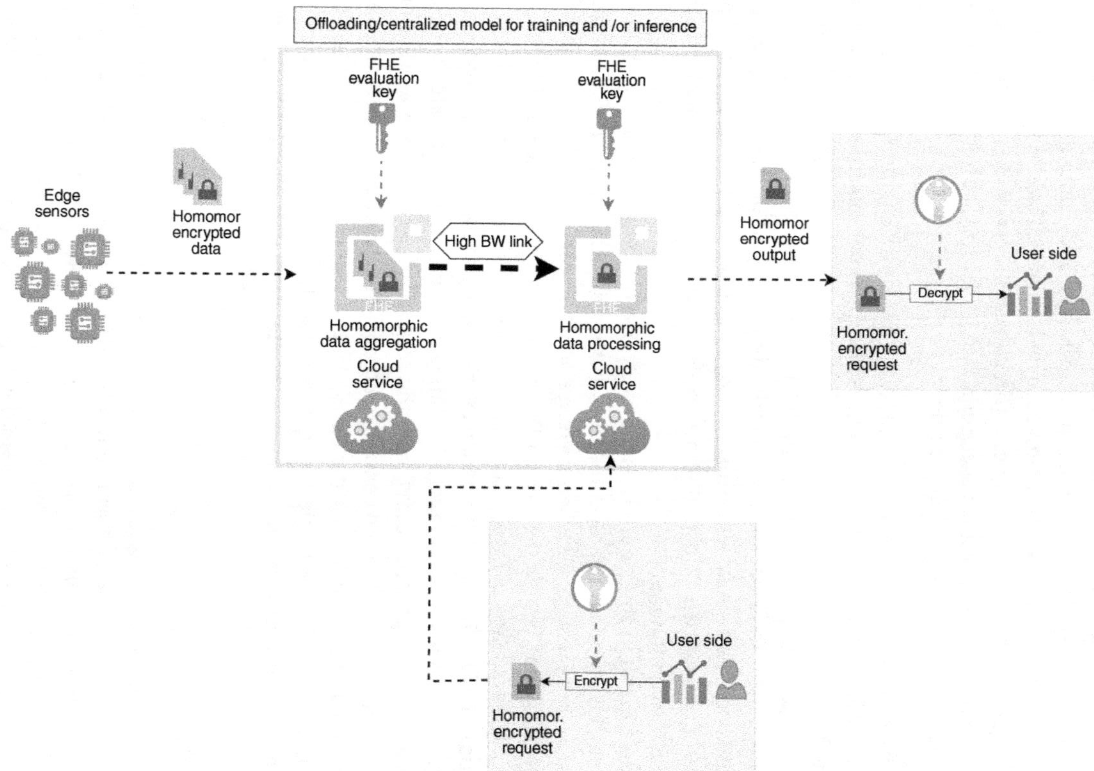

Figure 3.13 Typical scenario for offloading private computation to the cloud. Sensors transmit homomorphically encrypted data that is aggregated by a cloud or edge service, and then privately processed by another cloud service. The request is also sent encrypted by the user to the cloud service.

the area is how they address them. With time, techniques converged into three main stages or steps to evaluate a DT. Briefly speaking, they are:

1) **Feature Selection**: This is the first challenge an oblivious protocol for DT evaluation would have to face. How to correctly identify which features are evaluated at every node in the DT? Depending on the protocol, the question can be expanded to, *which features and in what order?* The privacy challenge here is clear, the model owner might not want to disclose what features belong to a DT and where.
2) **Comparison**: Once we got the features, it is time to compare the features versus the weights. As it is clearly explained in [100]. Many works aim to be constant round, and since the path should be kept secret, they tend to simply execute all comparisons in parallel all at once.
3) **Path Evaluation**: Finally, contributions consider how to actually explore the DT. Once we have all comparisons in place, how can we adapt to the cryptographic constraints imposed by previous stages to get to the leaves of the tree? It is worth noting that the path being explored needs to remain secret. Some works even consider simply exploring all paths in the tree and aggregating the results on the leaves to achieve this.

3.6.5 Relevant Contributions

3.6.5.1 On Passive Security

As with training, passive security is a popular approach among different works in the area. This is especially true for the offload model and 2-party setups. The common scenario, in this case, assumes a server holding the weights and tree structure of a DT or Random Forest and, a client holding some labeled features.

On the cryptographic side, we see two paradigms commonly used by researchers and algorithm designers: (i) Homomorphic Encryption and (ii) Garbling Circuits, typically referred to as Yao. It is worth commenting that Kiss et al. [100] systematized the general approach that most relevant works on passive security adopted, combining both GCs and HE. In this section, we give a brief intuition on how each of the three phases of the evaluation would work on both paradigms.

- **Homomorphic Encryption**: The setup is the same as we described above, a client C holds features and a server S some DT model (i.e. the weights or thresholds used). Both entities, C and S, need to evaluate the DT without learning each others' private inputs. The intuition is as follows:
  1) **Selection Phase**: Homomorphic approaches tend to rely on blinding factors: C would send public key encryptions of all her features to S in a given order. Then S can simply select the features it wants, add a blinding factor, and send them back to the client who then decrypts them and now holds a subset of features, all of which are blinded.
  2) **Comparison**: This can be realized via one of the popular protocols for comparisons that are out there; one name that you are going to find in the literature is DGK by Damgard et al. [105] and its subsequent performance improvements by other authors.
  3) **Path Evaluation Phase**: How the path evaluation works depends greatly on the output of the previous stage. Typically, here, the server is responsible for the exploration, touching every node and adding up the outcomes of the paths along the way.

- **Garbled Circuits**: Same setup as before and same general objective. Because of the nature of the underlying cryptographic technique in here, the main difference is our dependence on symmetric techniques rather than asymmetric techniques. The intuition is as follows:
  1) **Selection Phase**: The approach here is to find a function that is then modeled as a binary circuit, capable to have as an input a set of features and then output the selection needed by the server. As it described by Kiss et al. [100], some authors envision the use of Walksman's permutation network, or any other permutation network for that matter. The server then would input the specific selection bits for the switch gates so that they obtain the required client features.
  2) **Comparison**: Similarly, comparison is also implemented through Boolean circuit evaluation. Bear in mind that inputs in the case of GCs are expressed bit-wise, which makes comparisons easier. Recently, there has been the circuit proposed by Makri et al. [106], and the benchmarking of the same circuit by Aly et al. [98]. The latter measures two versions of the circuits using a different gate composition (excluding and including XOR gates) over a distributed implementation of Garbled Circuits.
  3) **Path Evaluation Phase**: Typically, it is based on the server permuting the tree and the client evaluating it. A common way to do it is to rely on a technique called Garbled Decision Tree (GDT) that borrows some common conceptions from Garbled Circuits themselves.

3.6.5.2 On Active Security

Albeit the subset of protocols that tackle the problem of private DT evaluation and provide active security is smaller, it is still significant enough to explore. The literature here points strongly to MPC protocols based on secret sharing.[4]

When we discuss active security, we typically refer to protocols involving several parties (more than two parties). Moreover, algorithms are organized in the same way as in this section, targeting the same three phases. Also, contributions in this area tend to cater for constant asymptotic complexity in terms of communication rounds. A good example of these are the contributions by Tsuchida and Nishide [107].

Here, the model is relatively different. An MPC cloud holds a vector of secret features, and a secret tree structure. We tend to assume complete trees and avoid leakage on the composition of the trees. Security comes from the underlying cryptographic constructions. Note that such cryptographic constructions can be at times slightly adapted to the needs of the DT's functionality.

However, there is a big difference between theoretical and practical efficiency. Many works, as the one above, target constant rounds given the theoretical challenge and advantage of having such bounds. In practice, however, it might turn to be much more effective to have different path exploration techniques, despite having sublinear communication round complexity. For example, if such exploration technique reduces the nonlinear functions while evaluating a DT. This is still an open line of work.

---

4 We note that this is not the only exclusive way to achieve active security and that other approaches working on different computational models might exist, however, they are not as prevalent in the literature.

3.7 Software and Frameworks

It all started with VIFF [108]. Indeed, Martin Geisler's thesis dating back to 2009, initiated a trend that continues to this day, the development of modular frameworks and compilers for MPC. That is not to say there were not frameworks that predated or were released at the same time as VIFF, like SEPIA [109] or FairplayMP [110]. But the impact of VIFF, a framework made fully over Python, opened the door to the construction of frameworks and compilers that can be extended with new protocols, could have their own Domain Specific Language (DSL), and can be more than proof of concepts. We stand far from those old days, and today we have several tools available, even dockerized, that can be adopted for PPML. We identify some of the most popular in Table 3.4. They all have different capabilities, and can offer active (A), passive (P), or covert (C) security. We consider also important to mention if they can let the user code using Arithmetic[5] (A), Boolean,[6] and/or Mixed[7] (M) circuits.

Many of the frameworks we have been discussing so far and are present on Table 3.4 rely on the online/offline model. Namely, they are capable of offloading the computation of the MPC preprocessing to an offline phase. This is the case of SCALE-MAMBA and MP-SPDZ. However, we have excluded from the list the tools that work on the somewhat similar to the Trusted Third Party (TTP) model or do not have a preprocessing phase. This is because

Table 3.4 MPC frameworks.

		Security			Protocols			
Framework	**Parties**	**A**	**C**	**P**	**A**	**B**	**M**	**Open source**
FANNG-MPC	n	✓	—	—	✓	✓	✓	✓
MPyC	n	—	—	✓	✓	—	—	✓
SCALE-MAMBA	n	✓	—	—	✓	✓	✓	✓
MP-SPDZ	n	✓	—	✓	✓	✓	✓	✓
Sharemind	n	—	✓	✓	✓	✓	—	—
ABY^3	3	—	—	✓	✓	✓	✓	✓
Obliv-C	2	—	—	✓	—	✓	—	✓
ObliVM	2	—	—	✓	✓	—	—	✓
EMP-toolkit	n	✓	—	✓	✓	—	—	✓
JIFF	n	—	—	✓	✓	—	—	✓
EzPC	2–3	—	—	✓	✓	—	—	✓

Source: Adapted from [113–115].

5 Via a member of the SPDZ protocol family for dishonest majorities, or replicated and/or Shamir secret sharing variations for honest majorities.

6 Via Garbled Circuits, colloquially referred to as Yao Circuits, and their multiparty generalizations or its Boolean sibling, GMW.

7 Via share conversion techniques, namely daBits by Rotaru and Wood [88], the evolution of daBits by Aly et al. [111] and edaBits by Escudero et al. [112].

Table 3.5 TTP-based frameworks for MPC as presented by [113–115].

		Security			Protocols			
Framework	**Parties**	**A**	**C**	**P**	**A**	**B**	**M**	**Open source**
Fresco	n	—	—	✓	✓	—	—	✓
TF-Encrypted	n	—	—	✓	✓	—	—	✓
CrypTen	n	—	—	✓	✓	—	—	✓
PySyft	n	—	—	✓	✓	—	—	✓

Source: Adapted from [63, 70, 107].

real deployments might be limited, when the model assumes that a TTP can provide the preprocessing. They, of course, are *out there* and part of the MPC for ML ecosystem. We have mentioned what we consider to be the most relevant ones in Table 3.5.

Bear in mind, this is an active space, and the complementarities between blockchain and MPC are triggering new up and coming implementations that can be generalized for ML and optimization. A good example of this is HoneyBadgerMPC [116], albeit it currently only offers semi-honest security, this might change in the future.

3.8 Lightweight FHE and MPC for IoT

FHE and MPC offer a solution for the privacy problem in IoT, but not without cost. MPC is interactive, hence it breaks one of the most fundamental and useful axioms of cloud computing: "the user sends data and forgets." With MPC, the user is forced to interact with the cloud server and to perform comparative computational effort. This is especially deleterious in the IoT scenario due to the low bandwidth and limited computational power. On the other hand, FHE is not only interactive but it also presents a similar problem in terms of bandwidth due to the FHE's ciphertext expansion.

This section presents two approaches that can circumvent these limitations: (i) MPC outsourcing; and (ii) hybrid-FHE. The former has already been described in Sections 3.5.5.4 and in 3.5.5.7. The latter consists of mixing FHE with symmetric key primitives.

3.8.1 MPC Outsourcing

MPC outsourcing trades-off trust for interactivity. It allows a user to send data and forget, but it requires trust in a set of servers. The idea is to split the data into shares, and handover these shares to a set of servers that interact directly with the cloud service providing the ML model, Figure 3.14. The shares are transmitted between the sensors to the servers through secure channels, encrypted end-to-end, hence preventing leakage in the IoT hub or other external entities.

Any MPC protocol can be used between the servers and the cloud service. Thus, if full threshold setting is chosen, the user only needs to trust one of the servers. Relaxing the level

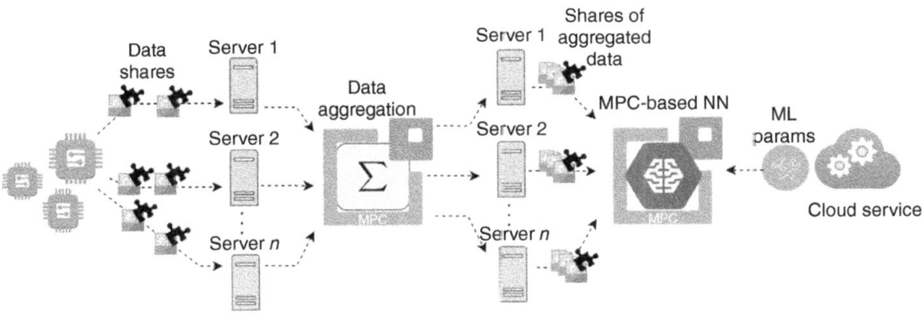

Figure 3.14 MPC outsourcing to a set of servers.

of trust also allows for less complex protocols in the honest majority setting. Moreover, the servers may also receive shares of several devices and perform data aggregation steps before engaging into private ML algorithms with the cloud service.

3.8.2 Hybrid-FHE

FHE can solve the privacy problem in the IoT scenario. However, its applicability is limited by the low hardware capabilities of IoT edge devices (in terms of computational power and energy budget) and the bandwidth constraints of current IoT communication standards. A potential solution to circumvent these limitations is Hybrid Fully Homomorphic Encryption (HFHE), as shown in Figure 3.15, which combines HE and symmetric key encryption (SKE). With HFHE, the IoT device can encrypt data using an SKE scheme. In principle, SKE schemes are much simpler than FHE schemes: (i) they require significantly lower computational effort; and (ii) are not affected by ciphertext expansion. Thus, the IoT edge devices can select a randomly generated key, and then encrypt this key with an HE scheme using the users' public key of the HE scheme. The IoT hub can convert SK-encrypted data into HE-encrypted data by performing an homomorphic evaluation of the SKE decryption circuit. This is, the IoT hub decrypts the data using the SKE key. Such key is homomorphically encrypted, thus the decrypt algorithm can only be done homomorphically, yielding homomorphically encrypted data. Thus, this process converts SK-encrypted data into HE-encrypted data. The IoT device saves bandwidth and computation. The cost is extra effort in the IoT hub, and potentially larger parameters for the HE scheme. Additionally, the IoT hub can also perform homomorphic data aggregation in order to pre-process data and reduce bandwidth consumption between the IoT hub and the cloud service, as shown in Figure 3.16.

The AES SKE scheme has already been proposed as a candidate for HHE, [117]. The reason is that IoT devices are commonly equipped with hardware acceleration for AES. Unfortunately, the AES decryption circuit requires a multiplicative depth more than 40, which must be evaluated homomorphically in the IoT hub. Adopting Public Key Encryption (PKE) can reduce multiplicative depth [118]. However, PKE is more complex than SKE, and it requires considerably more bandwidth. The latest trend is low-depth symmetric key ciphers [119, 120].

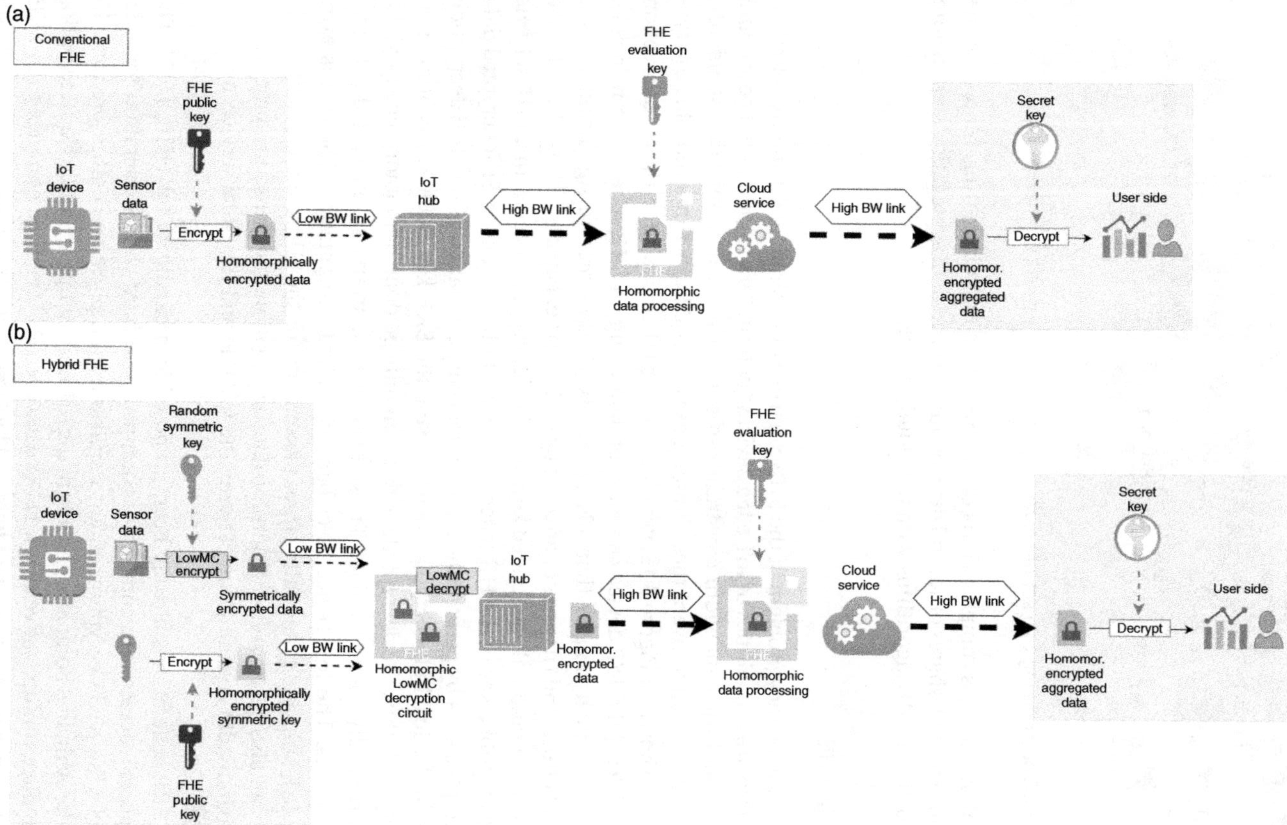

Figure 3.15 Conventional FHE (a) and hybrid-FHE (b). In hybrid-FHE, homomorphically encrypted data is transmitted through high bandwidth links, except for a short symmetric key which is transmitted only once.

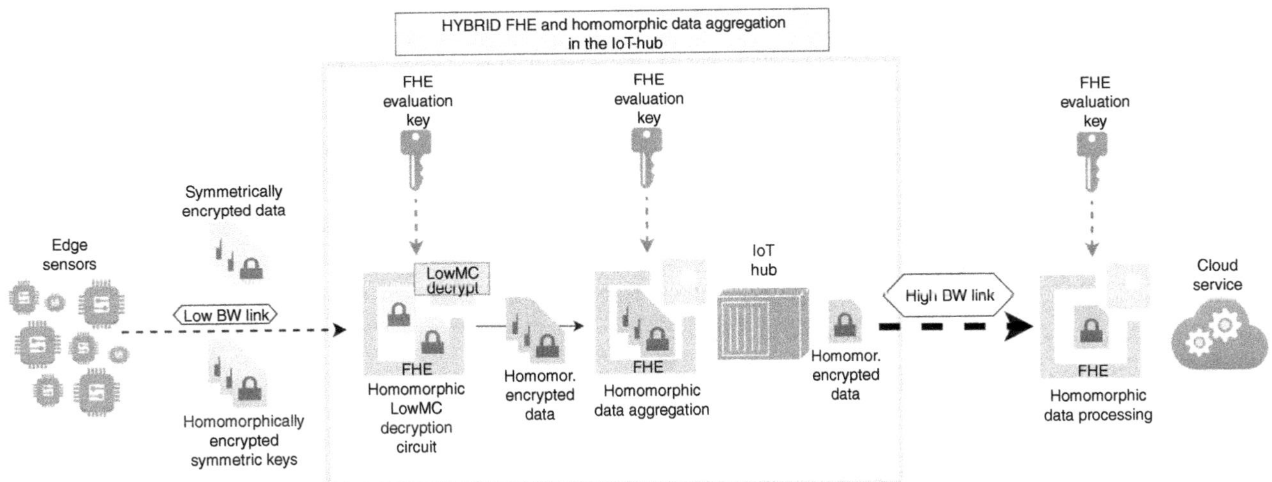

Figure 3.16 Hybrid-FHE with homomorphic data aggregation in the IoT hub.

3.9 Alternative Solutions

In addition to cryptographic primitives, other techniques have been proposed to provide privacy. Namely, model partition [121] and generative adversarial networks (GANs) [3, 122]. Model partition is based on sending the first layers of the model to the edge. The user's private data is preprocessed in the user side, and only the outcome of the first layers, specially designed to be irreversible, is delivered to the cloud service. This approach is convenient in the IoT scenario due to its simplicity. However, it does not provide the same privacy guarantees as confidential computing techniques based on MPC and FHE. GANs follow a different approach than MPC and FHE, an NN deployed in the user side converts the user's data into new random-looking data that outputs the same classification when processed by the NN in the cloud service side.

The name, adversarial network, stems from the fact that the GAN is trained against the cloud service's NN, and it tries to trick such network into producing equivalent results with fake data. However, this solution requires deployment of complex NN in the edge, and the privacy guarantees are also questionable since: (i) the user needs to trust the provider of such GAN network; and (ii) GANs cannot deliver formal proofs of privacy. Also, some results have recently shown empirically that synthetic data produced does not improve conventional anonymization techniques in terms of privacy or data usability [123].

Also, private data aggregation [1] is frequently suggested, mainly in the smart grid scenario [124, 125]. The idea is to preprocess and aggregate information from several end users to prevent leaking individual data in the cloud service. This, however, requires trust in the aggregator. It is also worth highlighting some recent works that cater for signal processing techniques based on quantization and noise addition [126, 127]. These works tackle the communication load problem and privacy issues in federated learning simultaneously. They cannot provide formal proofs of security, but they can measure privacy in terms of ϵ-differential privacy.

3.10 Conclusions

The combination of massive IoT and AI solutions will open up new possibilities in various domains, revolutionizing industries such as healthcare, finance, and transportation. AI and IoT will have the potential to gather and process vast amounts of data to enhance efficiency and decision-making, and to enable task automation in strategic fields. However, this has also raised privacy concerns which have fueled research on PPML for IoT. This chapter provides a holistic view of the main advances and limitations of MPC-based and FHE-based PPML solutions. It also highlights the need for techniques such as hybrid-FHE or MPC outsourcing to overcome the communication and computational limitations of IoT. The chapter also points out the existence of alternative approaches that do not require cryptography, such as generative models or signal processing.

References

1 Kardash, A. and Morin, S. (2022). The practices and challenges of generating nonidentifiable data. *IEEE Security and Privacy* 20 (1): 113–118. https://doi.org/10.1109/MSEC.2021.3126185.

2 Liu, Y., Guo, W., Fan, C.-I. et al. (2019). A practical privacy-preserving data aggregation (3PDA) scheme for smart grid. *IEEE Transactions on Industrial Informatics* 15 (3): 1767–1774. https://doi.org/10.1109/TII.2018.2809672.

3 Yoon, J., Drumright, L.N., and van der Schaar, M. (2020). Anonymization through data synthesis using generative adversarial networks (ADS-GAN). *IEEE Journal of Biomedical and Health Informatics* 24 (8): 2378–2388. https://doi.org/10.1109/JBHI.2020.2980262.

4 Yao, A.C. (1982). Protocols for secure computations. 23rd Annual Symposium on Foundations of Computer Science (SFCS 1982), 160–164. https://doi.org/10.1109/SFCS.1982.38.

5 Asharov, G., Lindell, Y., Schneider, T., and Zohner, M. (2013). More efficient oblivious transfer and extensions for faster secure computation. In: *Proceedings of the 2013 ACM SIGSAC Conference on Computer & Communications Security*, CCS '13, 535–548. New York, NY, USA: Association for Computing Machinery. ISBN: 9781450324779. https://doi.org/10.1145/2508859.2516738.

6 Chou, T. and Orlandi, C. (2015). The simplest protocol for oblivious transfer. In: *Progress in Cryptology – LATINCRYPT 2015. LATINCRYPT 2015*, Lecture Notes in Computer Science, vol. 9230 (ed. K. Lauter and F. Rodríguez-Henríquez), 40–58. Berlin, Heidelberg: Springer-Verlag. ISBN: 9783319221731 https://doi.org/10.1007/978-3-319-22174-8_3.

7 Hazay, C., Scholl, P., and Soria-Vazquez, E. (2017). Low cost constant round MPC combining BMR and oblivious transfer. In: *Advances in Cryptology –ASIACRYPT 2017* (ed. T. Takagi and T. Peyrin), 598–628. Cham: Springer International Publishing. ISBN: 978-3-319-70694-8.

8 Lindell, Y., Smart, N.P., and Soria-Vazquez, E. (2016). More efficient constant-round multi-party computation from BMR and SHE. In: *Theory of Cryptography* (ed. M. Hirt and A. Smith), 554–581. Berlin, Heidelberg: Springer-Verlag.

9 Ben-Or, M., Goldwasser, S., and Wigderson, A. (1988). Completeness theorems for non-cryptographic fault-tolerant distributed computation. In: *Proceedings of the 20th Annual ACM Symposium on Theory of Computing*, STOC '88, 1–10. New York, NY, USA: Association for Computing Machinery. ISBN: 0897912640. https://doi.org/10.1145/62212.62213.

10 Keller, M., Pastro, V., and Rotaru, D. (2018). Overdrive: making SPDZ great again. In: *Advances in Cryptology –EUROCRYPT 2018* (ed. J.B. Nielsen and V. Rijmen), 158–189. Cham: Springer International Publishing. ISBN: 978-3-319-78372-7.

11 Kolesnikov, V. (2005). Gate evaluation secret sharing and secure one-round two-party computation. In: *ASIACRYPT'05*, 136–155. Berlin, Heidelberg: Springer-Verlag. ISBN: 3540306846. https://doi.org/10.1007/11593447_8.

12 Zahur, S., Rosulek, M., and Evans, D. (2014). Two halves make a whole: reducing data transfer in garbled circuits using half gates. Cryptology ePrint Archive, Paper 2014/756. https://eprint.iacr.org/2014/756 (accessed 14 March 2024).

13 Beaver, D., Micali, S., and Rogaway, P. (1990). The round complexity of secure protocols. In: *Proceedings of the 22nd Annual ACM Symposium on Theory of Computing*, STOC '90, 503–513. New York, NY, USA: Association for Computing Machinery. ISBN: 0897913612 https://doi.org/10.1145/100216.100287.

14 Goldreich, O., Micali, S., and Wigderson, A. (1987). How to play any mental game. In: *Proceedings of the 19th Annual ACM Symposium on Theory of Computing*, STOC '87, –218, 229. Association for Computing Machinery. ISBN: 0897912217. https://doi.org/10.1145/28395.28420.

15 Shamir, A. (1979). How to share a secret. *Communications of the ACM* 22 (11): 612–613. https://doi.org/10.1145/359168.359176.

16 Beaver, D. (1992). Efficient multiparty protocols using circuit randomization. In: *Advances in Cryptology —CRYPTO '91* (ed. J. Feigenbaum), 420–432. Berlin, Heidelberg, Berlin, Heidelberg: Springer-Verlag.

17 Keller, M., Orsini, E., and Scholl, P. (2016). MASCOT: Faster malicious arithmetic secure computation with oblivious transfer. Cryptology ePrint Archive, Paper 2016/505. https://eprint.iacr.org/2016/505 (accessed 14 March 2024).

18 Mohassel, P. and Zhang, Y. (2017). SecureML: A system for scalable privacy-preserving machine learning. 2017 IEEE Symposium on Security and Privacy (SP), 19–38. https://doi.org/10.1109/SP.2017.12.

19 Kaviani, D. and Popa, R.A. (2023). MPC deployments. A hub for real-world MPC deployments. https://mpc.cs.berkeley.edu (accessed 14 March 2024).

20 Aumann, Y. and Lindell, Y. (2007). Security against covert adversaries: efficient protocols for realistic adversaries. In: *Theory of Cryptography* (ed. S.P. Vadhan), 137–156. Berlin, Heidelberg: Springer-Verlag. ISBN: 978-3-540-70936-7.

21 Aliasgari, M., Blanton, M., Zhang, Y., and Steele, A. (2012). Secure computation on floating point numbers. Cryptology ePrint Archive, Paper 2012/405. https://eprint.iacr.org/2012/405 (accessed 13 March 2024).

22 Archer, D.W., Atapoor, S., and Smart, N.P. (2021). The cost of IEEE arithmetic in secure computation. Cryptology ePrint Archive, Paper 2021/054. https://eprint.iacr.org/2021/054 (accessed 13 March 2024).

23 Catrina, O. and Saxena, A. (2010). Secure computation with fixed-point numbers. In: *Financial Cryptography and Data Security* (ed. R. Sion), 35–50. Berlin, Heidelberg: Springer-Verlag.

24 Liu, J., Juuti, M., Lu, Y., and Asokan, N. (2017). Oblivious neural network predictions via MiniONN transformations. In: *Proceedings of the 2017 ACM SIGSAC Conference on Computer and Communications Security*, CCS '17, 619–631. New York, NY, USA: Association for Computing Machinery. ISBN: 9781450349468. https://doi.org/10.1145/3133956.3134056.

25 Chen, H., Kim, M., Razenshteyn, I. et al. (2020). Maliciously secure matrix multiplication with applications to private deep learning. Cryptology ePrint Archive, Paper 2020/451. https://eprint.iacr.org/2020/451 (accessed 14 March 2024).

26 Rivest, R.L., Adleman, L., and Dertouzos, M.L. (1978). On data banks and privacy homomorphisms. *Foundations of Secure Computation* 4 (11): 169–180.

27 Rivest, R.L., Shamir, A., and Adleman, L. (1978). A method for obtaining digital signatures and public-key cryptosystems. *Communications of the ACM* 21 (2): 120–126.

28 ElGamal, T. (1985). A public key cryptosystem and a signature scheme based on discrete logarithms. *IEEE Transactions on Information Theory* 31 (4): 469–472.

29 Paillier, P. (1999). Public-key cryptosystems based on composite degree residuosity classes. International Conference on the Theory and Applications of Cryptographic Techniques, 223–238. Springer.

30 Boneh, D., Goh, E.-J., and Nissim, K. (2005). Evaluating 2-DNF formulas on ciphertexts. In: *Theory of Cryptography*. Theory of Cryptography Conference, TCC 2005, Lecture Notes in Computer Science, vol. 3378 (ed. J. Kilian), 325–341. Springer.

31 Gentry, C. (2009). *A Fully Homomorphic Encryption Scheme*, vol. 20. Stanford, CA: Stanford University.

32 Brakerski, Z., Gentry, C., and Vaikuntanathan, V. (2014). (Leveled) fully homomorphic encryption without bootstrapping. *ACM Transactions on Computation Theory (TOCT)* 6 (3): 1–36.

33 Brakerski, Z. (2012). Fully homomorphic encryption without modulus switching from classical GapSVP. In: *Advances in Cryptology –CRYPTO 2012* (ed. R. Safavi-Naini and R. Canetti), 868–886. Berlin, Heidelberg: Springer-Verlag.

34 Fan, J. and Vercauteren, F. (2012). Somewhat practical fully homomorphic encryption. IACR Cryptology ePrint Archive.

35 Ducas, L. and Micciancio, D. (2015). FHEW: Bootstrapping homomorphic encryption in less than a second. In: *Advances in Cryptology –EUROCRYPT 2015*, 617–640. Berlin, Heidelberg: Springer-Verlag.

36 Chillotti, I., Gama, N., Georgieva, M., and Izabachène, M. (2019). TFHE: Fast fully homomorphic encryption over the torus. *Journal of Cryptology* 33: 34–91.

37 Cheon, J.H., Kim, A., Kim, M., and Song, Y. (2017). Homomorphic encryption for arithmetic of approximate numbers. In: *Advances in Cryptology –ASIACRYPT 2017* (ed. T. Takagi and T. Peyrin), 409–437. Cham: Springer International Publishing.

38 Albrecht, M.R., Player, R., and Scott, S. (2015). On the concrete hardness of learning with errors. *Journal of Mathematical Cryptology* 9 (3): 169–203.

39 Gentry, C. (2010). Computing arbitrary functions of encrypted data. *Communications of the ACM* 53 (3): 97–105.

40 Bergerat, L., Boudi, A., Bourgerie, Q. et al. (2022). Parameter optimization & larger precision for (T) FHE. Cryptology ePrint Archive.

41 Biasioli, B., Marcolla, C., Calderini, M., and Mono, J. (2023). Improving and automating BFV parameters selection: an average-case approach. Cryptology ePrint Archive, Paper 2023/600.

42 Mono, J., Marcolla, C., Land, G. et al. (2023). Finding and evaluating parameters for BGV. International Conference on Cryptology in Africa - AFRICACRYPT 2023.

43 Albrecht, M.R., Chase, M., Chen, H. et al. (2018). Homomorphic Encryption Security Standard. *Technical Report*. Toronto, Canada: HomomorphicEncryption.org.

44 Al Badawi, A., Bates, J., Bergamaschi, F. et al. (2022). OpenFHE: Open-source fully homomorphic encryption library. Cryptology ePrint Archive, Paper 2022/915. https://eprint.iacr.org/2022/915 (accessed 13 March 2024).

45 PALISADE. https://palisade-crypto.org (accessed 14 March 2024).

46 SEAL (2019). *Microsoft SEAL (Release 3.4)*. Redmond, WA: Microsoft Research. https://github.com/Microsoft/SEAL.

47 Halevi, S. and Shoup, V. HElib. https://github.com/homenc/HElib (accessed 14 March 2014).

48 Encrypt-Everything-Everywhere. https://github.com/momalab/e3 (accessed 14 March 2024).

49 EVA. https://github.com/microsoft/EVA (accessed 14 March 2014).

50 nGraph-HE2. https://github.com/IntelAI/he-transformer (accessed 14 March 2024).

51 Transpiler. https://github.com/google/fully-homomorphic-encryption (accessed 14 March 2024).

52 Viand, A., Jattke, P., and Hithnawi, A. (2021). SoK: Fully homomorphic encryption compilers. 2021 IEEE Symposium on Security and Privacy (SP), 1092–1108. IEEE Computer Society.

53 Feldmann, A., Samardzic, N., Krastev, A. et al. (2021). F1: A fast and programmable accelerator for fully homomorphic encryption (extended version). arXiv preprint arXiv:2109.05371.

54 Reagen, B., Choi, W., Ko, Y. et al. (2021). Cheetah: Optimizing and accelerating homomorphic encryption for private inference. Proceedings of the 27th IEEE International Symposium on High Performance Computer Architecture (HPCA-27).

55 Riazi, M.S., Laine, K., Pelton, B., and Dai, W. (2020). HEAX: An architecture for computing on encrypted data. Proceedings of the 25th International Conference on Architectural Support for Programming Languages and Operating Systems, 1295–1309.

56 Roy, S.S., Turan, F., Jarvinen, K. et al. (2019). FPGA-based high-performance parallel architecture for homomorphic computing on encrypted data. 2019 IEEE International Symposium on High Performance Computer Architecture (HPCA), 387–398. IEEE.

57 Turan, F., Roy, S.S., and Verbauwhede, I. (2020). HEAWS: An accelerator for homomorphic encryption on the Amazon AWS FPGA. *IEEE Transactions on Computers* 69 (8): 1185–1196.

58 Bois, A., Cascudo, I., Fiore, D., and Kim, D. (2021). Flexible and efficient verifiable computation on encrypted data. In: *Public-Key Cryptography –PKC 2021*. PKC 2021 IACR International Conference on Public-Key Cryptography, Lecture Notes in Computer Science, vol. 12711 (ed. J.A. Garay), 528–558. Springer.

59 Fiore, D., Nitulescu, A., and Pointcheval, D. (2020). Boosting verifiable computation on encrypted data. In: *Public-Key Cryptography –PKC 2020*, IACR International Conference on Public-Key Cryptography, Lecture Notes in Computer Science, vol. 12111 (ed. A. Kiayias, M. Kohlweiss, P. Wallden, and V. Zikas), 124–154. Springer.

60 Ganeshand, C., Nitulescu, A., and Soria-Vazquez, E. (2021). Rinocchio: SNARKs for ring arithmetic. Cryptology ePrint Archive, Report 2021/322.

61 Marcolla, C., Sucasas, V., Manzano, M. et al. (2022). Survey on fully homomorphic encryption, theory, and applications. *Proceedings of the IEEE* 110 (10): 1572–1609.

62 Technology Innovation Institute (2023). Fanng-mpc. https://github.com/Crypto-TII/FANNG-MPC (accessed 14 March 2024).

63 Dowlin, N., Gilad-Bachrach, R., Laine, K. et al. (2016). CryptoNets: Applying neural networks to encrypted data with high throughput and accuracy. In: *Proceedings of*

the 33rd International Conference on International Conference on Machine Learning, vol. 48, ICML'16, 201–210. JMLR.org.

64 Rouhani, B.D., Riazi, M.S., and Koushanfar, F. (2018). Deepsecure: scalable provably-secure deep learning. In: *Proceedings of the 55th Annual Design Automation Conference*, DAC '18. New York, NY, USA: Association for Computing Machinery. ISBN: 9781450357005 https://doi.org/10.1145/3195970.3196023.

65 Juvekar, C., Vaikuntanathan, V., and Chandrakasan, A.P. (2018). Gazelle: A low latency framework for secure neural network inference. CoRR, abs/1801.05507. http://arxiv.org/abs/1801.05507 (accessed 14 March 2024).

66 Riazi, M.S., Samragh, M., Chen, H. et al. (2019). XONN: XNOR-based oblivious deep neural network inference. In: *28th USENIX Security Symposium (USENIX Security 19)*, 1501–1518. Santa Clara, CA, USA: USENIX Association. ISBN: 978-1-939133-04-5. https://www.usenix.org/conference/usenixsecurity19/presentation/riazi.

67 Rathee, D., Rathee, M., Kumar, N. et al. (2020). CrypTFlow2: Practical 2-party secure inference. In: *Proceedings of the 2020 ACM SIGSAC Conference on Computer and Communications Security*, CCS '20, 325–342. New York, NY, USA: Association for Computing Machinery. ISBN: 9781450370899. https://doi.org/10.1145/3372297.3417274.

68 Mishra, P., Lehmkuhl, R., Srinivasan, A. et al. (2020). Delphi: A cryptographic inference service for neural networks. 29th USENIX Security Symposium (USENIX Security 20). Boston, MA: USENIX Association. https://www.usenix.org/conference/usenixsecurity20/presentation/mishra (accessed 14 March 2024).

69 Patra, A., Schneider, T., Suresh, A., and Yalame, H. (2020). Aby2.0: Improved mixed-protocol secure two-party computation. Cryptology ePrint Archive, Paper 2020/1225. https://eprint.iacr.org/2020/1225 (accessed 14 March 2024).

70 Huang, Z., Lu, W.-j., Hong, C., and Ding, J. (2022). Cheetah: Lean and fast secure Two-Party deep neural network inference. In: *31st USENIX Security Symposium (USENIX Security 22)*, 809–826. Boston, MA: USENIX Association. ISBN: 978-1-939133-31-1. https://www.usenix.org/conference/usenixsecurity22/presentation/huang-zhicong.

71 Riazi, M.S., Weinert, C., Tkachenko, O. et al. (2018). Chameleon: A hybrid secure computation framework for machine learning applications. In: *Proceedings of the 2018 on Asia Conference on Computer and Communications Security*, ASIACCS '18, 707–721. New York, NY, USA: Association for Computing Machinery. ISBN: 9781450355766. https://doi.org/10.1145/3196494.3196522.

72 Knott, B., Venkataraman, S., Hannun, A. et al. (2022). CrypTen: Secure multi-party computation meets machine learning. In: *Advances in Neural Information Processing Systems 34 (NeurIPS 2021)*.

73 Gupta, K., Kumaraswamy, D., Chandran, N., and Gupta, D. (2022). LLAMA: A low latency math library for secure inference. Cryptology ePrint Archive, Paper 2022/793. https://eprint.iacr.org/2022/793 (accessed 14 March 2024).

74 Mohassel, P. and Rindal, P. (2018). ABY3: A mixed protocol framework for machine learning. In: *Proceedings of the 2018 ACM SIGSAC Conference on Computer and Communications Security*, CCS '18, 35–52. New York, NY, USA: Association for Computing Machinery. ISBN: 9781450356930. https://doi.org/10.1145/3243734.3243760.

75 Chaudhari, H., Choudhury, A., Patra, A., and Suresh, A. (2019). ASTRA: High throughput 3PC over rings with application to secure prediction. Cryptology ePrint Archive, Paper 2019/429. https://eprint.iacr.org/2019/429 (accessed 14 March 2024).

76 Wagh, S., Gupta, D., and Chandran, N. (2018). SecureNN: Efficient and private neural network training. IACR Cryptology ePrint Archive, 2018/442.

77 Koti, N., Pancholi, M., Patra, A., and Suresh, A. (2020). SWIFT: Super-fast and robust privacy-preserving machine learning. CoRR, abs/2005.10296. https://arxiv.org/abs/2005.10296 (accessed 14 March 2024).

78 Wagh, S., Tople, S., Benhamouda, F. et al. (2020). FALCON: Honest-majority maliciously secure framework for private deep learning. CoRR, abs/2004.02229. https://arxiv.org/abs/2004.02229 (accessed 14 March 2024).

79 Rachuri, R. and Suresh, A. (2019). Trident: Efficient 4PC framework for privacy preserving machine learning. CoRR, abs/1912.02631. http://arxiv.org/abs/1912.02631 (accessed 14 March 2024).

80 Byali, M., Chaudhari, H., Patra, A., and Suresh, A. (2019). FLASH: Fast and robust framework for privacy-preserving machine learning. Cryptology ePrint Archive, Paper 2019/1365. https://eprint.iacr.org/2019/1365 (accessed 14 March 2024).

81 Dalskov, A., Escudero, D., and Keller, M. (2021). Fantastic four: Honest-Majority Four-Party secure computation with malicious security. In: *30th USENIX Security Symposium (USENIX Security 21)*, 2183–2200. USENIX Association. ISBN: 978-1-939133-24-3. https://www.usenix.org/conference/usenixsecurity21/presentation/dalskov.

82 Koti, N., Patra, A., Rachuri, R., and Suresh, A. (2021). Tetrad: Actively secure 4PC for secure training and inference. CoRR, abs/2106.02850. https://arxiv.org/abs/2106.02850 (accessed 14 March 2024).

83 Zheng, W., Deng, R., Chen, W. et al. (2021). Cerebro: A platform for multi-party cryptographic collaborative learning. IACR Cryptology ePrint Archive.

84 Koti, N., Patil, S., Patra, A., and Suresh, A. (2022). MPClan: Protocol suite for privacy-conscious computations. Cryptology ePrint Archive, Paper 2022/675. https://eprint.iacr.org/2022/675 (accessed 14 March 2024).

85 Smart, N.P. and Tanguy, T. (2019). TaaS: Commodity MPC via triples-as-a-service. Cryptology ePrint Archive, Paper 2019/957. https://eprint.iacr.org/2019/957 (accessed 14 March 2024).

86 Bos, J.W., Lauter, K., Loftus, J., and Naehrig, M. (2013). Improved security for a ring-based fully homomorphic encryption scheme. In: *Cryptography and Coding* (ed. M. Stam), 45–64. Berlin, Heidelberg: Springer-Verlag. ISBN: 978-3-642-45239-0.

87 Dierckx, P. (1993). *Curve and Surface Fitting with Splines*. USA: Oxford University Press, Inc.

88 Rotaru, D. and Wood, T. (2019). Marbled circuits: mixing arithmetic and Boolean circuits with active security. In: *Progress in Cryptology –INDOCRYPT 2019* (ed. F. Hao, S. Ruj, and S.S. Gupta), 227–249. Cham: Springer International Publishing. ISBN: 978-3-030-35423-7.

89 Songhori, E.M., Hussain, S.U., Sadeghi, A. et al. (2015). TinyGarble: Highly compressed and scalable sequential garbled circuits. In: *2015 IEEE Symposium on Security* *and Privacy (SP)*, 411–428. Los Alamitos, CA, USA: IEEE Computer Society. https://doi.org/10.1109/SP.2015.32. https://doi.ieeecomputersociety.org/10.1109/SP.2015.32.

90 Kolesnikov, V. and Schneider, T. (2008). Improved garbled circuit: free XOR gates and applications. In: *Proceedings of the 35th International Colloquium on Automata, Languages and Programming, Part II*, ICALP '08, 486–498. Berlin, Heidelberg: Springer-Verlag. ISBN: 9783540705826 https://doi.org/10.1007/978-3-540-70583-3_40.

91 Dyer, E.L., Sankaranarayanan, A.C., and Baraniuk, R.G. (2013). Greedy feature selection for subspace clustering. *Journal of Machine Learning Research* 14 (1): 2487–2517.

92 Halevi, S. and Shoup, V. (2014). Algorithms in HElib. Cryptology ePrint Archive, Paper 2014/106. https://eprint.iacr.org/2014/106 (accessed 14 March 2024).

93 Halevi, S. and Shoup, V. (2018). Faster homomorphic linear transformations in HElib. Cryptology ePrint Archive, Paper 2018/244. https://eprint.iacr.org/2018/244 (accessed 14 March 2024).

94 Mishra, P., Lehmkuhl, R., Srinivasan, A. et al. (2020). Delphi: A cryptographic inference system for neural networks. In: *Proceedings of the 2020 Workshop on Privacy-Preserving Machine Learning in Practice*, PPMLP'20, 27–30. New York, NY, USA: Association for Computing Machinery. ISBN: 9781450380881 https://doi.org/10.1145/3411501.3419418.

95 Araki, T., Furukawa, J., Lindell, Y. et al. (2016). High-throughput semi-honest secure three-party computation with an honest majority. In: *Proceedings of the 2016 ACM SIGSAC Conference on Computer and Communications Security*, CCS '16, 805–817. New York, NY, USA: Association for Computing Machinery. ISBN: 9781450341394. https://doi.org/10.1145/2976749.2978331.

96 Keller, M. and Sun, K. (2021). Secure quantized training for deep learning. CoRR, abs/2107.00501, https://arxiv.org/abs/2107.00501 (accessed 14 March 2024).

97 Aly, A., Cong, K., Keller, M. et al. (2021). SCALE and MAMBA v1.14: documentation. https://nigelsmart.github.io/SCALE/.

98 Aly, A., Nawaz, K., Salazar, E., and Sucasas, V. (2022). Through the looking-glass: benchmarking secure multi-party computation comparisons for ReLU's. In: *Cryptology and Network Security: 21st International Conference*, CANS 2022, Abu Dhabi, United Arab Emirates, (13–16 November 2022), Proceedings, 44–57. Berlin, Heidelberg: Springer-Verlag. ISBN: 978-3-031-20973-4. https://doi.org/10.1007/978-3-031-20974-1_3.

99 Chatel, S., Pyrgelis, A., Troncoso-Pastoriza, J.R., and Hubaux, J.-P. (2021). SoK: Privacy-preserving collaborative tree-based model learning. *Proceedings on Privacy Enhancing Technologies* 3: 182–203.

100 Kiss, Á., Naderpour, M., Liu, J. et al. (2019). SoK: Modular and efficient private decision tree evaluation. *Proceedings on Privacy Enhancing Technologies* 2: 187–208.

101 Fang, W., Zhao, D., Tan, J. et al. (2021). Large-scale secure XGB for vertical federated learning. In: *Proceedings of the 30th ACM International Conference on Information and Knowledge Management*, CIKM '21, 443–452. New York, NY, USA: Association for Computing Machinery. ISBN: 9781450384469. https://doi.org/10.1145/3459637.3482361.

102 Leung, C., Law, A., and Sima, O. (2019). *Towards Privacy-Preserving Collaborative Gradient Boosted Decision Trees*. UC Berkeley.

103 de Hoogh, S., Schoenmakers, B., Chen, P., and op den Akker, H. (2014). Practical secure decision tree learning in a teletreatment application. In: *Financial Cryptography* *and Data Security* (ed. N. Christin and R. Safavi-Naini), 179–194. Berlin, Heidelberg: Springer-Verlag. ISBN: 978-3-662-45472-5.

104 Lindell, Y. and Pinkas, B. (2002). Privacy preserving data mining. *Journal of Cryptology* 15 (3): 177–206. https://doi.org/10.1007/s00145-001-0019-2.

105 Damgard, I., Geisler, M., and Kroigard, M. (2009). A correction to 'efficient and secure comparison for on-line auctions'. *International Journal of Applied Cryptography* 1 (4): 323–324. https://doi.org/10.1504/IJACT.2009.028031.

106 Makri, E., Rotaru, D., Vercauteren, F., and Wagh, S. (2021). Rabbit: efficient comparison for secure multi-party computation. In: *Financial Cryptography and Data Security* (ed. N. Borisov and C. Diaz), 249–270. Berlin, Heidelberg: Springer-Verlag. ISBN: 978-3-662-64322-8.

107 Tsuchida, H. and Nishide, T. (2021). Private decision tree evaluation with constant rounds via (only) fair SS-4PC. In: *Information Security and Privacy* (ed. J. Baek and S. Ruj), 309–329. Cham: Springer International Publishing.

108 Geisler, M. (2010). Cryptographic protocols: theory and implementation. PhD thesis. Aarhus University Denmark, Department of Computer Science.

109 Burkhart, M., Strasser, M., Many, D., and Dimitropoulos, X. (2010). SEPIA: Privacy-preserving aggregation of multi-domain network events and statistics. In: *Proceedings of the 19th USENIX Conference on Security*, USENIX Security'10, 15. Washington, DC, USA: USENIX Association. ISBN: 8887666655554.

110 Ben-David, A., Nisan, N., and Pinkas, B. (2008). FairplayMP: A system for secure multi-party computation. In: *Proceedings of the 15th ACM Conference on Computer and Communications Security*, CCS '08, 257–266. New York, NY, USA: Association for Computing Machinery. ISBN: 9781595938107. https://doi.org/10.1145/1455770.1455804.

111 Aly, A., Orsini, E., Rotaru, D. et al. (2019). Zaphod: Efficiently combining LSSS and garbled circuits in SCALE. In: *Proceedings of the 7th ACM Workshop on Encrypted Computing & Applied Homomorphic Cryptography*, WAHC'19, 33–44. New York, NY, USA: Association for Computing Machinery. ISBN: 9781450368292. https://doi.org/10.1145/3338469.3358943.

112 Escudero, D., Ghosh, S., Keller, M. et al. (2020). Improved primitives for MPC over mixed arithmetic-binary circuits. In: *Advances in Cryptology –CRYPTO 2020* (ed. D. Micciancio and T. Ristenpart), 823–852. Cham: Springer International Publishing. ISBN: 978-3-030-56880-1.

113 Hastings, M., Hemenway, B., Noble, Daniel, and Zdancewic, S. (2019). SoK: General purpose compilers for secure multi-party computation. 2019 IEEE Symposium on Security and Privacy (SP), 1220–1237. https://doi.org/10.1109/SP.2019.00028.

114 Keller, M. (2020). MP-SPDZ: A versatile framework for multi-party computation. In: *Proceedings of the 2020 ACM SIGSAC Conference on Computer and Communications Security*, CCS '20, 1575–1590. New York, NY, USA: Association for Computing Machinery. ISBN: 9781450370899 https://doi.org/10.1145/3372297.3417872.

115 Rotaru, D. (2023). Awesome-mpc. https://github.com/rdragos/awesome-mpc (accessed 14 March 2024).

116 GitHub, Inc. (2023). HoneybadgerMPC. https://github.com/initc3/HoneyBadgerMPC (accessed 13 March 2024).

117 Gentry, C., Halevi, S., and Smart, N.P. (2015). Homomorphic evaluation of the AES circuit (updated implementation).

118 Cheon, J.H. and Kim, J. (2015). A hybrid scheme of public-key encryption and somewhat homomorphic encryption. *IEEE Transactions on Information Forensics and Security* 10 (5): 1052–1063. https://doi.org/10.1109/TIFS.2015.2398359.

119 Méaux, P., Carlet, C., Journault, A., and Standaert, F.-X. (2019). *Improved Filter Permutators for Efficient FHE: Better Instances and Implementations*, Lecture Notes in Computer Science, vol. 11898. Springer International Publishing https://doi.org/10.1007/978-3-030-35423-7_4. https://hal.science/hal-03966901.

120 Dobraunig, C., Grassi, L., Helminger, L. et al. (2021). Pasta: a case for hybrid homomorphic encryption. Cryptology ePrint Archive, Paper 2021/731. https://eprint.iacr.org/2021/731 (accessed 14 March 2024).

121 Yu, C.-H., Chou, C.-N., and Chang, E. (2019). Distributed layer-partitioned training for privacy-preserved deep learning. 2019 IEEE Conference on Multimedia Information Processing and Retrieval (MIPR), 343–346. https://doi.org/10.1109/MIPR.2019.00068.

122 Boulemtafes, A., Derhab, A., and Challal, Y. (2020). A review of privacy-preserving techniques for deep learning. *Neurocomputing* 384: 21–45. https://doi.org/10.1016/j.neucom.2019.11.041.

123 Stadler, T., Oprisanu, B., and Troncoso, C. (2022). Synthetic data–anonymisation groundhog day. 31st USENIX Security Symposium.

124 Dias, L. and Rizzetti, T.A. (2021). A review of privacy-preserving aggregation schemes for smart grid. *IEEE Latin America Transactions* 19 (7): 1109–1120. https://doi.org/10.1109/TLA.2021.9461839.

125 Zhu, L., Li, M., Zhang, Z. et al. (2019). Privacy-preserving authentication and data aggregation for fog-based smart grid. *IEEE Communications Magazine* 57 (6): 80–85. https://doi.org/10.1109/MCOM.2019.1700859.

126 Gafni, T., Shlezinger, N., Cohen, K. et al. (2022). Federated learning: a signal processing perspective. *IEEE Signal Processing Magazine* 39 (3): 14–41. https://doi.org/10.1109/MSP.2021.3125282.

127 Lang, N. and Shlezinger, N. (2022). Joint privacy enhancement and quantization in federated learning. 2022 IEEE International Symposium on Information Theory (ISIT), 2040–2045. https://doi.org/10.1109/ISIT50566.2022.9834551.

4

Federated Learning-Based Intrusion Detection Systems for Massive IoT

Filippos Pelekoudas-Oikonomou[1], Parya H. Mirzaee[2], Waleed Hathal[2], Georgios Mantas[1,3], Jonathan Rodriguez[4], Haitham Cruickshank[2], and Zhili Sun[2]

[1] *Faculty of Engineering and Science, University of Greenwich, Chatham Maritime, UK*
[2] *Institute for Communication Systems (ICS), 5G&6G Innovation Centre (5G&6GIC), University of Surrey, Guildford, UK*
[3] *Instituto de Telecomunicações, Aveiro, Portugal*
[4] *Faculty of Computing, Engineering and Science, University of South Wales, Pontypridd, UK*

4.1 Introduction

As the development of 6G technology progresses, marking a significant shift in communications, its implications for both economic and societal aspects are substantial. 6G is characterized by its advanced technological features, and while it is expected to be commercially available in the 2030s, it represents a significant leap in wireless technology, breaking new ground in terms of connectivity [1]. 6G aims to integrate various environments such as space, air, ground, and sea, providing connectivity speeds over 1 Tbps – 10 to 100 times faster than 5G – and reducing latency to an impressive 10 to 100 μs [2].

6G technology is set to introduce a wide range of digital services, including wearable displays, implantable medical devices, mixed reality, and telepresence applications, leading to the enhancement of the Internet and the Internet of Things (IoT) [2]. However, the expansion of the IoT ecosystem with 6G, marked by a sharp increase in the number and diversity of IoT devices and connections, leading to massive IoT, presents both opportunities and challenges. The integration of these connected devices into various aspects of human life, particularly in sensitive areas like healthcare, poses significant risks and raises critical concerns in terms of data security and privacy [3]. For instance, the potential for personal data breaches from medical implants and the vulnerability of autonomous vehicles to security threats are serious concerns that need to be addressed [1]. Therefore, the development of innovative security solutions is essential to safeguard massive IoT ecosystem against future threats. These solutions should account for the limited resources typical of the massive IoT ecosystem and adhere to its demanding network performance and privacy requirements.

Security and Privacy for 6G Massive IoT, First Edition.
Edited by Georgios Mantas, Firooz B. Saghezchi, Jonathan Rodriguez, and Victor Sucasas.

Such advancements in security are crucial for the massive IoT to gain acceptance and ensure widespread adoption in the impending era of 6G technology [2].

In addressing the security challenges in the IoT and the forthcoming 6G networks, Intrusion Detection Systems (IDSs) and anomaly detection techniques have been widely adopted to safeguard networks from malicious activities. However, current centralized IDSs and anomaly detection methods face limitations in addressing the complexities and evolving nature of modern cyberattacks. Federated Learning (FL) emerges as a promising paradigm for addressing these challenges by enabling collaborative machine learning (ML) across diverse devices without compromising data privacy and without the need for direct data sharing with a centralized server [4]. This approach stands in contrast to traditional central learning models, where devices must upload their data to a server for centralized processing and learning. FL's ability to maintain data on the local devices. while only sharing model updates significantly enhances privacy and reduces the risks associated with centralized data storage [5]. The decentralized nature of FL not only alleviates privacy concerns but also contributes to reducing latency and increasing the scalability and reliability of network systems. In FL, the computational and learning tasks are distributed across numerous devices within the network while it also contains the learning server, which has the role of aggregating and broadcasting the updated learning model. This distributed approach to learning and computing is particularly advantageous in environments with a high density of devices, such as the massive IoT and those anticipated in 6G networks [6].

FL has already demonstrated its utility in various domains, including mobile and wireless networking, healthcare, and vehicle communications, and is increasingly being recognized for its potential in IoT security leading to the creation of Federated Learning-based Intrusion Detection Systems (FL-based IDSs) [4, 7]. Conventional IDSs often rely on centralized data collection and processing, which can be inefficient due to high communication costs, latency, unwillingness to share sensitive information, and the burden placed on central servers [6]. FL addresses these limitations effectively, offering a more efficient and secure alternative for tackling prevalent cyber threats such as malware and Denial of Service (DoS) or Distributed Denial of Service (DDoS) attacks [8]. For the aforementioned reasons, FL-based IDSs can be considered as a possible security solution for massive IoT.

Consequently, the aim of this chapter is to conduct an analysis of current FL-based IDSs in order to provide a roadmap to support the design and development of effective, efficient, and privacy-preserving IDS for protecting the emerging disruptive massive IoT applications in the 6G era. Following the introduction, the chapter is organized as follows: In Section 4.2, we provide the fundamentals of IDSs and ML techniques for IDSs in IoT, as well as the limitations that arise from their centralized nature. In Section 4.3, we introduce the concept of FL and explore its different categories and architectures. In Section 4.4, we present use cases, definitive examples as well as state-of-the-art research works of FL-based IDS for IoT that can be expanded into a more massive network scale, that of massive IoT. In Section 4.5, we explore the concept of model aggregation in the FL in detail, including model aggregation approaches and model aggregation algorithms. In Section 4.6, we present challenges and open research directions that require further exploration so that FL-based IDSs will become widely adopted for securing massive IoT environments. Finally, Section 4.7 concludes the chapter.

4.2 Intrusion Detection Systems (IDSs) in IoT

IDSs represent a critical component in the realm of IoT security, serving as security mechanism against unauthorized and malicious activities within IoT networks and systems. At its core, an IDS is designed to monitor and analyze the diverse array of activities occurring within a network or system, seeking to identify any anomalies or patterns indicative of potential security threats. These threats may include unauthorized access attempts, malware infections, or abnormal behavior that deviates from established rules.

4.2.1 Fundamentals of IDSs

The function of an IDS involves real-time monitoring and analysis of network or system traffic, often employing two primary methodologies [9]: signature-based detection and anomaly based detection. Signature-based detection relies on a database of known attack patterns or signatures, comparing the incoming data against this repository to identify matches. In contrast, anomaly based detection focuses on establishing a baseline of normal behavior and flagging any deviations from this baseline as potential threats. Modern IDSs often integrate both methodologies to enhance accuracy and comprehensiveness, leading to hybrid-based detection techniques.

Signature-based detection techniques operate by maintaining a repository of predefined attack signatures and comparing incoming network traffic or system actions against this signature database. Upon identifying a match, the system triggers a detection alert. While effective in opposing known attacks with established signatures, this method falls short in detecting novel or zero-day attacks, and it struggles against mutations of existing attack patterns [10]. Addressing this limitation, innovative research works have proposed alternative approaches to enhance signature-based techniques [11].

Anomaly based detection techniques operate by establishing a baseline profile of normal behavior within the monitored environment. This baseline serves as a reference for comparing system actions in real-time, with any deviations beyond the defined threshold triggering an alert without specifying the type of detected attack. While some endeavors involve the use of ML models to learn both normal and attack events for behavioral detection, the construction of normal profiles is deemed more effective as it allows for the inclusion of new attack events in real-world networks. Compared to signature-based detection techniques, anomaly based methods prove more adept at discovering novel attacks. However, a notable drawback lies in the challenge of constructing an accurate normal behavior baseline, leading to elevated false-positive rates [12]. Anomaly based detection techniques heavily rely on ML algorithms to develop these baseline profiles for monitored systems. Yet, the application of such ML techniques in resource- and energy-constrained IoT environments remains a challenge due to the considerable computational resources required for training and validating these models [13].

Hybrid-based detection techniques represent a strategic amalgamation of the previously mentioned methods, aiming to capitalize on their strengths while mitigating the weaknesses associated with detecting both known and novel attacks. In a notable example, researchers in [14] introduced SVELTE, an IDS tailored for IP-connected IoT systems utilizing RPL (Routing Protocol for Low-Power and Lossy Networks) as a routing protocol in 6LoWPAN networks. SVELTE adopts a hybrid approach, integrating both anomaly and signature-based detection techniques. This hybridization is carefully orchestrated to strike a balance between the storage and processing requirements inherent in each method. By seeking to optimize the storage cost associated with signature-based detection and the computational cost of anomaly based techniques, SVELTE exemplifies the effectiveness of hybrid-based strategies in achieving a comprehensive and resource-efficient approach to intrusion detection in IoT environments.

The implementation of IDS in IoT encompasses three essential modules. Initially, the data gathering module collects and analyzes data from various parts of IoT systems, aiming to recognize normal patterns of interaction. This early stage analysis is crucial for detecting potentially malicious behavior. Subsequently, the analysis module processes the collected data, employing various techniques and methods. The final module includes a mechanism for reporting the attacks.

ML and Deep Learning (DL) methods are well-suited for thoroughly analyzing data and acquiring knowledge about both normal and unusual behaviors within IoT ecosystems. ML/DL methods, deployed in the analysis module of an IDS, exceed in predicting new and evolving attack patterns by training and learning from existing legitimate samples. This predictive capability is especially valuable in the dynamic landscape of IoT security, where emerging threats may differ significantly from known attack patterns.

In the context of the massive IoT, where a multitude of interconnected devices communicate and share data, the implementation of IDS holds immense value. An IDS tailored for massive IoT environments can provide enhanced security by actively monitoring and analyzing the vast amount of data exchanged between IoT devices. By identifying and responding to anomalous behavior or potential security incidents, an IDS contributes to the overall resilience of massive IoT ecosystems, safeguarding sensitive data and ensuring the integrity and functionality of connected devices. In essence, the integration of IDS in massive IoT serves as a proactive measure, fortifying the security posture of interconnected systems and mitigating the risks associated with cyber threats in this dynamically evolving landscape.

4.2.2 Machine Learning Techniques for IDSs in IoT

In this section, we present the most popular ML techniques for IDSs in IoT. We start with Naive Bayes (NB) classifier, based on Bayes' theorem, that leverages previous observations of similar events to predict the probability of the occurrence of an event. In IDSs, this algorithm proves valuable for classifying normal and abnormal behaviors based on historical observations in supervised learning scenarios. Renowned for its simplicity and ease of implementation, the NB classifier calculates posterior probabilities to make labeling decisions, distinguishing between normal and anomalous traffic using independent features such as status flags, protocol, and latency. Despite its simplicity and efficiency, the NB classifier has limitations, particularly in its failure to account for interdependencies between features during classification, impacting its accuracy [9].

The K-Nearest Neighbor (KNN) algorithm is known for its simplicity, relying on a distance function, typically the Euclidean distance function, to determine the class of an object based on its proximity to its closest neighbors. The parameter "K" represents the number of nearest neighbors considered during the classification process. The choice of "K" can impact the classification outcome, making it crucial to find the optimal value for enhanced accuracy. This optimal value is determined through testing, although this process can be notably time-consuming in some cases [15].

Decision Trees (DTs) function by extracting features from dataset samples and organizing them into an ordered tree structure based on the values of these features. Each feature is represented by a node in the tree, with its corresponding values depicted by branches stemming from that node. The optimal feature node for dividing the tree is identified as the origin node. The DT algorithm involves two key processes: induction and inference. During induction, the model is constructed by adding nodes and branches through a process of feature selection based on information gain. This process aims to minimize overlap between different classes in the training dataset, enhancing the classifier's accuracy in identifying distinct instances. The leaves of each sub-DT are then classified according to their corresponding classes. The inference process involves classifying unknown instances by iteratively comparing them to the constructed DT. In the context of intrusion detection, DTs have shown potential as classifiers, with research utilizing them to detect DDoS attacks in IoT environments through the analysis of network traffic for identifying malicious sources [16].

Random Forests (RFs) represent a supervised ML algorithm characterized by the integration of multiple DTs to provide more accurate and resilient classification outcomes [17]. These DTs are constructed randomly and are trained to provide classification results through a majority voting mechanism. While DTs and RF share similarities, they are distinct in their approaches; DTs build a rule-set during training for subsequent classification, whereas RF constructs a rule-subset utilizing all member DTs. This distinction contributes to a more robust and accurate output, is resistant to overfitting, and requires substantially fewer inputs without the need for feature selection. Studies emphasize the suitability of RF for anomaly and intrusion detection in IoT networks [18, 19]. Notably, research has demonstrated the superiority of RF over methods like KNN and Support Vector Machines (SVMs) in detecting DDoS attacks in IoT networks [18, 19]. This advantage is attributed to RF's ability to operate with fewer input features and circumvent the heavy computational demands associated with real-time feature selection in IDS.

SVMs represent a formidable classifier, operating by establishing a hyperplane within the feature set of two or more classes. This splitting hyperplane is strategically positioned to maximize the distance from the nearest data point of each compared class. SVMs are suitable for scenarios where classification is required for classes with large feature sets based on a limited number of data samples. Particularly suited for anomaly detection tasks involving the classification of normal and abnormal classes, SVMs exhibit scalability and real-time capabilities, including online learning. Numerous research studies evaluating SVM-based IDSs in IoT systems have consistently demonstrated superior accuracy compared to other ML algorithms, including DTs, NB, and RFs [20].

K-means clustering stands as an unsupervised algorithm designed for the discovery of k clusters within data samples, where each instance is allocated to a specific cluster based on its features. The distribution of samples across k clusters relies on the estimation of centroids using squared Euclidean distance. The iterative process involves the recalibration of cluster centroids by calculating the mean of data points assigned to each cluster. Iterations persist until no further modifications to the clusters can be made. Noteworthy limitations include the necessity to appropriately select the value of k and the assumption of an equal distribution of the sample dataset across these clusters.

Ensemble Learning (EL) is a powerful approach that capitalizes on the strengths of multiple classifiers, combining their results to achieve a majority vote for classification. This technique significantly enhances classification accuracy by leveraging outputs from diverse homogeneous or heterogeneous classifiers. The foundation of EL lies in recognizing that no single ML algorithm can serve as a universally effective solution across all applications and datasets [21]. For generalized applications, EL becomes particularly advantageous, maximizing accuracy through variance reduction and mitigating the risk of overfitting. While EL's accuracy is a clear advantage, it comes at the cost of increased time complexity, attributable to the parallel use of multiple classifiers. In the case of IDSs, EL's effectiveness has been explored in various studies [22, 23].

4.2.3 Limitations of ML-Based IDSs in IoT

ML-based IDSs for IoT have gained significant traction in recent years due to their ability to effectively detect and prevent IoT attacks. However, ML-based IDSs, characterized by a centralized nature, face the following challenges in the IoT context, which can hinder their effectiveness [24]:

- **Data Privacy Concerns**: Centralized ML-based IDSs typically require centralized data collection and processing, raising concerns about data privacy and security [25]. IoT devices often generate sensitive data, and centralized storage can expose this data to unauthorized access and potential breaches. Moreover, sharing sensitive data with a central server can violate data privacy regulations and erode trust among users.
- **Scalability and Communication Overhead**: Centralized ML-based IDSs often struggle to scale effectively in large-scale IoT deployments. With the expansion of IoT to massive IoT and a large number of IoT devices generating data, centralized processing can become computationally expensive and lead to bottlenecks. Additionally, the communication overhead for sending raw data to a central server can be significant, especially in bandwidth-constrained environments [26].
- **Heterogeneity and Label Scarcity**: IoT devices often have diverse data collection capabilities and generate data with varying features and characteristics. ML-based IDSs may struggle to handle this heterogeneity, leading to suboptimal performance and reduced detection accuracy. Moreover, in some IoT applications, labeled data for training ML models may be scarce, making it difficult to train accurate models.

FL emerges as a promising solution to address the limitations of centralized ML-based IDSs for massive IoT. FL enables collaborative model training without compromising data privacy by keeping raw data on local devices; it also enhances the scalability and communication efficiency of ML-based IDSs for IoT and can leverage unlabeled data from multiple participants to improve model training [27]. Participants train a shared model locally on their own data and only exchange aggregated model parameters with a central server. In the next section, we will examine FL further.

4.3 Federated Learning: A Decentralized ML Approach

In this section, we discuss FL and explain its basic properties, different categories, and architectures. FL, also referred to as collaborative learning, emerges as an innovative ML approach that diverges from traditional centralized methods. This approach involves training algorithms across multiple independent sessions, each utilizing its distinct dataset. In contrast to centralized methodologies, where local datasets are merged into a single training session, FL facilitates the training of ML models across distributed nodes without the explicit exchange of raw data samples. This novel approach effectively addresses crucial concerns related to data privacy, access rights, and the integration of heterogeneous datasets [28]. The capability of FL to construct a robust and shared model without compromising individual datasets positions it as a promising solution in the pursuit of privacy-preserving and secure ML paradigms. The FL framework operates through iterative interactions between clients and a central server. In each iteration:

- **Model Distribution**: The central server coordinates the training process by distributing a global model to a subset of clients selected using various strategies (e.g. random sampling, weighted selection based on device capabilities). This global model can be a pretrained model or the result of previous training rounds.
- **Local Training**: Each selected client trains the received global model locally on its own private dataset. This local training leverages techniques like Stochastic Gradient Descent (SGD) or its variants to update the model parameters based on the client's data. Importantly, the raw data itself never leaves the client device, ensuring data privacy.
- **Parameter Aggregation**: After local training, clients upload their updated model parameters, not the raw data, back to the central server. These updates can be further processed through techniques like differential privacy to provide additional privacy guarantees.
- **Global Model Update**: The central server aggregates the received model updates from all participating clients. This aggregation can be performed using various methods, like averaging or weighted averaging, which consider factors like client data quality or device capabilities. The aggregated update is then applied to the global model, resulting in a new, improved model.
- **Model Re-distribution**: The updated global model is then distributed back to the clients, initiating the next iteration of the training process. This iterative process continues until a predefined convergence criterion is met, signifying that the model has achieved the desired level of performance.

4.3.1 Definition of Federated Learning

FL involves building ML models using datasets distributed across multiple devices, with a focus on preventing data leakage [29, 30]. To be more specific, we can provide a mathematical definition of FL. FL is a collaborative ML approach involving multiple data owners, denoted as $\{F_1, \ldots, F_N\}$. Each owner possesses their own dataset $\{D_1, \ldots, D_N\}$. Unlike conventional methods where all data sets are combined into a single dataset $D = D_1 \cup \ldots \cup D_N$ to train a model (denoted as M_{SUM}), FL allows each data owner F_i to contribute to the training of a collective model M_{FED} without exposing their individual data D_i to other participants. The key aspect of this system is maintaining the confidentiality of each participant's data while collaboratively training a model. The effectiveness of the FL model, M_{FED}, is measured by its accuracy (V_{FED}), which should be comparable to the accuracy of the conventionally trained model, M_{SUM} (V_{SUM}). Formally, FL is considered successful in terms of accuracy if the difference between V_{FED} and V_{SUM} is less than a small, nonnegative real number δ. This condition, represented as $| V_{FED} - V_{SUM} | < \delta$, is referred to as δ-accuracy loss, indicating a minimal loss in accuracy despite the decentralized nature of the learning process [7].

4.3.2 Categories of Federated Learning

To effectively apply FL, the distribution of data across participating parties is considered. This is why FL is categorized into the following three primary types based on the distribution of data [31]:

- **Horizontal Federated Learning (HFL)**: In HFL, data is distributed across parties with the same feature space but distinct sample IDs. Each party possesses a unique collection of samples, but the features they represent are identical. This scenario is prevalent in collaborative learning settings, where data owners contribute anonymized or pseudonymized data from their respective datasets.
- **Vertical Federated Learning (VFL)**: VFL emerges when parties share the same sample IDs but possess different feature spaces. Each party contributes distinct features to the same samples, potentially reflecting specialized data domains or access restrictions. This approach is particularly useful when data is sensitive or regulated, enabling collaboration without compromising privacy.
- **Federated Transfer Learning (FTL)**: FTL tackles the challenge of data distribution disparities, where parties hold data with distinct feature spaces and sample IDs. This scenario arises when data originates from heterogeneous sources with varying data formats or collection methods. FTL aims to leverage the strengths of each party's data to train a unified model, bridging the gaps in data representation.

Figure 4.1 illustrates the fundamental concepts of HFL, VFL, and FTL in a two-party scenario. HFL involves parties contributing separate data slices with the same feature structure. VFL depicts parties sharing the same sample data but contributing different feature sets. FTL represents a more complex scenario, where parties possess distinct feature spaces and sample IDs.

The selection of FL approach depends on the specific characteristics of the distributed data. HFL is suitable for scenarios with shared feature spaces and independent sample IDs, while VFL is applicable when parties share the same sample IDs but contribute unique feature sets. FTL is particularly useful in situations where data distribution disparities hinder traditional FL methods.

4.3.3 Federated Learning Architectures

While we have explained the three categories of FL based on the distribution of data, in this section we focus on their corresponding architecture and their advantages.

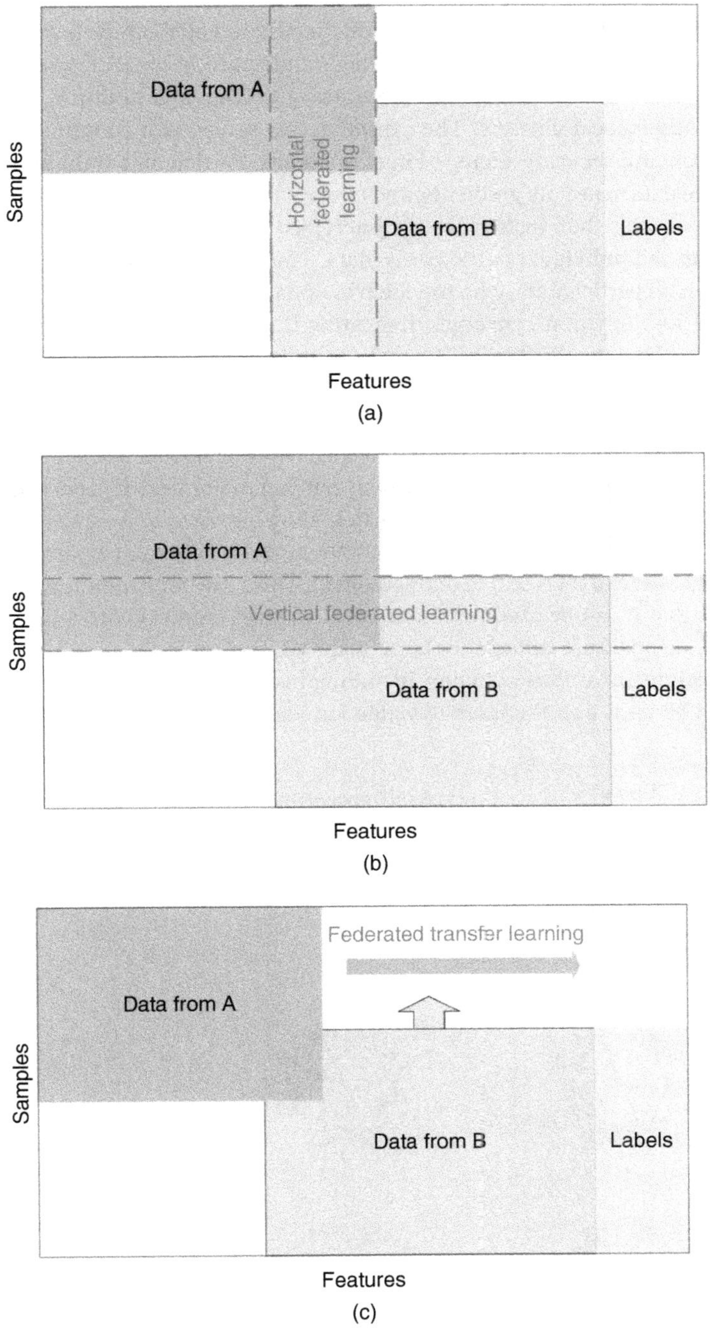

Figure 4.1 FL is categorized into three primary types based on the distribution of data: (a) horizontal federated learning (HFL), (b) vertical federated learning (VFL), and (c) federated transfer learning (FTL). Source: Filippos Pelekoudas-Oikonomou.

4.3.3.1 Horizontal Federated Learning (HFL)

HFL is a distributed ML paradigm that enables multiple parties to collaboratively train an ML model without revealing their sensitive data. The architecture of an HFL system typically involves a central server that coordinates the training process and multiple participants who contribute their local datasets. The central server serves as a coordinator, facilitating communication and securing aggregation of gradients. Participants train local models on their respective data, compute gradients, and encrypt them before sending them to the central server. The server then aggregates the encrypted gradients without learning any information about the individual participants' data. The aggregated gradients are decrypted and shared with all participants, who use them to update their local models. This process iterates until the loss function converges, indicating that the model has learned the underlying patterns in the data effectively. The training process of HFL is depicted in Figure 4.2.

HFL offers several advantages over traditional ML approaches [32]. First, it preserves data privacy by keeping sensitive data on the participants' devices. This means that participants do not have to share their data with a central server, which can be a major security concern. Second, HFL enables collaboration among multiple parties, allowing them to leverage the collective knowledge of their datasets to train more accurate models. This can be particularly beneficial in domains where data is scarce or fragmented. Third, the HFL architecture is scalable, which means that it can be used to train models on large datasets with a large number of participants. This makes it suitable for large-scale ML applications.

HFL has found applications in various domains, including healthcare, finance, and the IoT. In the IoT, HFL can be used to train anomaly detection and predictive maintenance

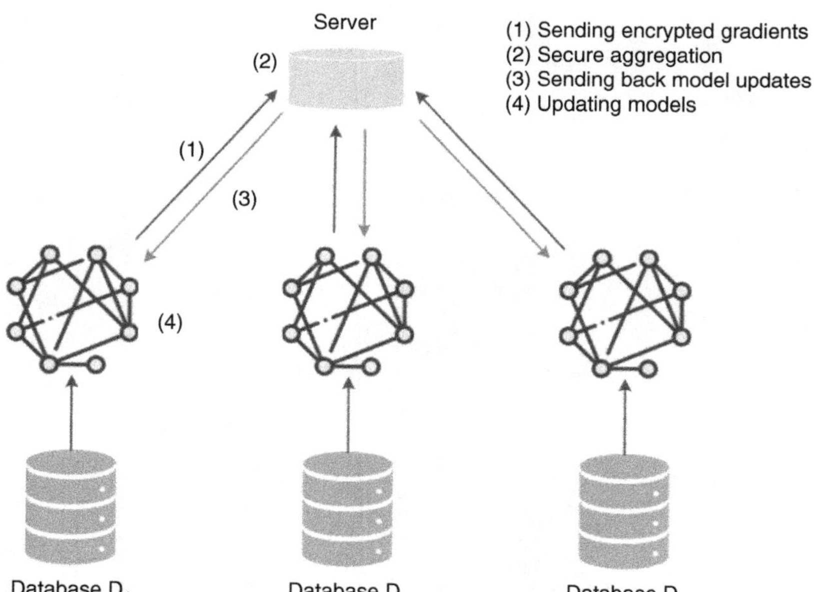

Figure 4.2 HFL architecture. Source: Filippos Pelekoudas-Oikonomou.

models. This can help to prevent equipment failures and downtime, while also improving the overall efficiency and reliability of IoT systems.

4.3.3.2 Vertical Federated Learning (VFL)

VFL is a privacy-preserving collaborative ML approach that enables multiple parties to train an ML model together without sharing their raw data. The architecture of a VFL system involves a central server and multiple participants, each contributing their own data with distinct feature spaces representing specific aspects or attributes of the target domain [33]. Participants retain ownership and control over their data throughout the training process. The central server acts as a trusted intermediary, facilitating secure communication and aggregation of partial information from participants. The server does not have access to participants' raw data, ensuring privacy.

The training process of VFL involves several steps to ensure privacy and collaboration among participants with distinct datasets. As a preliminary step, before training the model, the system performs encrypted entity alignment to identify the common entities shared by participants. This process utilizes encryption techniques to align user IDs without revealing sensitive information. The training process starts with key distribution in which collaborator C (central server) generates encryption pairs and distributes the public key to participants A and B. As a second step, participants A and B exchange encrypted intermediate results for gradient and loss calculations. The following step includes encrypted gradient and loss calculation in which participants A and B compute encrypted gradients and loss values while adding additional privacy-preserving masks. Participant B also computes encrypted loss values using its label data. Following the encrypted gradient and loss calculation, participants A and B send their encrypted gradients, loss, and masks to collaborator C in the encrypted gradient and losses aggregation step. Then, the collaborator C decrypts the encrypted gradients and loss, applies additional privacy-preserving transformations, and re-encrypts the information before sending it back to participants A and B. After this step, participants A and B receive the re-encrypted gradients and unmask them to obtain the original gradients. They then update their local models based on the decrypted gradients and masks. Finally, once the model has converged, participants can share the final model parameters with each other, allowing them to collaborate further or deploy the model for inference tasks. The training process of VFL is depicted in Figure 4.3.

VFL offers several advantages over traditional ML approaches. By employing encrypted entity alignment and secure aggregation techniques, VFL can train models on heterogeneous data without revealing sensitive information or compromising privacy [34]. This collaborative approach leverages the collective knowledge of distributed data, leading to more accurate and robust models [35]. Finally, VFL is designed to handle a large number of participants and datasets, making it more scalable and suitable for large-scale ML applications. The architecture efficiently manages distributed data and collaborative training, enabling the development of models with broader applicability.

4.3.3.3 Federated Transfer Learning (FTL)

FTL extends the capabilities of VFL by enabling collaborative training on data with both overlapping and nonoverlapping features. This approach is particularly useful when the

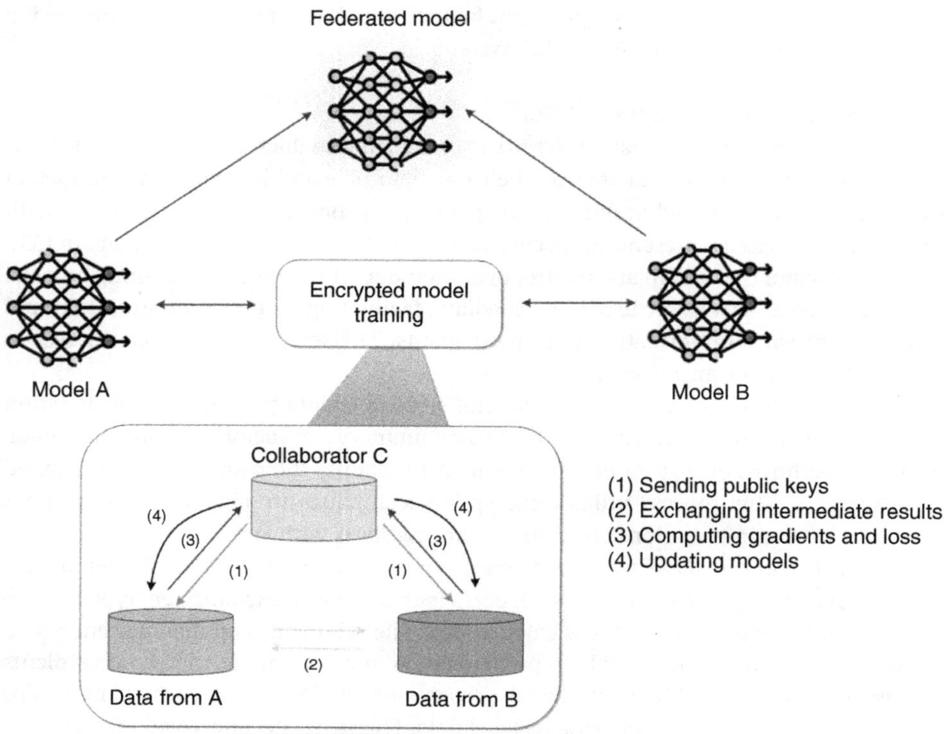

Figure 4.3 Training process of VFL. Source: Filippos Pelekoudas-Oikonomou.

overlap between datasets is limited but the parties possess complementary knowledge that can be leveraged to improve model performance. The FTL architecture incorporates the core components of VFL while introducing additional mechanisms to handle data heterogeneity across parties. Similar to VFL, FTL involves multiple participants with unique datasets. However, these datasets may have limited overlap in terms of features and/or labels. The central server acts as a secure intermediary, facilitating communication and aggregation of partial information from participants. As in VFL, the server does not have access to participants' raw data.

During the training process, similar to VFL, FTL performs encrypted entity alignment. Parties A and B exchange not only encrypted gradients and loss values but also intermediate representations of their data. This enables them to learn a common representation that captures the shared knowledge across domains. Gradient computations for parties A and B differ from VFL due to the incorporation of transfer learning. Party A focuses on predicting labels for its nonoverlapping data, while Party B leverages its labeled data to guide the learning process and refine the common representation. Collaborator C aggregates the encrypted gradients and intermediate representations, ensuring privacy and confidentiality during the aggregation process. The aggregated information is then used to update the model parameters for both parties. Inference in FTL requires both parties to participate. Party A uses the updated model parameters to predict labels for its nonoverlapping data, while Party B utilizes its labeled data to provide additional guidance.

FTL offers several advantages over traditional ML approaches and VFL. Unlike VFL, which is limited to scenarios with overlapping data, FTL enables training on entire datasets, even with limited overlap in features and labels. This is achieved by leveraging transfer learning techniques, which allow participants to share knowledge across domains and improve model performance without compromising data privacy. FTL's hybrid approach seamlessly integrates VFL and transfer learning, making it suitable for a wide range of data scenarios, including those with limited overlap. Additionally, FTL facilitates collaboration among multiple parties and can accommodate large-scale datasets, making it a scalable and versatile tool for collaborative ML [36].

4.4 Federated Learning-Based IDSs for IoT

In this section, we present use cases, definitive examples as well as state-of-the-art research works of FL-based IDS for IoT that can be expanded into a more massive network scale, that of massive IoT. Figure 4.4 illustrates IoT-relevant use cases where FL-based IDSs can be applied.

4.4.1 FL-Based IDSs in IoT Use Cases

In this section, we present FL-based IDSs that have been applied to IoT use cases.

4.4.1.1 FL-Based IDSs in Smart Homes

Smart homes are becoming increasingly popular, filled with connected devices like thermostats, lights, and cameras. These devices generate a constant stream of sensing data,

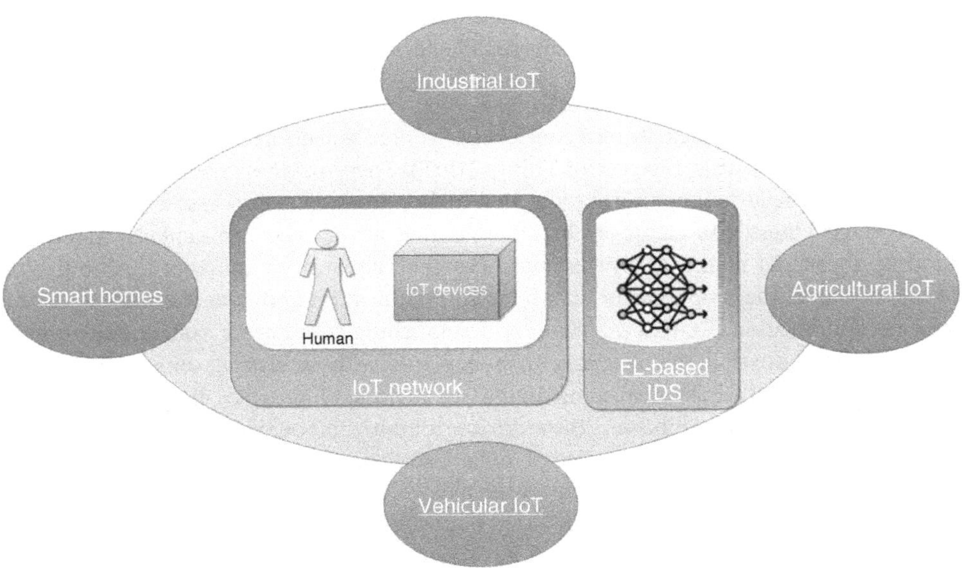

Figure 4.4 FL-based IDS in IoT use case. Source: Filippos Pelekoudas-Oikonomou.

offering convenience and automation. However, this data can also reveal a lot about the habits and routines of the residents, raising privacy concerns. Traditional IDSs in smart homes might require sending this raw data to a central server for analysis. While this approach can be effective for detecting security threats, it comes at the cost of user privacy. FL offers a new approach that balances security and privacy in smart home environments.

FL can offer local anomaly detection with Gate Recurrent Units (GRUs) suggested in the research by Mothukuri et al. [37] for the purpose of creating models that are trained on historical sensor data collected by the device itself. GRUs are a type of artificial neural network particularly adept at handling sequential data, like the continuous stream of readings from sensors. Over time, the GRU model on a sensing device learns to identify normal patterns in temperature fluctuations specific to each environment or smart home. The devices periodically share only updates to their local models with a coordinator. The key advantage of FL is that these updates essentially capture the anomaly detected without revealing the specific sensor data that triggered it, hence ensuring privacy and anonymity. The coordinator then aggregates these model updates from all devices in the network.

This aggregated information creates a global intrusion detection model. This model incorporates learnings from various sensors, making it more robust and adaptable to diverse threats. The updated global model is then sent back to individual devices. Each local anomaly detection model is not just based on its own data but also incorporates knowledge from all the participating devices, strengthening its ability to detect novel or evolving threats.

In FL-based IDSs for smart homes, user privacy is protected because raw sensor data stays local. Additionally, the system benefits from collaborative learning, resulting in improved intrusion detection capabilities. However, there are challenges to consider. Firstly, data patterns can vary significantly across different smart homes, impacting the effectiveness of the FL model. Secondly, strategies are needed to ensure the integrity of model updates and prevent malicious actors from manipulating the system. Finally, running an FL model might require more processing power on individual devices compared to simpler methods.

4.4.1.2 FL-Based IDSs in Industrial IoT

Industrial facilities rely heavily on a complex network of sensors and machines, collectively known as the Industrial Internet of Things (IIoT). These devices generate a plethora of data that is crucial for monitoring and optimizing production processes. However, this interconnectedness also creates vulnerabilities that adversaries can exploit. Traditional security measures might involve isolating the IIoT network or utilizing a central IDS. While effective, these approaches can limit communication and data sharing, hindering process optimization. FL-based IDSs offer a compelling solution for collaborative threat detection in industrial IoT settings. In a network with industrial sensors and other network-connected mechanical components and devices, FL provides distributed learning with model updates. FL empowers these devices to train their own local anomaly detection models based on the data they collect. Unlike traditional methods, these models are not built on a central server with access to all the data.

Authors introduce MV-FLID, an FL-based IDS offering a multifaceted approach to attack detection, classification, and mitigation [38]. This means each device can learn from different aspects of the data (temperature, pressure, etc.), creating a more comprehensive understanding of normal operations. Crucially, these devices do not share their raw production data. Instead, they periodically share only updates to their local models with a central coordinator.

The coordinator then aggregates these model updates from all the devices in the factory. By combining the learnings from various devices, the coordinator builds a more robust global intrusion detection model. This model incorporates knowledge about normal operations across different parts of the production line, making it more adept at detecting subtle deviations that might indicate a potential attack. The updated global model is then distributed back to individual devices. Each device's local model is not just based on its own data but also incorporates knowledge from all the other devices in the network. This collaborative learning significantly improves the overall ability to detect novel or evolving threats.

Evaluation results demonstrate MV-FLID's superior performance compared to traditional, non-decentralized approaches. Beyond its accuracy, MV-FLID boasts several advantages:

1. **Applicability**: Adaptable to various network architectures and communication protocols, including Message Queuing Telemetry Transfer protocol (MQTT), Transmission Control Protocol (TCP), and Modbus. It extends to different device types by focusing on network traffic analysis instead of relying solely on device specifications.
2. **Scalability**: FL empowers efficient training even with limited data, making it suitable for diverse scenarios irrespective of the communication protocol or specific end devices.
3. **Security Enhancements**: By aligning with secure multiparty computation principles, MV-FLID protects individual device data from exposure and promotes differential privacy.

FL-based IDSs offer substantial advantages for industrial IoT security. Firstly, sensitive production data remains confidential as raw data is never shared. Secondly, the collaborative learning approach leads to a more comprehensive and adaptable IDS. However, challenges exist. FL models can be impacted by variations in data collected by different devices. Additionally, ensuring the integrity of model updates and preventing malicious actors within the network are crucial considerations.

4.4.1.3 FL-Based IDSs in Agricultural IoT

Agricultural IoT, with its diverse stakeholders and distributed data across the supply chain, poses unique challenges for IDS. Centralized learning approaches raise privacy concerns and struggle with massive amounts of data and varying device capabilities. To address these issues, authors in [39] propose FELIDS, an FL-based IDS specifically designed for the security of Agricultural IoT networks.

FELIDS prioritizes data privacy by employing local learning on individual devices. Each device trains its own model on its local data and only shares the model updates with an aggregation server. This server combines these updates to create an improved global model without ever accessing the raw data from individual devices. To effectively combat a wide range of attacks, FELIDS leverages three DL techniques: Deep Neural Networks (DNNs), Convolutional Neural Networks (CNNs), and Recurrent Neural Networks (RNNs). Each of these approaches has its own strengths and weaknesses, and by combining them, FELIDS aims to achieve comprehensive and accurate intrusion detection.

The effectiveness of FELIDS is rigorously evaluated using three real-world datasets: CSE-CIC-IDS2018 [40], MQTTset [41], and InSDN [42]. These datasets represent diverse real-world scenarios and challenges faced by Agricultural IoT networks. The results demonstrate that FELIDS achieves competitive accuracy compared to centralized models, even surpassing them in some cases. This highlights its ability to protect data privacy while maintaining high detection performance.

Future work of the authors based on FELIDS focuses on enhancing its robustness against adversarial attacks, where malicious actors attempt to manipulate the learning process. Additionally, the system's efficiency will be studied using the recently released Edge-IIoTset dataset [43], ensuring its scalability and practicality for large-scale Agricultural IoT deployments.

4.4.1.4 FL-Based IDS in Vehicular IoT Networks

Traditional IDS for Vehicular Ad-hoc Networks (VANETs) rely on centralized training of ML models [44]. This approach faces challenges in terms of privacy concerns, as raw data needs to be uploaded to a central server and limited computation power on individual vehicles. FL offers a promising solution to address these limitations.

Vehicles in a VANET can act as client nodes in the FL framework. FL as a distributed learning technique can train models on these individual nodes without sharing raw data. The FL server coordinates the training process by sending a global model to selected vehicles. These vehicles train the model locally using their own data and upload the updated model parameters back to the server. The server aggregates these updates to create a new, improved global model, which is then distributed back to the vehicles for further training. This iterative process continues until a desired level of accuracy is achieved.

There are several advantages to using FL for intrusion detection in VANETs:

- **Enhanced Privacy**: Raw data remains on individual vehicles, protecting sensitive information about driving habits and vehicle performance.
- **Reduced Communication Overhead**: Only model updates, significantly smaller than raw data, are transmitted, minimizing data transfer and bandwidth usage.
- **Improved Scalability**: The workload is distributed across vehicles, making the system highly scalable to a large number of vehicles in the vehicular IoT network.
- **Strengthened Accuracy**: Collaborative training leverages data from all vehicles, leading to a more robust intrusion detection model that can adapt to evolving threats.

By enabling privacy-preserving and collaborative learning, FL empowers vehicular IoT networks to effectively detect intrusions while ensuring data privacy and efficient resource utilization. This paves the way for a more secure and reliable transportation ecosystem.

4.4.2 Cross-layer FL for Lightweight IoT Privacy-Preserving IDS

In [45], the authors propose a lightweight and privacy-preserving federated IDS, named cross-layer federated IDS. This system is able to address challenges at two layers:

1. **Sampling**: Utilizing cluster-based sampling with K-means reduces memory and processing overhead by extracting representative subsets of network traffic.

2. **Detection**: A lightweight IDS based on the baseline K-means algorithm provides efficient anomaly detection with minimal computational requirements.

To preserve privacy, only model statistics and anonymized summaries of local data are shared with a central coordinator for collaborative learning without exposing sensitive information. Additionally, limited information exchange between nodes makes the system robust against adversarial attacks and network disruptions, even allowing local functionality in isolation.

Evaluation results using the NSL-KDD dataset [46] demonstrate improvements in detection accuracy for both coordinator and worker nodes through collaborative learning. The study also reveals trade-offs between precision and recall based on statistic-sharing levels and the impact of sampling rate adjustments on performance trends. This cross-layer federated IDS offers a promising solution for intrusion detection in resource-constrained IoT networks by effectively balancing efficient operation, accurate detection, and data privacy. Its adaptability and robustness position it as a valuable tool for securing the ever-expanding massive IoT ecosystem.

4.5 Model Aggregation Approaches and Algorithms in FL

In this section, we explore the concept of model aggregation in the FL in detail, including model aggregation approaches and model aggregation algorithms. Model aggregation is a technique in ML that combines multiple models into a single, more powerful model. This can be done in a variety of ways, with the ultimate goal of creating a model that can make more accurate predictions than any of the individual models separately. Similarly to ML, in FL, the process of model aggregation involves merging the trained models from various clients to form a single global model [47]. This process ensures that the collective knowledge of the network is effectively utilized for training and updating the shared model. This procedure aims to accomplish two primary objectives:

1. **Enhanced Accuracy**: By combining the knowledge from multiple clients, the global model can learn from a larger and more diverse dataset, which can lead to improved accuracy on the target task.
2. **Privacy Preservation**: Unlike conventional methods, FL does not require the clients to disclose their raw data with the server. Instead, the clients only share their local model updates, which is much less sensitive than sharing their raw data. This makes FL a promising approach for applications where data privacy is a significant concern.

4.5.1 Model Aggregation Approaches in FL

Several aggregation approaches have been proposed for FL, each with its own set of advantages and disadvantages [48]. Average aggregation is a popular approach for model aggregation due to its simplicity and efficiency. It works by averaging the parameter updates from all participating devices to create the new global model. This straightforward approach offers ease of implementation and low computational overhead. However, it has limitations. It can be sensitive to outliers and updates from malicious clients, potentially compromising the model's accuracy. Additionally, average aggregation might not perform well in scenarios with non-IID (non-independent and identically distributed) data, where data distributions differ significantly across participating devices. While average aggregation offers a baseline approach, addressing these limitations is crucial for robust FL applications. We will explain FedAvg, the algorithm for average aggregation, in the model aggregation algorithms. An alteration of average aggregation is clipped average aggregation, which addresses the issue of outliers by clipping the model updates to a predefined range before taking the average. This reduces the effect of outliers and malicious clients, but it can be more computationally intensive than average aggregation.

FL offers privacy advantages by keeping data on local devices. However, ensuring strong privacy guarantees during the model aggregation process remains crucial. Secure aggregation techniques, such as homomorphic encryption or secure multiparty computation, offer the highest level of privacy protection by preventing any information about the raw data from being revealed. These techniques operate on encrypted data, ensuring sensitive information remains confidential. However, they can be computationally expensive and slower than simpler methods. Additionally, secure aggregation protocols require careful implementation and management to maintain their security properties. Moreover, an alternative approach is differential privacy aggregation, which injects controlled noise into the model updates before averaging. This approach provides a strong level of privacy by obscuring the contribution of any individual participant. However, the added noise can also introduce a trade-off in model accuracy. The amount of noise needs to be carefully calibrated to achieve a balance between protecting privacy and maintaining the effectiveness of the trained model. Differential Privacy Federated Averaging (DP-FedAvg) is the algorithm that leverages the differential privacy aggregation.

Beyond basic averaging techniques, FL offers more sophisticated aggregation methods to address various challenges. Momentum aggregation tackles the issue of slow or noisy client updates by incorporating a momentum term. This term leverages the previous update direction, accelerating convergence and smoothing out fluctuations in the learning process. However, the choice of momentum term requires careful tuning based on the noise level in the updates. Weighted aggregation addresses data heterogeneity by assigning weights to client contributions during the aggregation process. These weights can be based on factors like a client's past performance, data quality, or resource availability. This approach allows the model to prioritize updates from more reliable or representative clients, potentially improving overall accuracy. However, careful calibration of weights is crucial to avoid bias introduced by skewed performance metrics or noisy data. In addition, Bayesian aggregation utilizes Bayesian inference to incorporate uncertainty into the aggregation process. This approach can enhance model generalization and reduce overfitting by considering the uncertainty associated with each client's update. Nevertheless, it comes at the cost of higher computational demands and potentially larger data requirements. Additionally, the underlying assumptions of the Bayesian model might not be suitable for all types of data.

Moreover, other aggregation approaches address challenges like malicious actors and communication overhead. Adversarial aggregation tackles the issue of malicious clients or outliers by employing detection and mitigation methods. These techniques can identify and downplay the influence of misleading updates, safeguarding model accuracy. As a limitation, adversarial aggregation can be computationally expensive due to the added detection and mitigation steps, requiring sophisticated algorithms. For scenarios with bandwidth limitations, quantization reduces the bit representation of model updates before transmission. This approach improves communication efficiency by sending smaller data packets. However, it introduces quantization error, potentially impacting model accuracy. Finding the optimal level of quantization involves a trade-off between communication efficiency and maintaining model effectiveness.

Besides, ensemble-based aggregation takes a different approach by combining the updates from multiple models, each trained on a distinct subset of the data. This technique leverages the diversity of perspectives offered by these models to enhance overall accuracy. Nonetheless, ensemble-based aggregation can be computationally expensive due to the need to train and manage multiple models. Additionally, careful management of the ensemble models is crucial to ensure their effectiveness. Hierarchical aggregation tackles communication efficiency from a structural perspective. It introduces a tiered architecture where local aggregation occurs at lower levels, significantly reducing the amount of data transmitted to the central server. This approach requires a well-defined hierarchical structure and careful management of data flow and aggregation protocols across different levels. Furthermore, FL extends its capabilities beyond basic aggregation techniques to encompass methods that personalize the learning process and leverage model diversity. Personalized aggregation incorporates information about each client's unique data characteristics during the update aggregation step. This approach allows the model to better adapt to individual client distributions, potentially leading to improved overall performance. A deficit of personalized aggregation is that it might necessitate additional communication and computational overhead to accommodate the client-specific adjustments.

Finally, by incorporating these advanced aggregation methods, FL applications can achieve improved robustness against malicious actors, enhanced communication efficiency, and overall better performance in real-world scenarios. Choosing the most suitable aggregation method depends on the specific application's requirements for privacy, accuracy, and computational efficiency. These advanced aggregation methods showcase the versatility of FL in catering to diverse learning scenarios.

4.5.2 Model Aggregation Algorithms in FL

Various model aggregation algorithms have been developed to address the challenges associated with FL based on the aforementioned approaches. Federated Averaging (FedAvg) is one of the most widely used aggregation algorithms. In its function, it simply averages the parameter updates from all participating clients to create the new global model [49]. This approach is straightforward and computationally efficient, making it a popular choice for FL. However, FedAvg may not adequately account for the unequal contributions from different clients, particularly those with smaller datasets. To address these issues, VarAvg (Variance Normalized Federated Averaging) normalizes the parameter updates according to their variance, effectively reducing the influence of noisy or unreliable updates.

While FedAvg remains a popular approach for FL, it can face challenges when dealing with large numbers of participants or high communication costs. Hierarchical Federated Averaging (HFA) addresses these limitations by introducing a tiered structure [50]. In HFA, models are aggregated at intermediate servers before being sent to the central server. This hierarchical approach can significantly improve convergence speed by allowing for more frequent local updates within smaller groups of devices. Additionally, it reduces communication overhead by aggregating updates locally before transmitting them to the central server, which is particularly beneficial in settings with large client populations or high bandwidth limitations. This hierarchical structure makes HFA a valuable alternative for FL applications requiring scalability and efficient communication.

DP-FedAvg incorporates differential privacy techniques into the model aggregation process [51]. These techniques add controlled noise to the updates exchanged between clients and the server. This noise ensures that the contribution of any single client has minimal impact on the overall model update. As a result, DP-FedAvg preserves the privacy of individual client data while still enabling collaborative learning across participants. This approach fosters trust and encourages participation in FL applications where data privacy is paramount.

FL can be vulnerable to malicious actors injecting corrupted data to manipulate the model training process. Robust Federated Averaging (RFA) addresses this challenge by employing a robust aggregation mechanism based on the geometric median [48]. This approach helps to identify and discard outliers, effectively filtering out misleading updates from potentially malicious clients. By protecting the integrity of the global model from such attacks, RFA ensures the accuracy and reliability of the trained model in FL applications. This robust aggregation technique is particularly suitable for scenarios where the security of the training data is a major concern.

FL typically focuses on training a model for a single task. However, real-world applications often involve scenarios where a model needs to perform multiple related tasks. Federated Multitask Learning (MTL) addresses this need by extending FL to handle these complexities [52]. In MTL, a single model is trained collaboratively across participants, but it learns to perform multiple tasks simultaneously. To achieve optimal performance across all tasks, federated MTL utilizes task-specific weights during the model aggregation process. These weights ensure that the contributions from different tasks are balanced, allowing the model to effectively learn the nuances of each task while leveraging the underlying relationships between them. This approach makes federated MTL a valuable tool for FL applications where a single model needs to handle diverse but interrelated tasks.

Beyond standard model training, FL can be extended for even more advanced functionalities. Federated Neural Architecture Search (FNAS) integrates FL with Neural Architecture Search (NAS) techniques [53]. NAS automates the process of finding the optimal architecture for a DL model. In FNAS, NAS is performed in a federated manner, where different network designs are explored on participating devices. This distributed approach leverages the computational power of the network while preserving data privacy. Ultimately, FNAS selects the best-performing architecture for the global model, leading to potentially more efficient and accurate models for FL applications. This integration can lead to optimizing model design within the FL framework.

The above-mentioned examples highlight the diverse range of model aggregation algorithms being developed for FL. Model aggregation stands as a pivotal element in FL, playing a crucial role in realizing the objectives of enhanced accuracy and privacy protection within this decentralized ML framework. As the field continues to evolve, we can expect to see even more innovative and effective approaches to combine local models and train robust privacy-preserving federated models. The selection of a model aggregation algorithm depends on the specific FL application and the desired balance between accuracy and privacy.

4.6 Challenges and Future Directions in FL-Based IDS for IoT

FL offers a promising paradigm for IDS in the IoT landscape. However, alongside its advantages, several challenges and open research directions require further exploration to unlock the full potential of FL-based IDSs for securing massive IoT environments.

4.6.1 Validation and Standardization

One of the significant challenges in deploying FL-based IDSs for IoT security lies in the lack of standardized mechanisms for validation [9]. Evaluating the effectiveness and generalizability of these systems remains an open issue. The absence of standardized datasets encompassing diverse attack scenarios and heterogeneous data distributions representative of real-world IoT environments hinders comprehensive evaluation. Additionally, ethical and practical considerations limit the feasibility of replicating real-world attacks for testing purposes. Developing standardized benchmarks, privacy-preserving evaluation metrics, and open-source testing environments are crucial for promoting reproducibility and facilitating a fair comparison of different FL-based IDS approaches.

4.6.2 Data Heterogeneity and Non-IID Data

A fundamental challenge in FL for IDS is the presence of non-IID data across participating IoT devices [54]. This data disparity arises due to the inherent heterogeneity of IoT networks, where devices collect and store data with varying features and characteristics. Traditional FL approaches often suffer from performance degradation when dealing with non-IID data distributions. Existing works primarily focus on mitigating this challenge through algorithmic and architecture-level improvements aimed at enhancing the training effectiveness and prediction accuracy of FL models [55]. These advancements involve techniques like FTL and federated meta-learning, which enable models to adapt to diverse data distributions and learn from past learning experiences across different devices.

4.6.3 Security and Privacy Enhancements

While FL offers advantages in data privacy compared to centralized approaches, further advancements are necessary to ensure robust security and privacy guarantees. The reliance on a central parameter server in traditional FL introduces a single point of failure and raises concerns about potential security vulnerabilities. Malicious actors might exploit these vulnerabilities to launch poisoning attacks, compromising the integrity of the trained model. In such attacks, malicious participants intentionally inject misleading gradients or manipulated data points into the training process. These poisoned updates aim to steer the global model toward misclassifying legitimate traffic as malicious or vice versa, compromising the overall effectiveness of the IDS. Mitigating poisoning attacks in FL-based IDSs requires a multipronged approach [56]. Future research directions include incorporating differential privacy techniques to prevent unauthorized model reconstruction and enhance the system's resilience against inference attacks, model inversion attacks, and other privacy breaches. Additionally, exploring the application of redactable blockchain technology within the FL framework can offer promising avenues for secure and transparent model updates while preserving data privacy.

4.6.4 Communication Efficiency and Scalability

Scalability and communication efficiency are critical aspects for successful deployment in large-scale IoT networks. The distributed nature of FL can lead to significant communication overhead due to the exchange of model parameters between devices and the central server [57]. To address this challenge, future research efforts should focus on developing techniques for adaptive learning rates, efficient model compression, and federated knowledge distillation. These approaches can minimize communication costs and improve the scalability of FL-based IDSs, especially in resource-constrained IoT environments.

4.6.5 Explainability and Interpretability

Enhancing the explainability and interpretability of FL-based IDSs is crucial for building trust and transparency in deployed security solutions. Traditional black-box models often lack transparency in their decision-making processes, making it difficult to understand how they identify and classify threats. Future research directions should explore techniques that can explain the rationale behind the model's predictions and identify potential biases or vulnerabilities. This allows for improved user trust and facilitates the identification and rectification of any issues within the FL-based IDS.

By addressing the above-mentioned challenges and pursuing the promising future directions described above, the field of FL-based IDSs for IoT can move toward a more standardized and reliable validation process. This will ultimately lead to the development of more trustworthy and effective IDSs for securing the ever-expanding world of massive IoT.

4.7 Conclusion

In this chapter, we conducted an analysis of current FL-based IDSs in order to provide a roadmap to support the design and development of effective, efficient, and privacy-preserving IDS for protecting the emerging disruptive massive IoT applications in the 6G era. In particular, we provided the fundamentals of IDSs and ML techniques for IDSs in IoT, as well as the limitations that arise from their centralized nature. In addition, we introduced the concept of FL, and explored its different categories and architectures. Furthermore, we presented use cases, definitive examples as well as state-of-the-art research works of FL-based IDS for IoT that can be expanded into a more massive network scale, that of massive IoT. Moreover, we explored the concept of model aggregation in the FL in detail, including model aggregation approaches and model aggregation algorithms. Finally, we presented challenges and open research directions that require further exploration so that FL-based IDSs will become widely adopted for securing massive IoT environments.

List of Abbreviations

IoT	internet of things
FL	federated learning
IDS	intrusion detection system
FL-based IDS	federated learning -based intrusion detection system
ML	machine learning
DoS	denial of service
DDoS	distributed denial of service
RPL	routing protocol for low-power and lossy networks
DL	deep learning
NB	naïve bayes
KNN	K-nearest neighbor
DT	decision trees
RF	random forest
SVM	support vector machines
EL	ensemble learning
SGD	stochastic gradient descent
HFL	horizontal federated learning
VFL	vertical federated learning
FTL	federated transfer learning
GRU	gate recurrent units
IIoT	industrial internet of things
MQTT	message queuing telemetry transfer protocol
TCP	transmission control protocol
DNN	deep neural network
CNN	convolutional neural network
RNN	recurrent neural networks
VANET	vehicular Ad-hoc network
IID	independent and identically distributed
FedAvg	federated averaging
DP-FedAvg	differential privacy federated averaging
VarAvg	variance normalized federated averaging
HFA	hierarchical federated averaging
RFA	robust federated averaging
MTL	multi-task learning
FNAS	federated neural architecture search
NAS	neural architecture search

References

1 Abdel Hakeem, S.A., Hussein, H.H., and Kim, H. (2022). Security requirements and challenges of 6G technologies and applications. *Sensors* 22 (5): 1–43. https://doi.org/10.3390/s22051969.

2 Nguyen, V.L., Lin, P.C., Cheng, B.C. et al. (2021). Security and privacy for 6G: a survey on prospective technologies and challenges. *IEEE Communication Surveys and Tutorials* 23 (4): 2384–2428. https://doi.org/10.1109/COMST.2021.3108618.

3 Guo, F., Yu, F.R., Zhang, H. et al. (2021). Enabling massive IoT toward 6G: a comprehensive survey. *IEEE Internet of Things Journal* 8 (15): 11891–11915. https://doi.org/10.1109/JIOT.2021.3063686.

4 Alazab, M., Rm, S.P., Parimala, M. et al. (2022). Federated learning for cybersecurity: concepts, challenges, and future directions. *IEEE Transactions on Industrial Informatics* 18 (5): 3501–3509. https://doi.org/10.1109/TII.2021.3119038.

5 Li, D., Han, D., Weng, T.-H. et al. (2022). Blockchain for federated learning toward secure distributed machine learning systems: a systemic survey. *Soft Computing* 26 (9): 4423–4440. https://doi.org/10.1007/s00500-021-06496-5.

6 Jayasinghe, S., Siriwardhana, Y., Porambage, P. et al. (2022). Federated learning based anomaly detection as an enabler for securing network and service management automation in beyond 5G networks. 2022 Joint European Conference on Networks and Communications & 6G Summit (EuCNC/6G Summit), Grenoble, France (7–10 June 2022), pp. 345–350. https://doi.org/10.1109/EuCNC/6GSummit54941.2022.9815754.

7 Yang, Q., Liu, Y., Chen, T., and Tong, Y. (2019). Federated machine learning: concept and applications. *ACM Transactions on Intelligent Systems and Technology* 10 (2): 1–19. https://doi.org/10.1145/3298981.

8 Cheema, M.A., Qureshi, H.K., Chrysostomou, C., and Lestas, M. (2020). Utilizing blockchain for distributed machine learning based intrusion detection in Internet of Things. Proceedings – 2020 16th International Conference on Distributed Computing in Sensor Systems (DCOSS), Marina del Rey, CA, USA (25–27 May 2020), pp. 429–435. https://doi.org/10.1109/DCOSS49796.2020.00074.

9 Asharf, J., Moustafa, N., Khurshid, H. et al. (2020). A review of intrusion detection systems using machine and deep learning in Internet of Things: challenges, solutions and future directions. *Electronics (Switzerland)* 9 (7) MDPI AG, 1177: https://doi.org/10.3390/electronics9071177.

10 Keshk, M., Sitnikova, E., Moustafa, N. et al. (2021). An integrated framework for privacy-preserving based anomaly detection for cyber-physical systems. *IEEE Transactions on Sustainable Computing* 6 (1): 66–79. https://doi.org/10.1109/TSUSC.2019.2906657.

11 Liu, C., Yang, J., Chen, R. et al. (2011). Research on immunity-based intrusion detection technology for the Internet of Things. Proceedings – 2011 Seventh International Conference on Natural Computation, Shanghai, China (26–28 July 2011), pp. 212–216. https://doi.org/10.1109/ICNC.2011.6022060.

12 Mitchell, R. and Chen, I.R. (2014). A survey of intrusion detection techniques for cyber-physical systems. *ACM Computing Surveys* 46 (4): https://doi.org/10.1145/2542049.

13 El Mokadem, R., Ben Maissa, Y., and El Akkaoui, Z. (2023). Federated learning for energy constrained devices: a systematic mapping study. *Cluster Computing* 26 (2): 1685–1708. https://doi.org/10.1007/S10586-022-03763-4/TABLES/5.

14 Raza, S., Wallgren, L., and Voigt, T. (2013). SVELTE: real-time intrusion detection in the Internet of Things. *Ad Hoc Networks* 11 (8): 2661–2674. https://doi.org/10.1016/J.ADHOC.2013.04.014.

15 Zachos, G., Essop, I., Mantas, G. et al. (2021). An anomaly-based intrusion detection system for Internet of Medical Things networks. *Electronics* 10 (21): 2562.

16 Alharbi, S., Rodriguez, P., Maharaja, R. et al. (2018). Secure the Internet of Things with challenge response authentication in fog computing. 2017 IEEE 36th International Performance Computing and Communications Conference (IPCCC), San Diego, CA, USA (10–12 Dec 2017), pp. 1–2. https://doi.org/10.1109/PCCC.2017.8280489.

17 Breiman, L. (2001). Random forests. *Machine Learning* 45 (1): 5–32. https://doi.org/10.1023/A:1010933404324/METRICS.

18 Zhang, J. and Zulkernine, M. (2006). A hybrid network intrusion detection technique using random forests. Proceedings – First International Conference on Availability, Reliability and Security (ARES'06), Vienna, Austria (20–22 April 2006), pp. 262–269. https://doi.org/10.1109/ARES.2006.7.

19 Chang, Y., Li, W., and Yang, Z. (2017). Network intrusion detection based on random forest and support vector machine. Proceedings – 2017 IEEE International Conference on Computational Science and Engineering (CSE) and IEEE International Conference on Embedded and Ubiquitous Computing (EUC), Guangzhou, China (21–24 July 2017), pp. 635–638. https://doi.org/10.1109/CSE-EUC.2017.118.

20 Torres, J.M., Comesaña, C.I., and García-Nieto, P.J. Review: machine learning techniques applied to cybersecurity. *International Journal of Machine Learning and Cybernetics*. https://doi.org/10.1007/s13042-018-00906-1.

21 Domingos, P. (2012). A few useful things to know about machine learning. *Communications of the ACM* 55 (10): 78–87. https://doi.org/10.1145/2347736.2347755.

22 Gaikwad, D.P. and Thool, R.C. (2015). Intrusion detection system using bagging ensemble method of machine learning. Proceedings – 2015 International Conference on Computing Communication Control and Automation, Pune, India (26–27 February 2015), pp. 291–295. https://doi.org/10.1109/ICCUBEA.2015.61.

23 Reddy, R.R., Ramadevi, Y., and Sunitha, K.V.N. (2017). Enhanced anomaly detection using ensemble support vector machine. Proceedings – 2017 International Conference on Big Data Analytics and Computational Intelligence (ICBDACI), Chirala, Andhra Pradesh, India (23–25 March 2017), pp. 107–111. https://doi.org/10.1109/ICBDACI.2017.8070818.

24 Ullah, S., Ahmad, J., Khan, M.A. et al. (2023). TNN-IDS: transformer neural network-based intrusion detection system for MQTT-enabled IoT networks. *Computer Networks* 237: 110072. https://doi.org/10.1016/J.COMNET.2023.110072.

25 Vadigi, S., Sethi, K., Mohanty, D. et al. (2023). Federated reinforcement learning based intrusion detection system using dynamic attention mechanism. *Journal of Information Security and Applications* 78: 103608. https://doi.org/10.1016/J.JISA.2023.103608.

26 Duan, Q., Lu, Z., Alsamiri, J., and Alsubhi, K. (2023). Federated learning for intrusion detection systems in internet of vehicles: a general taxonomy, applications, and future directions. *Future Internet* 15 (12): 403. https://doi.org/10.3390/FI15120403.

27 Alazab, A., Khraisat, A., Singh, S. et al. (2023). Enhancing privacy-preserving intrusion detection through federated learning. *Electronics* 12 (16): 3382. https://doi.org/10.3390/ELECTRONICS12163382.

28 Kairouz, P., Brendan McMahan, H., Avent, B. et al. (2019). Advances and open problems in federated learning. *Foundations and Trends in Machine Learning* 14 (1–2): 1–210. https://doi.org/10.1561/2200000083.

29 Konečn, J., Brendan McMahan, H., Yu, F.X. et al. (2016). Federated learning: strategies for improving communication efficiency [Online]. Available: https://arxiv.org/abs/1610.05492v2 (accessed 14 Dec 2023).

30 Konečn, J., Brendan McMahan, H., Ramage, D., and Richtárik, P. (2016). Federated optimization: distributed machine learning for on-device intelligence [Online]. Available: https://arxiv.org/abs/1610.02527v1 (accessed 14 Dec 2023).

31 Yu, L., Xin, L., and Guo, M. (2022). A safe architecture of 5G network intelligence based on federated learning and NWDAF. AISS '22: Proceedings of the 4th International Conference on Advanced Information Science and System. https://doi.org/10.1145/3573834.3574529.

32 Li, J., Cui, T., Yang, K. et al. (2021). Demand forecasting of e-commerce enterprises based on horizontal federated learning from the perspective of sustainable development. *Sustainability* 13 (23): https://doi.org/10.3390/su132313050.

33 Li, S., Yao, D., and Liu, J. (2023). FedVS: straggler-resilient and privacy-preserving vertical federated learning for split models. International Conference on Machine Learning. PMLR.

34 Feng, S., Yu, H., Zhu, Y. et al. (2024). MMVFL: a simple vertical federated learning framework for multi-class multi-participant scenarios. *Sensors* 24 (2): 619. https://doi.org/10.3390/S24020619.

35 Xu, Y., Hu, X., Wei, J. et al. (2023). VF-CART: a communication-efficient vertical federated framework for the CART algorithm. *Journal of King Saud University, Computer and Information Sciences* 35 (1): 237–249. https://doi.org/10.1016/J.JKSUCI.2022.11.013.

36 Liu, Y., Kang, Y., Xing, C. et al. (2020). A secure federated transfer learning framework. *IEEE Intelligent Systems* 35 (4): 70–82. https://doi.org/10.1109/MIS.2020.2988525.

37 Mothukuri, V., Khare, P., Parizi, R.M. et al. (2022). Federated-learning-based anomaly detection for IoT security attacks. *IEEE Internet of Things Journal* 9 (4): 2545–2554. https://doi.org/10.1109/JIOT.2021.3077803.

38 Attota, D.C., Mothukuri, V., Parizi, R.M., and Pouriyeh, S. (2021). An ensemble multi-view federated learning intrusion detection for IoT. *IEEE Access* 9: 117734–117745. https://doi.org/10.1109/ACCESS.2021.3107337.

39 Friha, O., Ferrag, M.A., Shu, L. et al. (2022). FELIDS: federated learning-based intrusion detection system for agricultural Internet of Things. *Journal of Parallel and Distributed Computing* 165: 17–31. https://doi.org/10.1016/J.JPDC.2022.03.003.

40 CSE-CIC-IDS2018. https://www.kaggle.com/datasets/solarmainframe/ids-intrusion-csv (accessed 15 Jul 2024).

41 Vaccari, I., Chiola, G., Aiello, M. et al. (2020). MQTTset, a new dataset for machine learning techniques on MQTT. *Sensors* 20 (22): 6578. https://doi.org/10.3390/S20226578.

42 Elsayed, M.S., Le-Khac, N.A., and Jurcut, A.D. (2020). InSDN: a novel SDN intrusion dataset. *IEEE Access* 8: 165263–165284. https://doi.org/10.1109/ACCESS.2020.3022633.

43 Ferrag, M.A., Friha, O., Hamouda, D. et al. (2022). Edge-IIoTset: a new comprehensive realistic cyber security dataset of IoT and IIoT applications for centralized and federated learning. *IEEE Access* 10: 40281–40306. https://doi.org/10.1109/ACCESS.2022.3165809.

44 Driss, M., Almomani, I., Huma, Z., and Ahmad, J. (2022). A federated learning framework for cyberattack detection in vehicular sensor networks. *Complex & Intelligent Systems* 8 (5): 4221–4235. https://doi.org/10.1007/S40747-022-00705-W/TABLES/5.

45 Hajj, S., Azar, J., Bou Abdo, J. et al. (2023). Cross-layer federated learning for lightweight IoT intrusion detection systems. *Sensors* 23 (16): 7038. https://doi.org/10.3390/S23167038.

46 Revathi, S. and Malathi, A. (2013). A detailed analysis on NSL-KDD dataset using various machine learning techniques for intrusion detection. *International Journal of Engineering Research and Technology.*

47 Ek, S., Portet, F., Lalanda, P., and Vega, G. (2021). A federated learning aggregation algorithm for pervasive computing: evaluation and comparison. 2021 IEEE International Conference on Pervasive Computing and Communications (PerCom), Kassel, Germany (22–26 March 2021). https://doi.org/10.1109/PERCOM50583.2021.9439129.

48 Moshawrab, M., Adda, M., Bouzouane, A. et al. (2023). Reviewing federated learning aggregation algorithms; strategies, contributions, limitations and future perspectives. *Electronics* 12 (10): 2287. https://doi.org/10.3390/ELECTRONICS12102287.

49 Nilsson, A., Smith, S., Ulm, G. et al. (2018). A performance evaluation of federated learning algorithms. DIDL '18: Proceedings of the Second Workshop on Distributed Infrastructures for Deep Learning, Rennes, France (10–11 Dec 2018), pp. 1–8. https://doi.org/10.1145/3286490.3286559.

50 Abdellatif, A.A., Mhaisen, N., Mohamed, A. et al. (2022). Communication-efficient hierarchical federated learning for IoT heterogeneous systems with imbalanced data. *Future Generation Computer Systems* 128: 406–419. https://doi.org/10.1016/j.future.2021.10.016.

51 Li, Y., Chang, T.H., and Chi, C.Y. (2020). Secure federated averaging algorithm with differential privacy. 2020 IEEE 30th International Workshop on Machine Learning for Signal Processing (MLSP), Espoo, Finland (21–24 Sept 2020). https://doi.org/10.1109/MLSP49062.2020.9231531.

52 Chen, J. and Zhang, A. (2022). FedMSplit: correlation-adaptive federated multi-task learning across multimodal split networks. Proceedings of the 28th ACM SIGKDD Conference on Knowledge Discovery and Data Mining (Aug 2022), pp. 87–96. https://doi.org/10.1145/3534678.3539384.

53 Zhu, H., Zhang, H., and Jin, Y. (2021). From federated learning to federated neural architecture search: a survey. *Complex & Intelligent Systems* 7 (2): 639–657. https://doi.org/10.1007/S40747-020-00247-Z/FIGURES/10.

54 Hsu, H.Y., Keoy, K.H., Chen, J.R. et al. (2023). Personalized federated learning algorithm with adaptive clustering for non-IID IoT data incorporating multi-task learning and neural network model characteristics. *Sensors (Basel)* 23 (22): https://doi.org/10.3390/S23229016.

55 Lian, Z., Wang, W., Han, Z., and Su, C. (2023). Blockchain-based personalized federated learning for Internet of Medical Things. *IEEE Transactions on Sustainable Computing* 8 (4): 694–702. https://doi.org/10.1109/TSUSC.2023.3279111.

56 Zhang, Z., Zhang, Y., Guo, D. et al. (2022). SecFedNIDS: robust defense for poisoning attack against federated learning-based network intrusion detection system. *Future Generation Computer Systems* 134: 154–169. https://doi.org/10.1016/J.FUTURE.2022.04.010.

57 Chen, M., Shlezinger, N., Vincent Poor, H. et al. (2021). Communication-efficient federated learning. *Proceedings of the National Academy of Sciences of the United States of America* 118 (17): e2024789118. https://doi.org/10.1073/PNAS.2024789118/SUPPL_FILE/PNAS.2024789118.SAPP.PDF.

5

Securing Massive IoT Using Network Slicing and Blockchain*,†

Shihan Bao[1], *Zhili Sun*[2], *and Haitham Cruickshank*[2]

[1] *State Key Laboratory of Wireless Mobile Communications, China Academy of Telecommunications Technology, Beijing, China*
[2] *Institute for Communication Systems (ICS), 5G&6G Innovation Centre (5G&6GIC), University of Surrey, Guildford, UK*

5.1 Introduction

The massive Internet of Things (IoT) is a prominent technology that can offer pervasive connectivity for a vast number of everyday items [1]. According to IoT Analytics, the number of connected IoT devices reached 14.4 billion [2]. Massive IoT use cases (e.g. e-health, smart grid, autonomous driving) will deal with massive amounts of data, while cloud servers will collect and manage all these data. Consequently, these servers will require high storage capacity and a reliable network for system maintenance, which will pose potential threats to data integrity and availability due to their singular points of vulnerability. In addition, the connected massive IoT devices often suffer from limited protection, easy physical access, and sometimes even insufficient security updates. Therefore, IoT systems are becoming an increasingly attractive target for attackers.

A promising solution for these security issues is network slicing and segmentation when IoT devices explode the attack surface. IoT use cases come with their own requirements, including data rate, coverage, latency, mobility, and connection density. Hence, only dedicated networks could meet these needs [3]. The network slicing of 5G networks enables multiple dedicated network services using Software Defined Network (SDN) and Network Function Virtualization (NFV). It decouples network function (NF) functionality from the physical infrastructure and relocates NFs from dedicated appliances to pools of resources. This significantly improves network efficiency and can adjust virtual networks without disturbing the overall service operation. Most notably, network slicing enables flexibility and modularity to establish multiple subnetworks. Each subnetwork can be specialized for a distinct use-case from one sheared network infrastructure to meet various network needs of diverse verticals.

*This chapter uses content originally published in the Journal of Communications and Information Networks, Vol.7, No.4, Dec. 2022 (Blockchain for Network Slicing in 5G and Beyond: Survey and Challenges, by Shihan Bao, Yacong Liang, Hui Xu).
†This chapter uses content originally published in the PhD thesis of Dr Shihan Bao (Dynamic pseudonym management for privacy preservation in vehicular communication systems – University of Surrey).

Security and Privacy for 6G Massive IoT, First Edition.
Edited by Georgios Mantas, Firooz B. Saghezchi, Jonathan Rodriguez, and Victor Sucasas.

Creating on-demand End-to-End (E2E) network slices based on different service requirements is a critical feature of the 5G network. An E2E Network Slice Instance (NSI) could contain various subnetworks over many administrative domains. Therefore, network slicing enables an E2E ecosystem to provide a consistent experience for distinct services [4]. Network slicing has also been in the spotlight of many standardization bodies (e.g. 3GPP [3rd Generation Partnership Project], ITU-T [International Telecommunication Union – Telecommunication Standardization Sector], and ETSI [European Telecommunications Standards Institute]). The motivation of network slicing is to facilitate a new business ecosystem. The industry and standardization bodies are expecting network slicing to enable innovative services and networking capabilities for premium consumers and industry stakeholders. As one of the key enablers of future mobile networks, network slicing can speed the development of network services by creating customized services for vertical industries and could create partnerships among network operators and verticals. The objectives of network slicing are the provision of enhanced Quality of Services (QoS) for consumers and cost-effective approaches for network operators and vertical industries.

The current 5G mobile network still inherits the traditional network security system of external border protection. While new security challenges are emerging, conventional protection methods cannot keep up with the rising pace of new issues. In addition, security and privacy threats of network slicing are still not getting enough attention, which might hinder the development of the next-generation mobile network. Hence, the International Mobile Telecommunications-2020 (IMT-2020) and 3GPP both have stated emerging requirements for enhanced security and privacy [5]. Researchers are starting to believe that independent and self-growing security capabilities could be a key enabler for the next-generation mobile network. It is envisioned that next-generation networks will face more challenging security and privacy issues while the network structure becomes more complex and intelligent among more mature network slicing mechanisms. In particular, the most benefits of network slicing may come from the cost-effectiveness and efficiency of aggregating services on common public infrastructure and resources, but the security, privacy, and reliability of public networks are considerable concerns not just for network operators but also for consumers. Isolation is one of the most fundamental features of network slicing. The security level of a network slice depends on assuring a required level of isolation. Furthermore, the reliability concern is another issue with network slicing. Having a single point of failure at any life cycle of network slicing could easily shut down the network slice and even affect the whole system.

On the other hand, blockchain has the characteristics of decentralization, distributed structure, immutability, robustness, traceability, and openness, which suit the needs of endogenous security design. Since the success of blockchain applications in various areas, more and more entities believe that blockchain will become a fundamental technology for future telecommunications and the Internet. Blockchain could achieve decentralized, dynamic network management. In addition, IoT systems can use blockchain technology to build a unified authentication system for better interoperability and security. Moreover, distributed data storage is enabled in massive IoT systems by using blockchain, which prevents a single point of failure. The inherent characteristics of the blockchain offer a great opportunity to tackle massive IoT challenges, which could lead to a secure distributed fabric for massive IoT systems. The following could be the potential benefits and motivations for developing a blockchain-based IoT framework: (i) Resilience: The blockchain technology offers a distributed ledger that stores copies of transaction records over blockchain peers, which helps maintain resilience and data immutability. IoT applications require data integrity and need to be resilient to data tampering and breakage. (ii) Fault tolerance: Network for massive IoT systems require high availability. Blockchain provides a Byzantine fault-tolerant mechanism that could avoid single-point failure problem. (iii) Immutability: One of the most distinguished features of blockchain is distributed ledger which has distributed consensus for record immutability. (iv) Trust: Blockchain eliminates centralized control and enables trust directly between peers. (v) Interoperability: Various blockchain applications have proven that blockchain can work over a heterogeneous network. Using blockchain for an IoT framework has a high potential to adapt to dynamic environments and scenarios of massive IoT systems.

The main contributions of this chapter can be summarized as follows:

- to shed light on key challenges inherent to massive IoT, specifically highlighting security, privacy, and location privacy concerns.
- to review standardization efforts related to network slicing from renowned entities such as ITU, 3GPP, and ETSI.
- to conduct an in-depth examination of employing blockchain and network slices to secure massive IoT systems.

The remainder of the chapter is structured as follows: Section 5.2 discusses background basics regarding massive IoT and blockchain technology. In Section 5 3, we present security and privacy threats for IoT systems. Section 5.4 provides research challenges and solutions using network slicing with blockchain to secure massive IoT. Section 5.5 follows open challenges that need more attention in the next-generation mobile networks. Finally, section 5.6 concludes this paper.

5.2 Background

5.2.1 Massive IoT

IoT is used as an umbrella keyword for covering the various combinations of the digital and physical objects [6]. Massive IoT is the evolution of the IoT system toward supermassiveness in number and convergence of communications, caching, computing, control, sensing, and localization [7]. Nowadays, IoT devices can be found in almost every field including utilities, agriculture, smart cities, connected transportation, and logistics. Various types of smart sensors, meters, monitors, and tracking devices are able to exchange messages about operations and automated processes to make more efficient and smart decisions. Moreover, new forms of IoT-enabled interactions and use cases are expected to come in the near future, i.e. holographic communications, extended reality, and brain-computers.

For current low-bandwidth scenarios, LTE or Low Power Wider Area (LPWA) is utilized in most IoT solutions. But future massive IoT systems may include one million connected devices per square meter, and thus 5G and beyond networks will be essential for supporting such scenarios [8].

5.2.2 Blockchain

Nowadays, Bitcoin attracts a lot of attention along with its blockchain concept, which was proposed in 2008 [9]. In simple terms, a blockchain is a synchronized and distributed ledger that stores a list of blocks. Each block records a set of validated transactions (e.g. user information and a receipt) and securely links to the previous block. Central managers are removed from the blockchain structure, and the public ledger is maintained by all the network participants instead. This is realized by a protocol that achieves a trustworthy consensus about the chain of blocks created. In other words, network nodes can agree (deterministically) on the history and order of blocks that were created and on which node is allowed to add the next block to the chain.

The leader election of the node that can add the next block may be performed through a variety of techniques. For example, Proof of Work (PoW) poses a cryptographic puzzle to nodes based on a cryptographic hash function, the last local block seen, and the pool of transactions to be processed at a local node. The node that solves this puzzle announces the solution on the network, and other nodes accept such solution only if all transactions in the new block are valid, the block correctly points to the last block, and no other such solution was received beforehand. Since solving a puzzle is hard but verifying a solution is easy, this system provides security and effective validation and does not have a single point of failure.

The network will reach eventual consistency since some regions may temporarily diverge in their opinion of who won the next block. Since nodes hold an entire block tree, such disputes get resolved eventually as all nodes consider the path in the local tree with the "biggest overall work" to be the genuine chain (and this choice may vary over time).

Blockchain offers a means of creating a trustworthy record of transaction histories in a network of nodes in which there exists mistrust. This is a conservative trust model for Vehicular Communication Systems (VCSs), where some parts of the network would be within trusted computing bases but other parts would be more open or even publicly accessible.

Blockchain security is achieved in a manner that reflects the design choices of the blockchain. For example, when PoW is used for consensus, then one would need to control more than 50% of the nodes in the network in order to rewrite the blockchain history and so corrupt data veracity. This high degree of resiliency is what makes blockchain attractive in settings in which faults and malicious manipulation may corrupt the integrity of data ledgers. All transactions on the blockchain are traceable, back to the very first "genesis" block. The transactions are tamperproof – any attempt to tamper with them is immediately broadcast to the entire network and detected. That also means that once transactions are conducted on the blockchain, there is no going back to edit or delete them. Blockchains can modify all that as they enable connected devices to be smart, independent agents that can not only identify themselves to other machines securely but also carry out microtransactions based on a set of rules or smart contracts that cannot be tampered with.

5.2.2.1 Blockchain Applications

Blockchains are beginning to be used not only for decentralized cryptocurrencies but also for a wide range of applications including IoT.

Danezis and Meiklejohn [10] proposed a new digital currency based on Bitcoin that provides better scalability and flexibility. Zyskind et al. [11] used blockchain to build a decentralized personal data management system. The blockchain acts as a monitor for the access control of personal data. When a user intends to access the database or retrieve data from it, an access transaction is generated. After getting permission to access, users will describe their needs in data transactions to complete the communication in the blockchain system.

Despite the fact that blockchain has received a lot of attention from the banking industry, people find that the use of blockchain can also improve other systems such as insurance, electric vehicle charging, and car-sharing services [14]. Papers by Mengelkamp et al. [12] and Pop et al. [13] aim to use blockchain to provide a decentralized platform for trading local energy generation between both energy prosumers and consumers.

5.2.2.2 Consensus Mechanisms

Consensus mechanisms are protocols that all participated nodes agree on, which are used to synchronize the time with each other and make sure transactions are legitimate. The consensus mechanism is an important part of the blockchain system in order to function rightly. Basically, the mechanism makes sure all nodes are using the same blockchain, all transactions are constantly checked, and the blockchain is constantly audited by all nodes. There are many ways to agree on a consensus, however, the following mechanisms are the most popular ones:

PoW is one of the most commonly used consensus mechanisms and was the first used by Bitcoin. For most cryptocurrencies, mining is the process of adding a new transaction to the blockchain (through securing and verifying) while releasing new currency/coins. These new individual blocks are added by miners and contain PoW. Many cryptocurrencies have used the adapted version of this consensus mechanism. The PoW process is also known as mining and all nodes that participate in the process are called miners. The mining process is a complex mathematical puzzle. The first node that solves the problem has the right to update the network with a new block and earn incentives. In summary, mining keeps all nodes in the Bitcoin network in sync. It enforces that the transactions are carried out in the same order by all nodes, thus preventing someone from double-spending. In the following, we briefly illustrate the process in detail. The blockchain consists of blocks that contain transactions. Each miner needs to hash the latest block header by using Double SHA 256 [15]. In order to make the mathematical puzzle artificially hard to solve for miners, the hash result of the solved block has to meet certain conditions. First, the hash functions are asymmetric, which means it would take a lot of time to find the answer, but it is very easy to verify if the answer is right. Second, the hash of a block represents a numerical value. To solve the puzzle, miners need to keep trying to get the hash result below a certain numerical value. The numerical value is known as the target hash. As the nature of the hash functions, there is only one hash for each block, as known as a digital fingerprint. In order to get the hash of the block to fall under the target hash, miners add a random number within their block, which is called the nonce. As a result, miners keep trying new nonce that is included in the block, until the hash result falls below the target hash or a new block has been solved by another miner. In this puzzle-solving process, every miner enjoys the same probability if they have the same computation power, or more specifically, the same hash rate. No matter how long this miner stays in the blockchain network, the miner has no advantage in each round compared with miners who just participated in the network. It also means that if one miner wants to solve the puzzle faster than others, it needs more computation power. Furthermore, the difficulty target of these mathematical puzzles changes periodically based on the block-solving time. For example, target hash rate will change every 2016 block in the Bitcoin network. The Bitcoin network sets the average mining time of each round at 10 minutes. If blocks are generated too fast, the puzzles get more difficult to solve, and if they are generated for longer than 10 minutes, the puzzle difficulty would change to a lower level. To solve the hash function, many types of hardware could be used for mining blocks, namely, central processing unit (CPU) mining, graphics processing unit (GPU), Application-specific integrated circuit (ASIC) mining, etc. CPU mining uses a personal computer's CPU to validate transactions in a PoW consensus mechanism. GPU mining makes use of GPU cards to mine blocks, which are more powerful and have less power consumption, as GPUs are more purpose-driven to do a singular task. Last but not least, the ASIC miner is a customized circuit for mining. It could easily outperform CPUs and GPUs in terms of speed and efficiency.

There are many research studies demonstrating that PoW has a major drawback, which is the massive resource consumption. With the total computation power improvement, miners need to consume much more resources, especially electricity, to mine blocks, which leads to a very unsustainable processing. As a result, some blockchains are changing to other consensus mechanisms.

Delegated Proof of Stake (DPoS) is designed as digital democracy, using the voting and election process to prevent blockchain from centralization and malicious use. In a DPoS system, users can stake their coins to vote for a certain amount of delegates. The weight of their vote depends on their stake; for example, if A stakes 10 coins for a delegate and B stakes one coin for a delegate, A's vote weighs 10 times heavier than B's vote. Since delegates want to obtain as many votes as possible, they are constantly incentivized to create things valuable to the community, as they are likely to receive extra votes for doing so.

Proof of Stake (PoS) is another consensus mechanism in blockchain making use of the premise that a person can maintain the blockchain network and validate transactions according to how many coins he or she holds. Those who the own most coins in the network are not only more capable of mining but also more responsible for keeping the value of their coins high. In the PoS system, users can stake their tokens to become validators who have the chance to generate blocks. Their tokens will be locked up for a certain time. Then a randomized process is used to select who gets to produce the next block among all validators.

Proof of Capacity (PoC) is another different consensus mechanism that makes use of plotting. Unlike PoW, in PoC, miners use computation power to solve the mathematical puzzle, and the solutions are prestored in digital storage (e.g. hard disks). Once storage has been plotted, which means it has been filled with solutions, it can take part in the block creation process.

Similar to the PoW, whoever has the fastest solution to the puzzle of a (new) block can produce the next block. The more storage capacity means, the more solutions that can be stored, which can finally lead to the higher possibilities of creating a block.

5.3 Challenges on Massive IoT

The evolution of massive IoT increases the threat surface, which lacks E2E security mitigation. Massive IoT applications will not only include mission-critical circumstances (e.g. e-health, emergency rescue, intelligent transportation systems, smart power grids) but could also contain business-oriented applications (e.g. banking systems, logistics, insurance). The need for E2E security mitigation is crucial, especially for mission-critical IoT applications. Many researchers have shown considerable interest in securing IoT using blockchain, as blockchain technology can ensure the integrity of the IoT data and transactions. All participants in the blockchain network are able to see these data and transactions, and each of them keeps a copy of these data as a distributed digital ledger. Information in the digital ledger that is stored in the blockchain is unalterable. Furthermore, smart contract in the blockchain system enables massive IoT networks to execute and manage nodes and systems autonomously. Therefore, blockchain technology has been seen as a key enabler to mitigate security and privacy threats in massive IoT.

5.3.1 Security Requirements in Massive IoT

The nature of massive IoT systems has expanded the threat surface. Each end device is a potential door for adversaries hacking into the system. The number of end devices in massive IoT is huge, as the name suggests. Hence, security and privacy of the system face tremendous challenges. Most current IoT devices lack security protection due to various reasons including device specification constraints, cost considerations, and lack of standardization. Moreover, many security issues have happened because of insecure defaults, such as simple administrative credentials, open access to the management system, and no security tracking after one-time authentication.

In an IoT communication network, security of communication is crucial for providing a secure channel of message transmission. There are six elements for security requirements.

- **Confidentiality**: Confidentiality could be achieved by encrypting the message. So the message would become ciphertext, which is unreadable by a human or a computer without the correct key to decrypt it. A common method to justify the confidentiality is the ciphertext length. Longer ciphertext means higher security level as it requires longer processing time for the eavesdropper to decrypt the data.
- **Integrity**: Integrity means that the message from sender to receiver has not been modified. For example, since the safety messages in VCS affects driver's decision, it is important to guarantee the consistency, trustworthiness, and accuracy of the message. An extra data block might be attached to the message, including Checksum or Message Authentication Code (MAC), to provide integrity.
- **Availability**: The data need to stay available consistently whenever the user requires to access it. To achieve availability, the security manager should ensure a suitable message distribution scheme that effectively uses the capacity of the communication network. In addition, the computing system should store and process the data and also provide robustness in case of hardware failures or attacks such as Denial of Service (DoS) or Distributed Denial of Service (DDoS) [16].

- **Authenticity**: To make the receiver trust the authenticity of messages from the sender, a receipt should be attached to the message. Due to the nature of the wireless communication, malicious users could launch a man-in-the-middle attack [17]. The message could be tampered with or deleted during the transmission. Therefore, the receiver needs to make sure that the message is sent from a legitimate source. Authentication receipts, such as digital certificates, are used to provide legitimacy of the message. Digital certificates are supposed to include information about the valid time of the sender. In addition, blockchain is a technology to achieve distributed authentication, which has been taken into account by many researchers recently.
- **Traceability and Revocation**: A user who abuses the network should be traceable by an authority. The revocation of the misbehaving user by the authority should be timely. This reduces the impact of the misbehaving user on the network.

5.3.2 Privacy Requirements in Massive IoT

Despite the promises of massive IoT, it will face huge challenges of privacy that might hinder its development [18]. Due to IoT features, communication among devices usually has private information about a user or event. Much of this data is sensitive or personal data, and this brings data privacy questions to the forefront. Many privacy violations are conducted by eavesdropping on communication networks. In addition, the massive data from IoT sensors not only contain geographical and environmental information but also sensitive human behavior information in some cases, like wearable-IoT devices. Even though it gives users/industry plenty of benefits, including efficiency, intelligence, and convenience, the privacy of massive IoT has concerned individuals, governmental departments, and national infrastructures. Malicious ownership of private data could allow attackers to easily explore individuals and have the possibility to hack them. However, the trends indicate that future technology is going to heavily rely on high degrees of interconnectivity and analyzing personal behavior to provide more accurate customized services. Eckhoff and Sommer [19] state that "if there is data, there is also abuse" [19]. Such lack of privacy and sense of security deter individuals from participating in massive IoT networks. As massive IoT is a crucial part of current 5G and beyond network applications, the development of massive IoT may face obstruction concerns. Therefore, providing reliable, secure, and private network services is extremely necessary.

In IoT networks, privacy tends to be specific for hiding device identity and position and preventing them from being tracked by malicious entities. In terms of IoT privacy, there are six different requirements: short-term linkability, long-term unlinkability, anonymity, pseudonymity, accountability (non-repudiation), and location privacy [20]. Next, these six requirements for massive IoT are explained, with a focus on Intelligent Transportation System (ITS) applications.

- **Short-term Linkability**: Short-term linkability is a desirable property in ITS applications [21]. Assuming a vehicle sends two or more messages within a small time frame, the receiving vehicle should be able to verify that these messages came from the same source. Enforcing short-term linkability can effectively prevent Sybil attacks by a malicious user or a compromised On-Board Unit (OBU) impersonating multiple vehicles [22].

In vehicular networks, vehicles frequently broadcast messages with their speed, current location, and acceleration. These messages are used to build up a trajectory of nearby OBUs. The location privacy of users is not affected by short-term linkability because the small time frame increment does not impact the location privacy of the vehicle [23].

- **Long-term Unlinkability**: A basic location privacy requirement for vehicular networks is long-term unlinkability. A malicious user must not be able to link messages sent by a vehicle to the attributes of that vehicle such as location, type of car and applications. Tracking protection must be implemented to protect the users from linkability of two or more successive positions.
- **Anonymity**: Anonymity describes that a target vehicle cannot be distinguished from a set of other vehicles. It is a term used to mean the ability of the user to access a resource or service without disclosing its user identity. This means that the subject may perform an action without disclosing its user identity to third parties. Anonymity techniques provide means to know the set of users that cannot see the identity of someone performing certain actions. Hence, when the action of a user is anonymous, another subject will not be able to determine either the identity or even a reference to the identity of the user.
- **Pseudonymity**: Pseudonymity is a technique used to ensure that a user may access a resource or service without disclosing its identity but can still be accountable for that use. Since the user is accountable for the use, it can be regarded as conditional pseudonymity or reversible pseudonymity. Conditional pseudonymity is required where the identity information of violators needs to be revealed by law enforcement authorities for liability purposes. Similarities exist between pseudonymity and anonymity. While both protect the identity of the user, pseudonymity uses a reference to the user's identity for accountability.
- **Non-repudiation**: Non-repudiation means that a vehicle cannot deny its malicious or illegal behavior. While anonymity protects the location privacy of users, a misbehaving or faulty vehicle that caused an accident needs to be identified for possible prosecution. Note that pseudonyms are linked to information that allows the certificate authority to establish a forensic evidence against a misbehaving user for the purpose of accountability.
- **Location Privacy**: The exact location information of the user must be protected from unauthorized entities. The user's trajectory, which contains position data about the user's present and past locations, including points of interest, must not be revealed to unauthorized entities.

Eckhoff and Sommer [19] introduce a classification of privacy based on the data privacy-enhanced techniques. Four types of data are introduced: published data, observable information, repurposed data, and leaked data. Published data means that the information was made available on purpose. This type of data could be a database of anonymized position information of IoT. Observable information is the data that malicious users try to obtain and use. For instance, a malicious user could eavesdrop on unencrypted messages sent by the target IoT device and use the information to track the relevant user. Repurposed data means privacy violations by the entity that was given confidential information by users, but he/she misuses it for unauthorized purposes – e.g. a location-based service provider that sells user location data to third parties. Leaked data are information that was intended to be confidential but was exposed using illegal measures or was accidentally made public.

5.3.3 Location Privacy Challenges

More and more individuals seem to accept the growing interest from industry to collect personal information as they will get personalized services. Many online services that seem to be free of charge require the user to uncover personal data to be used. In this case, users are able to decide whether the benefits they obtained outweigh the value of information they abandoned. Therefore, it is debated whether privacy or private information could be seen as a kind of currency [24, 25].

Location privacy is the ability to hinder third parties from obtaining the current location and location changes of IoT devices, such as vehicles [26]. However, some studies state that many test subjects could sell one month's location data for only US$35 for third-party commercial use [25, 27]. A violation of location privacy could cause severe effects. Exposed IoT devices that refer to individuals not only could have obtrusive advertisements but also have the implicit danger of being tracked by malicious third parties. Thus, unprotected location data could cause serious results from embarrassment to financial loss or loss of social standing. Obviously, different entities that attack users' location privacy can cause different levels of malicious influence. For example, when location privacy is compromised, adversaries could use this information to plan and act maliciously.

5.4 Securing IoT Using Network Slicing and Blockchain

IoT applications include a wide range of use cases from environmental monitoring to industrial automation. Each case may have distinct requirements in terms of networking, scalability, connectivity, and security. Hence, network slicing enables IoT systems to achieve network isolation and dedicated security protection in different use cases. For instance, a logistics industrial IoT scenario may have one slice for accurate, sensitive enterprise communications, one for cargo distributions, and another slice for factory-connected device lighting. While network slicing allows connected IoT devices to run on a shared network and security architecture, the segmentation keeps security isolated and matches various network needs based on different use cases.

5.4.1 Network Slicing Standardization

The third generation partnership project (3GPP) standards established the foundation of 5G network slicing. The 3GPP SA2 Release 15 [28] puts forward the basic concept of network slicing and formulates the basic functions, schemes, and procedures required for 5G networks to support network slicing. A network slice was defined as a logical E2E network that could be dynamically created to serve a purpose or service category or customers. Standard network slices are mainly divided into five categories in terms of service type, including enhanced Mobile Broadband (eMBB) slices, Vehicular-to-Everything (V2X) slices, Ultra Reliable Low Latency Communications (URLLC) slices, massive IoT slices, and mission-critical communication slices. A network slice contains the core network control plane and user plane network functions and 5G access network. When a user accesses the network, network slices will be selected based on user subscription data, slice selection strategies, user requests, etc. to access corresponding network slices. While the access and mobility management (AMF) instance logically belongs to all network slices, other network functions, such as the session management function (SMF) and user plane function (UPF), can be specific to a single network slice. In Release 16, the network slicing security mechanism has been further enhanced in TS 23.501. It mainly focuses on supporting third parties to authenticate and authorize terminal access to network slicing. Network-slicing-specific authentication and authorization mechanisms have been formulated. In Release 17, part of the Generic Network Slicing Template (GST) is studied and standardized within the scope of 3GPP, including the control of the number of UEs in the slice, the number of Protocol Data Unit (PDU) sessions, and the data rate. A simplified network slicing example based on TS 23.501 is shown in Figure 5.1. As can be seen from the figure, each slice has dedicated AMF, SMF, Network Repository Function (NRF), and UPF as its core network. End users, such as mobile phones, connected vehicles, and smart IoT devices, will have options to access various network slices based on their needs and scenarios.

The work group 3GPP SA5 has studied the Management and Orchestration (MANO) of network slicing in TR 28.801 [29] and normative specification works for Release 15 upon the report. Network slice concepts, use cases, and requirements were presented in TS 28.530 [30]. In addition, provisioning, MANO for network slices were presented in TS 28.531 [31]. In terms of management, a complete network slice should include all network function

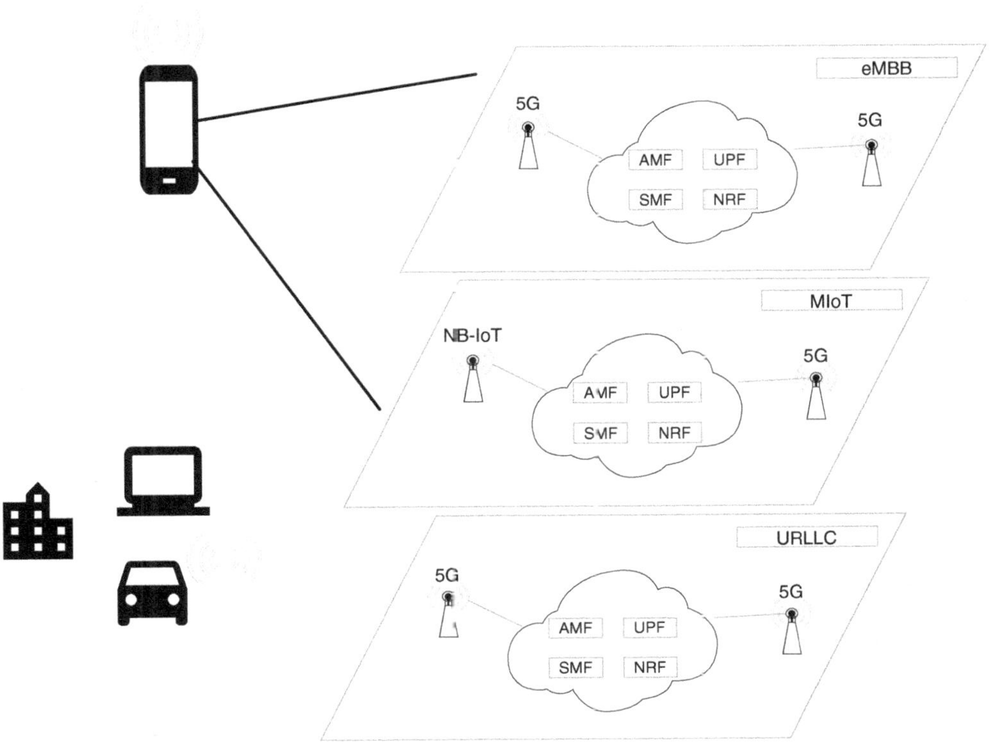

Figure 5.1 A simplified network slicing example based on TS 23.501.

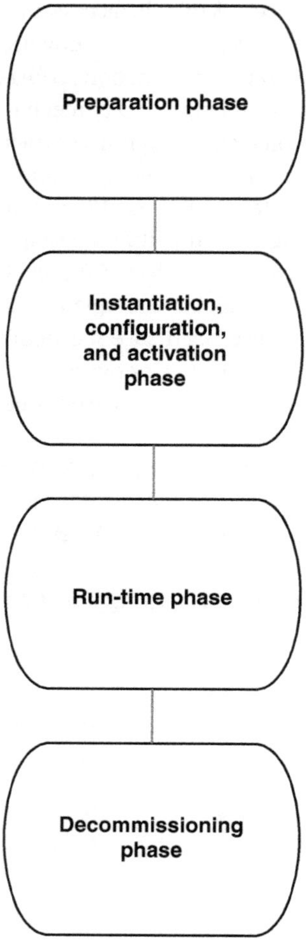

Figure 5.2 Lifecycle of network slicing.

instances and supporting resources to provide specific services for certain business purposes or operational efficiency purposes. TR 28.801 has studied a general network slice lifecycle, from the preparation phase, then the configuration and activation phase, to the run-time phase and decommissioning phase when the slice is no longer needed. Figure 5.2 depicts the lifecycle for the general network slicing procedure. 3GPP in TR 28.804 [32] studied network slice performance and fault monitoring in multiple tenant environments.

The research work on 3GPP for network slicing security is mainly carried out in SA3. In Release 14, 3GPP SA3 TR 33.899 [33] studied security isolation of network slices, security mechanism differentiation for network slices, the access security of network slices, and security architecture for network slices. Isolation requirements were defined to have limited influence on other slices. The technical report also pointed out that lack of security isolation could cause DoS attacks from one slice to another. In Release 15, SA3 launched the research and standardization work for the management security of network slicing in TS 33.501 [33]. The slice management requester needs to be authenticated and authorized based on Transport Layer Security (TLS) on the slice management interface. Different network slices could have different security policies, in terms of various types of authentication needs or because of different types of nodes (e.g. limited resource IoT devices vs. mobile equipment). Release 16 mainly focuses on open security issues to conduct security enhancement research, including network slice-specific authentication and authorization, network slicing-specific reauthentication and reauthorization, and authorization revocation process. In the Rel-16 stage, 3GPP SA3 began working on TR 33.813 in 2018 [34]. The enhanced network slicing security mainly studies the open issues left from the previous stage, including the authentication of access to specific network slices, key isolation, security features for Network Slice as a Service (NSaaS), and privacy protection. The technical report shows concern about the security and privacy of device access. Forged slice selection information and original selection information eavesdropping may lead to further damage to the network. In addition, slice management functions may be exposed through APIs that are vulnerable and need to be secured.

ETSI has set up a special security subgroup under NFV to conduct in-depth research on NFV security. Currently, it has established standards on the execution architecture of sensitive NFV components, NFV security management and monitoring, MANO component and interface security, and NFV security enhancement architecture [35, 36]. ETSI [37] studied the reliability and availability of network slicing. The allocated resources are concerned to be isolated during NSI creation and management operations from a security point of view. Moreover, ETSI [38] states the E2E network slicing management solutions including provisioning, performance, and fault management of a NSI across multiple management domains.

ITU-T describes security threats and potential attacks and defines security requirements for IMT-2020 network management and orchestration. In addition, ITU-T defines slice fault management, security management, and control plane security management in their proposed network management and orchestration framework [39].

Research projects have worked on extending or envisioning standardization landscapes into various schemes and proposals. Industry programs like Horizaon 2020 (H2020) and 5G Infrastructure Public Private Partnership (5GPPP) focus on the network slicing with diverse scenarios.

The EU project 5G Exchange (5GEx) [40] has proposed a bottom-up architecture further extending the concept of ETSI NFV architecture, on condition that physical and virtual resources of a network slice are instantiated over multiple domains or parties. The proposed architecture has 3 layers, including the resource domain at the bottom, the single-domain orchestration at the middle, and the multi-domain orchestration at the top. Each multi-domain orchestration controller interconnects with one or multiple single domain orchestrators via the orchestrator administrative domain.

5G-TRANSFORMER (5GT), which is H2020 project, proposes an E2E composite NFV network services architecture based on the combination technology between SDN and NFV. Multiple administrative domains help to manage the E2E deployment, requiring network slice federation functionalities [41]. The relevant project by Hortiguela et al. [42] proposes the service federation functionality of the 5GT service orchestrator and covers gaps that were identified in the ETSI NFV reports on the relevance of multi-domain resource orchestration. Based on the 5GT, the 5Growth project aims to enhance the 5GT architecture to achieve better performance, flexibility, automation, and security [43].

5.4.2 Network Slicing Security, Privacy, and Trust Threats for 5G and Beyond

Network slicing, as a network resource-sharing mechanism, refers to creating several virtual network slices using the same physical network resources. Each slice is tailored and optimized for a specific use and can be managed completely by the slice owner. According to ETSI [44], an NFV framework of network slice management is shown in Figure 5.3. The communication service management function is responsible for translating the communication service-related requirement to network slice-related requirements, and the network slice management function or network slice subnet management function is responsible for implementing the network slice related requirements by using VNFs or the connectivity to the Physical Network Functions (PNFs) to create network slices. The Network Functions Virtualization Orchestration (NFVO), Virtual Network Function Manager (VNFM), and Virtual Infrastructure Manager (VIM) are all management functions in the NFV framework that manage the life cycle and resource allocation of the VNF and NFV Infrastructure (NFVI).

Network slices that serve different types of services may have different levels of security and privacy policy requirements [45]. Combining the evolution of network architecture [46], the SDN and NFV are the main enablers for network slicing. Network softwarization consists of running network functions as software components to be hosted in the cloud, inside the virtual machines or containers. The management between network slices and inter-slices access in the 5G network is more complicated than in previous one-fit-all mobile networks because of the E2E slicing approach. The access management to slices, securing

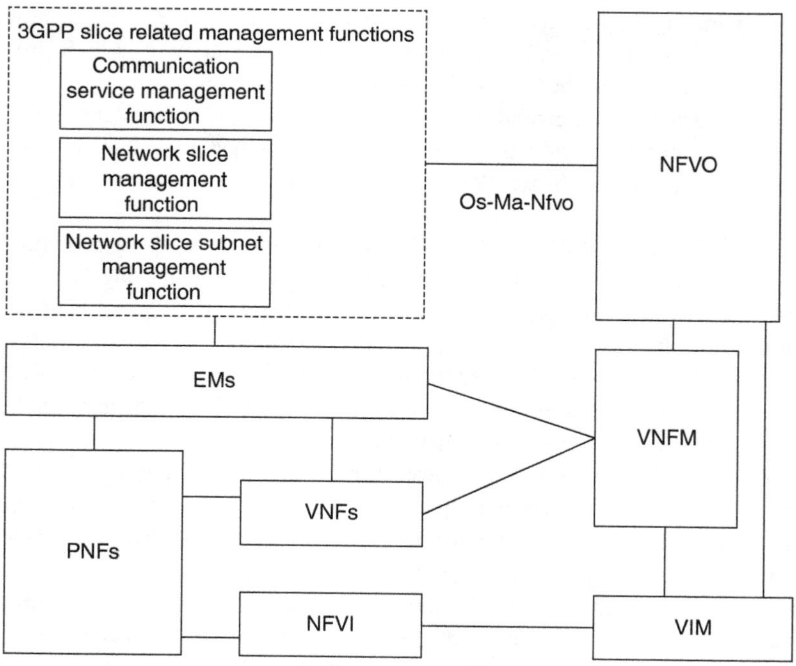

Figure 5.3 Network slice management in an NFV framework.

mutual access between the radio access network and the core network resources in the 5G network, and securing the connection between UE and NSIs are the main challenges for secure E2E network slicing management. Security threats can be caused by the concept of sharing resources among network slices. Reynaud et al. [47] and Sathi et al. [48] state that most security risks associated with network softwarization are the availability of controllers or orchestrators, isolation failure, compromised insider and NFV instances, and unauthorized data access. Moreover, the centralized slice manager also brings security issues such as network slice template, APIs, unauthorized access, and trust [49]. In addition, the next-generation mobile network may face more security and privacy challenges when the network slicing is associated with multi-domain infrastructures or multi-tenants. Cunha et al. [45] propose that the network slice security should follow traditional security principles including confidentiality, integrity, authenticity, and availability.

Regarding confidentiality threats, a compromised slice manager could monitor the traffic through both northbound and southbound interfaces, which may leak the slice configurations. A vulnerable point of attack in the configuration phase is the API. A compromised API could allow adversaries to interfere in the installation, configuration, or activation of a slice [33]. Similar to the slice manager, NFV Orchestrator (NFVO) also may breach confidentiality through interfaces. The confidentiality of inter-slice communications should be considered as well.

Regarding integrity threats, a poorly designed or improperly implemented network slice template may compromise the integrity of the template in the network slice preparation phase [50]. Injecting or forging traffic into slice manager's interfaces could compromise its integrity. Data exchange between slices may expose vulnerabilities as well. Another integrity threat is that an attack could change network slice configuration, leading to new threats such as slice deactivation at run-time phase [51]. In addition, the service level agreement (SLA) of network slicing faces the threat of being tampered or spoofed. Conventional SLA monitoring methods rely on third-party auditors, which could still temper the report for benefits [52].

Regarding authenticity threats, a compromised network function where credentials are locally stored would breach the system's authenticity. In addition, unauthorized access to the network slice may cause data leakage and lead to further damage to the network. Fan et al. [53] believe that rapid authentication needs to be innovated and rapidly adopted into the upcoming mobile communication network system since the network slice development requires UE to more frequently change in different slices.

Regarding availability threats, the availability issue of network slices may directly affect network capabilities and performance. Hence, the availability of a network slice is a major challenge. Launching DoS or DDoS attacks could break the availability of the network and delay the communication between entities. In addition, the attackers tend to launch DDoS attacks on a large number of UEs [53]. Compared with conventional single-user devices, beyond 5G networks, which will support extreme massive machine-type communication and extremely high-capacity communication may suffer from this threat more seriously than before.

Apart from the above security threats, privacy and trust issues are other critical concerns that have not been addressed well in current 5G networks [54, 55].

5.4.3 Blockchain-Enabled Secure Network Slicing

5.4.3.1 Integration of Distributed Ledger Technology (DLT) with Network Slicing

Recent studies have included distributed ledger technology (DLT), a key technology to assist automation and management in beyond 5G and future 6G networks [7, 56]. Network slices could also benefit from DLT, as DLT can provide a distributed framework, a secure and accountable digital ledger, and a fine-grained security indicator. It is a possible solution to resolve network slicing security challenges and maintain the confidentiality, integrity, and availability of the system. Bihannic et al. [57] believe that blockchain suits scenarios with multiple partners in the E2E value chain of any ecosystem. Multiple partners can not only be a horizontal chain between companies, suppliers, and end users but also a vertical chain among internal entities. In addition, many studies show that merging DLT/blockchain and network slicing could achieve multiple purposes. First, since the DLT with smart contract could create an automated marketplace or auction platform, network slices, and corresponding services can be negotiated among consumers, network operators, and industrial vertical players [58, 59]. Secondly, another challenge raised by multi-tenant and multi-domain network slicing scenarios is having much higher chances of attacks inside the network. However, the use of DLT can be a solution to establish a trust mechanism among network slicing parties and could also build an authentication layer for the multiple administrative domains to meet security requirements [60, 61]. Moreover, the distributed nature of DLT provides the capability of avoiding single-point failure because the system has high robustness even if one or a few nodes go out of reach. It also assures reliability and availability on an E2E NSI with multiparties. Due to these properties, blockchain has the potential to be integrated with network slicing [62, 63].

5.4.3.2 Blockchain-Powered Network Slicing in Verticals

Since the blockchain could provide trust and secure solutions, it has been noticed that research has considered using blockchain to tackle existing security, privacy, and trust challenges in network slicing. Xiao et al. [64] proposed a blockchain-based privacy-preserving network slicing SLA audit scheme, aiming to preserve data privacy and integrity. Conventional SLA audit process faces threats of data leakage and alterations in the audit report. This scheme introduces a blockchain-based audit strategy that can preserve data privacy in the audit process, and the smart contract inside is designed to operate audit tasks and carry out punishments. The blockchain in this model is a platform to provide a public and transparent way to store immutable data across multiple parties. Then a data encryption scheme called order-revealing encryption (ORE) is introduced and implemented in the smart contract of the proposed blockchain-based SLA auditing scheme. In their more recent research [65], this ORE scheme is more refined to be able to encrypt monitored parameters in the SLA and realize the comparison over auditing ciphertexts, resulting in the prevention of data leakage in SLA auditing. In addition to the integrity, an order-revealing encryption algorithm is used for encrypting the monitored data and parameters in the SLA audit. The smart contract in the blockchain can also be used to provide fine-grained security [57]. Future smart contracts could monitor security indicators and conduct in-depth analysis once any attack is detected in a given slice. The work of Camilo et al. [66] presents an SDN and DLT-based framework to provide secure, automatic, and distributed data. It uses the Hyperledger Fabric platform to show a secure and fast transaction performance through blockchain.

5.4.3.3 Multiple Participant Coordination and Trust Management by Blockchain for Network Slicing

Orchestration across multiple infrastructure providers can help simplify infrastructural operations and enable better scaling and faster deployment of network services. For the implementation of certain network services, it is required to create E2E NSI across multiple participants. In 3GPP TR 28.801, if a participant wants to create an E2E NSI across multiple participants, this participant should decompose the service request of the E2E network slice and provide its management data to the other participants. Preconditions of this connection are trust relationships that are assumed to exist between operators. This trust relationship makes the coordination between participants transparent and visible. However, the majority of participants worry about whether sensitive data could be exposed, and other participants might not follow the agreement on network slicing. With the above trust concern in mind, research such as [67] suggests that blockchain can be deployed to ensure trustworthiness between different telecom operators for multiple participants' coordination management in the network slice.

Rathi et al. [68] propose an architecture comprising multi-domain edge orchestration to achieve SLA autonomy management by smart contracts. In addition, the Hyperledger Fabric blockchain is used to improve the trustworthiness of the system by storing records in orchestrators. Samdanis et al. [69] introduces the concept of the 5G Network Slice Broker, which can facilitate on-demand resource allocation and perform admission control based on traffic monitoring and forecasting. Nour et al. [70] and Zanzi et al. [71] propose the use of blockchain to deploy the brokering mechanism through smart contracts. An example of a network slice broker using a blockchain system is shown in Figure 5.4. Mobile Virtual Network Operators (MVNOs), Over-the-Top (OTT) providers, and Infrastructure Providers (InPs) all participate in the blockchain network, and the slice broker mechanism will handle transactions between multiple traders. So the peer-to-peer system is able to replace the conventional centralized approach to slice brokering. Chendeb et al. [72] designed a multilayer blockchain-based secure network slicing architecture to be used in the medical field. The paper states that using decentralized storage platforms to perform as slice brokers could secure the network slicing. In addition, Gong et al. [73] propose an automated

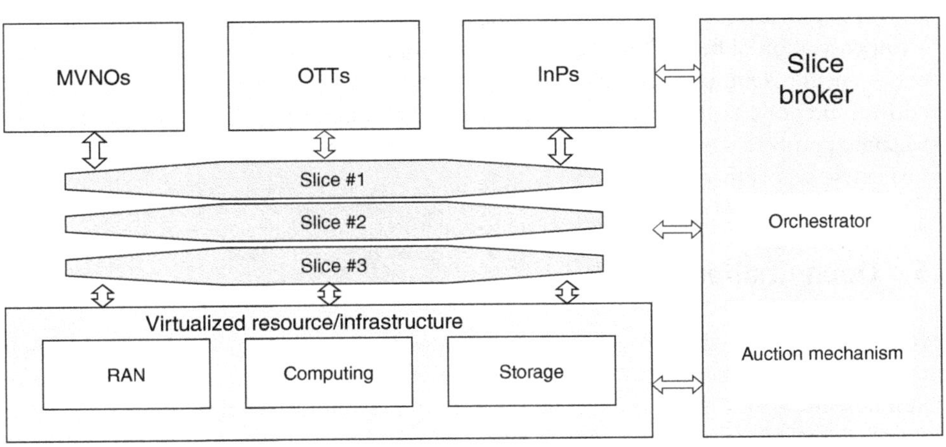

Figure 5.4 An example of blockchain-based slice-brokering system.

Category of threats	The role of blockchain	References
Integrity	Realize SLA trust automation management	[68, 72, 76–78]
Authenticity	Control the access to network slice	[74]
Availability	Resist DoS attacks	[75]
Trust	Establish trust relationship	[68, 70, 71, 73, 76, 77]
Privacy	Ensure privacy among network slices	[64, 74]

Figure 5.5 Categorization of blockchain-based solutions for threats in network slice.

resource allocation based on Deep Reinforcement Learning (DRL) and blockchain, which can ensure the security of transactions. However, these papers provide only a qualitative analysis of the security performance of the dynamic slicing endowed by the blockchain and do not elaborate on how to use the blockchain to establish a trust relationship.

The combination of network slicing and blockchain has become an inevitable trend, and it will be expected to establish a trust relationship between multiple participants. STewARD is a blockchain-based trust assessment framework that provides automated risk management in IoT devices [74]. The blockchain system will store trust scores of all devices so that the network controllers can leverage this information and then allow devices that meet the expected trust levels to connect with the network slice. Rebello et al. [75] develop a blockchain-based secure and isolated software infrastructure for virtual functions. In order to isolate these slices and avoid common attacks in shared infrastructures, they provide different categories of blockchains for different slice requirements, while all slices share the same infrastructure. Saad et al. [76] propose that DLT can be used to solve the trust problem. In addition, Theodorou et al. [77] present a zero-touch approach to cross-domain network slicing based on enterprise blockchain and AI-driven closed-loop automation architecture. It can detect or predict SLA violations with immediate action for mitigation without disrupting the service. Finally, a blockchain-based trusted architecture was proposed by Scheid et al. [78] to provide E2E security by utilizing smart contracts for SLA management.

A categorization of blockchain-based solutions for security, privacy, and trust threats in network slicing is summarized in Figure 5.5. The threats of network slicing are categorized into integrity, authenticity, availability, trust, and privacy. In the figure, the role of blockchain points to what benefits the blockchain system provides to each of the aforementioned categories of threats.

5.5 Open Challenges

Beyond 5G networks, the trios of trust, security, and privacy are interconnected to some extent. The challenges remain in multidisciplinary technologies, standardization, techno-economics, etc. [79]. Despite the fact that blockchain technology facilitates the security of network slicing, affecting the security of massive IoT, there are open challenges in terms of security, privacy, and trust. In this section, potential issues and challenges of endogenous security in network slicing beyond 5G networks will be reviewed.

5.5.1 Edge Intelligence in Network Slicing

A popular use case for edge intelligence is the automation of virtual resource management and orchestration in beyond 5G networks [80]. The management procedures of the RAN network slice subnet management function will be automated by edge intelligence to decrease the management and orchestration complexity. Therefore, the processed data among slices and users could be exposed. Secure and lightweight data-preserving technologies can be further investigated, aiming to handle data confidentiality and privacy for the future scalable and fast-changing next-generation mobile networks. In this regard, standardization and international recommendations play key roles in the development of automated network slicing to ensure privacy, security, and trust in massive IoT systems.

Moreover, the state-of-the-art security approaches for network slices as stated above are mostly human- or machine-centric. For instance, even automated anomaly detection still needs human intervention to address false negatives [81]. Since promising advanced telecommunication technologies and artificial intelligence techniques are emerging, the next-generation mobile networks are inevitably facing more and more automated and advanced attacks. So more sophisticated and intelligent security mechanisms are essential to overcome future massive-scale and automated attacks. Machine learning could play a crucial role in designing intelligent security solutions for network slicing. Machine learning algorithms and models, such as Markov models, neural networks, DRL, and genetic algorithms, could be utilized to find configuration errors in order to limit human intervention.

5.5.2 Post-quantum Security

Apart from conventional threats, many believe that the upcoming quantum computing will bring serious challenges to network security. The current cryptographic systems fall mainly under two categories: symmetric and asymmetric [82]. However, mathematicians state that either symmetric algorithms or asymmetric algorithms will be breakable once quantum computers come, as quantum computing is significantly more powerful than current computing systems. The research and development of quantum-resistant cryptographic algorithms will be a big milestone for the next-generation mobile networks. Future quantum-resistant cryptographic algorithms should be secure enough against both quantum computing threats and classic cryptography threats.

5.5.3 Blockchain Scalability

In the blockchain, there is a triangle, namely, decentralization, security, and scalability. In this triangle, most blockchain-based systems can only meet two subjects. Both generations of the blockchain (Bitcoin and Ethereum) have similar drawbacks, low transaction throughputs, small block size, and huge energy consumption. In terms of low transaction throughputs, Bitcoin can do 3–5 transactions per second (TPS), Ethereum can do 25 tps, and Visa can do 1500 tps. As a fundamental technology for future networks, DLT is eager to seek solutions and innovations to meet the trio of decentralization, security, and scalability together. Novel blockchain architectures and consensus algorithms are being researched to increase the throughput of today's blockchain networks.

5.5.4 RAN Slicing

The Next Generation Radio Access Network (NG-RAN) would support a great number of RAN slice subnets [80]. To satisfy the advanced service requirements of E2E slicing in the next-generation network, the future radio base stations are supposed to be more customizable and more dynamic. Maintaining a significant level of isolation will be a key to provide security and privacy on the E2E slicing. Hence, more reliable and resilient slice isolation technologies require further study. Moreover, the Open-RAN (O-RAN) has drawn a lot of attention recently. The main concept of openness and smartness in this new RAN framework is to let all participants build an open source, open hardware, software-driven, slicing-aware, and resource-efficient innovated radio network together [83]. However, O-RAN is still at early stage and needs further research to be fully realized in the next-generation mobile networks. Major problems, especially security related, can be summarized as the hijacking attack on open networking, the doubt of trustworthy open-RAN, interoperability, and standardization [84, 85].

5.6 Conclusion

In this chapter, we illustrate an intensive study on securing massive IoT with blockchain and network slices. It showed an overview of the basic background of massive IoT, IoT-relevant blockchain applications, and its consensus mechanisms. In this chapter, we presented security and privacy requirements and challenges and explored security countermeasures that could be applied to secure massive IoT. Moreover, we presented a review of network slicing security, privacy, and trust threats for 5G and beyond. Finally, we discussed potential blockchain-based approaches for secure network slicing in order to protect massive IoT.

References

1 Atzori, L., Iera, A., and Morabito, G. (2010). The Internet of Things: a survey. *Computer Networks* 54: 2787–2805.

2 IoT Analytics (2022). State of IoT–Spring 2022 Report. Technical Report. https://iot-analytics.com/number-connected-iot-devices (accessed 17 July 2024).

3 Foukas, X., Patounas, G., Elmokashfi, A., and Marina, M.K. (2017). Network slicing in 5G: survey and challenges. *IEEE Communications Magazine* 55 (5): 94–100. https://doi.org/10.1109/MCOM.2017.1600951.

4 Afolabi, I., Ksentini, A., Bagaa, M. et al. (2017). Towards 5G network slicing over multiple-domains. *IEICE Transactions on Communications* 100 (11): 1992–2006.

5 Čolaković, A. and Hadžialić, M. (2018). Internet of Things (IoT): a review of enabling technologies, challenges, and open research issues. *Computer Networks* 144: 17–39.

6 Ali, M.S., Vecchio, M., Pincheira, M. et al. (2019). Applications of blockchains in the Internet of Things: a comprehensive survey. *IEEE Communications Surveys & Tutorials* 21: 1676–1717.

7 Guo, F., Yu, F.R., Zhang, H. et al. (2021). Enabling massive IoT toward 6G: a comprehensive survey. *IEEE Internet of Things Journal* 8 (15): 11891–11915.

8 Ericsson (2022). Ericsson Mobility Report. Technical report.

9 Nakamoto, S. (2008). Bitcoin: A Peer-to-Peer Electronic Cash System.

10 Danezis, G. and Meiklejohn, S. (2015). Centrally banked cryptocurrencies. arXiv:1505.06895.

11 Zyskind, G., Nathan, O., and Pentland, A.S. (2015). Decentralizing privacy: using blockchain to protect personal data. In: *2015 IEEE Security and Privacy Workshops (SPW)*, 180–184. https://doi.org/10.1109/SPW.2015.27.

12 Mengelkamp, E., Notheisen, B., Beer, C. et al. (2018). A blockchain-based smart grid: towards sustainable local energy markets. *Computer Science - Research and Development* 33 (1–2): 207–214.

13 Pop, C., Cioara, T., Antal, M. et al. (2018). Blockchain based decentralized management of demand response programs in smart energy grids. *Sensors* 18 (1): 162.

14 Dorri, A., Steger, M., Kanhere, S.S., and Jurdak, R. (2017). Blockchain: a distributed solution to automotive security and privacy. *IEEE Communications Magazine* 55 (12): 119–125. https://doi.org/10.1109/MCOM.2017.1700879.

15 Penard, W. and van Werkhoven, T. (2008). On the secure hash algorithm family. *Cryptography in Context* 1–18.

16 Park, K. and Lee, H. (2001). On the effectiveness of route-based packet filtering for distributed dos attack prevention in power-law internets. In: *ACM SIGCOMM computer communication review*, vol. 31, 15–26. ACM.

17 Hubaux, J.P., Capkun, S., and Luo, J. (2004). The security and privacy of smart vehicles. *IEEE Security Privacy* 2 (3): 49–55. https://doi.org/10.1109/MSP.2004.26.

18 Raya, M. and Hubaux, J.-P. (2007). Securing vehicular ad hoc networks. *Journal of Computer Security* 15 (1): 39–68.

19 Eckhoff, D. and Sommer, C. (2014). Driving for big data? Privacy concerns in vehicular networking. *IEEE Security and Privacy* 12 (1): 77–79.

20 Asuquo, P., Cruickshank, H., Morley, J. et al. (2018). Security and privacy in location-based services for vehicular and mobile communications: an overview, challenges and countermeasures. *IEEE Internet of Things Journal* 5 (6): 4778–4802. https://doi.org/10.1109/JIOT.2018.2820039.

21 Golle, P., Greene, D., and Staddon, J. (2004). Detecting and correcting malicious data in VANETs. In: *Proceedings of the 1st ACM International Workshop on Vehicular ad Hoc Networks*, 29–37. ACM.

22 Douceur, J.R. (2002). The sybil attack. In: *International Workshop on Peer-to-Peer Systems*, 251–260. Springer.

23 Studer, A., Shi, E., Bai, F., and Perrig, A. (2009). Tacking together efficient authentication, revocation, and privacy in VANETs. In: *2009 6th Annual IEEE Communications*

Society Conference on Sensor, Mesh and Ad Hoc Communications and Networks, 1–9. IEEE.

24 Cvrcek, D., Kumpost, M., Matyas, V., and Danezis, G. (2006). A study on the value of location privacy. In: *Proceedings of the 5th ACM Workshop on Privacy in Electronic Society*, 109–118. ACM.

25 Leman-Langlois, S. (2013). Privacy as currency: crime, information and control in cyberspace. In: *Technocrime*, 135–161. Willan.

26 Sommer, C. and Dressler, F. (2014). *Vehicular Networking*. Cambridge University Press.

27 Danezis, G., Lewis, S., and Anderson, R.J. (2005). How much is location privacy worth? In: *WEIS*, vol. 5. Citeseer.

28 3GPP (2018). System architecture for the 5G system.

29 3GPP (2017). Study on management and orchestration of network slicing for next generation network.

30 3GPP (2022). Management and orchestration; concepts, use cases and requirements.

31 3GPP (2022). Management and orchestration; provisioning.

32 3GPP (2019). Study on tenancy concept in 5G networks and network slicing management V16.0.0.

33 3GPP (2017). Study on the security aspects of the next generation system.

34 3GPP (2020). Study on security aspects of network slicing enhancement.

35 European Telecommunications Standards Institute (ETSI) (2022). Network Functions Virtualisation (NFV) Release 4; Management and Orchestration; Functional Requirements Specification. Technical Report GS NFV-IFA 010, European Telecommunications Standards Institute(ETSI). Version 4.3.1.

36 Yuan, Q. (2019). Analysis on technologies and developments of 5G network slicing security. *Mobile communication* 10: 26–30.

37 ETSI (2019). Network Functions Virtualisation (NFV) Release 3; Reliability; Report on NFV Resiliency for the Support of Network Slicing. Technical Report GR NFV-REL 010. European Telecommunications Standards Institute(ETSI), Version 3.1.1.

38 ETSI (2021). Zero-Touch Network and Service Management (ZSM); End-to-end Management and Orchestration of Network Slicing. Technical Report GS ZSM 003. European Telecommunications Standards Institute(ETSI), Version 1.1.1.

39 ITU-T, 9 (2017). IMT-2020 network management and orchestration framework.

40 Guerzoni, R., Pérez-Caparrós, D., Monti, P. et al. (2016). Multi-domain orchestration and management of software defined infrastructures: a bottom-up approach.

41 De la Oliva, A., Li, X., Costa-Perez, X. et al. (2018). 5G-transformer: slicing and orchestrating transport networks for industry verticals. *IEEE Communications Magazine* 56 (8): 78–84.

42 Hortiguela, J.B., Mangues-Bafalluy, J., Martinez, R. et al. (2020). Realizing the network service federation vision: enabling automated multidomain orchestration of network services. *IEEE Vehicular Technology Magazine* 15 (2): 48–57. https://doi.org/10.1109/MVT.2020.2979558.

43 Li, X., Garcia-Saavedra, A., Costa-Perez, X. et al. (2021). 5Growth: An end-to-end service platform for automated deployment and management of vertical services over 5g networks. *IEEE Communications Magazine* 59 (3): 84–90. https://doi.org/10.1109/MCOM.001.2000730.

44 ETSI (2017). Network functions virtualisation (NFV) release 3; evolution and ecosystem; report on network slicing support with ETSI NFV architecture framework.

45 Cunha, V.A., Silva, E.D., Carvalho, M. et al. (2019). Network slicing security: challenges and directions. *Internet Technology Letters* 2 (5): e125.

46 Afolabi, I., Taleb, T., Samdanis, K. et al. (2018). Network slicing and softwarization: a survey on principles, enabling technologies and solutions. *IEEE Communications Surveys & Tutorials* 20 (3): 2429–2453.

47 Reynaud, F., Aguessy, F.-X., Bettan, O. et al. (2016). Attacks against network functions virtualization and software-defined networking: state-of-the-art. In: *Workshop on Security in Virtualized Networks (Sec-Virtnet 2016), workshop of 2nd IEEE Conference on Network Softwarization (NetSoft 2016)*, 471–476. Seoul, South Korea. https://doi.org/10.1109/NETSOFT.2016.7502487. https://hal.archives-ouvertes.fr/hal-01393740.

48 Sathi, V.N., Srinivasan, M., Thiruvasagam, P.K., and Chebiyyam, S.R.M. (2018). A novel protocol for securing network slice component association and slice isolation in 5G networks. In: *Proceedings of the 21st ACM International Conference on Modeling, Analysis and Simulation of Wireless and Mobile Systems, MSWIM'18*, 249–253. New York: Association for Computing Machinery. ISBN: 9781450359603. https://doi.org/10.1145/3242102.3242135.

49 NMNG Alliance (2018). Security aspects of network capabilities exposure in 5G. Final deliverable (approved-P Public).

50 Olimid, R.F. and Nencioni, G. (2020). 5G network slicing a security overview. *IEEE Access* 8: 99999–100009.

51 Khan, R., Kumar, P., Jayakody, D.N.K., and Liyanage, M. (2019). A survey on security and privacy of 5G technologies: potential solutions, recent advancements, and future directions. *IEEE Communications Surveys & Tutorials* 22 (1): 196–248.

52 Zhou, S., Wu, L., and Jin, C. (2017). A privacy-based SLA violation detection model for the security of cloud computing. *China Communications* 14 (9): 155–165.

53 Fan, C.-I., Shih, Y.-T., Huang, J.-J., and Chiu, W.-R. (2021). Cross-Network-Slice Authentication Scheme for the 5th Generation Mobile Communication System. *IEEE Transactions on Network and Service Management* 18 (1): 701–712. https://doi.org/10.1109/TNSM.2021.3052208.

54 Ni, J., Lin, X., and Shen, X.S. (2018). Efficient and secure service-oriented authentication supporting network slicing for 5G-enabled IoT. *IEEE Journal on Selected Areas in Communications* 36 (3): 644–657. https://doi.org/10.1109/JSAC.2018.2815418.

55 Ni, J., Zhang, K., Lin, X., and Shen, X. (2017). Securing fog computing for Internet of Things applications: challenges and solutions. *IEEE Communications Surveys & Tutorials* 20 (1): 601–628.

56 Akhtar, M.W., Hassan, S.A., Ghaffar, R. et al. (2020). The shift to 6G communications: vision and requirements. *Human-centric Computing and Information Sciences* 10 (1): 1–27.

57 Bihannic, N., Lejkin, T., Finkler, I., and Frerejean, A. (2018). Network slicing and blockchain to support the transformation of connectivity services in the manufacturing industry. https://sdn.ieee.org/newsletter/march-2018/network-slicing-and-blockchain-in-the-manufacturing-industry (accessed 01 September 2021).

58 Kabi, O.R. and Franqueira, V.N.L. (2018). Blockchain-based distributed marketplace. In: *International Conference on Business Information Systems*, 197–210. Springer.

59 Kapassa, E., Touloupos, M., Kyriazis, D., and Themistocleous, M. (2019). A smart distributed marketplace. In: *European, Mediterranean, and Middle Eastern Conference on Information Systems*, 458–468. Springer.

60 Chaer, A., Salah, K., Lima, C. et al. (2019). Blockchain for 5G: opportunities and challenges. In: *2019 IEEE Globecom Workshops (GC Wkshps)*, 1–6. IEEE.

61 De Filippi, P., Mannan, M., and Reijers, W. (2020). Blockchain as a confidence machine: the problem of trust & challenges of governance. *Technology in Society* 62: 101284.

62 He, G., Su, W., Gao, S. et al. (2021). NetChain: A blockchain-enabled privacy-preserving multi-domain network slice orchestration architecture. *IEEE Transactions on Network and Service Management* 19 (1): 188–202.

63 Hewa, T., Kalla, A., Porambage, P. et al. (2021). How dos attacks can be mounted on network slice broker and can they be mitigated using blockchain? In: *2021 IEEE 32nd Annual International Symposium on Personal, Indoor and Mobile Radio Communications (PIMRC)*, 1525–1531. IEEE.

64 Xiao, K., Geng, Z., He, Y. et al. (2020). A blockchain based privacy-preserving cloud service level agreement auditing scheme. *International Conference on Wireless Algorithms, Systems, and Applications.*

65 Xiao, K., Geng, Z., He, Y. et al. (2021). A blockchain-based privacy-preserving 5G network slicing service level agreement audit scheme. *EURASIP Journal on Wireless Communications and Networking* 2021: 1–16.

66 Camilo, G.F., Rebello, G.A.F., de Souza, L.A.C., and Duarte, O.C.M.B. (2020). AutAvailChain: Automatic and secure data availability through blockchain. *GLOBECOM 2020–2020 IEEE Global Communications Conference*, 1–6. https://doi.org/10.1109/GLOBECOM42002.2020.9322396.

67 Gorla, P., Chamola, V., Hassija, V., and Niyato, D. (2020). Network slicing for 5G with UE state based allocation and blockchain approach. *IEEE Network* 35 (3): 184–190.

68 Rathi, V.K., Chaudhary, V., Rajput, N.K. et al. (2020). A blockchain-enabled multi domain edge computing orchestrator. *IEEE Internet of Things Magazine* 3 (2): 30–36.

69 Samdanis, K., Costa-Perez, X., and Sciancalepore, V. (2016). From network sharing to multi-tenancy: the 5G network slice broker. *IEEE Communications Magazine* 54 (7): 32–39.

70 Nour, B., Ksentini, A., Herbaut, N. et al. (2019). A blockchain-based network slice broker for 5G services. *IEEE Networking Letters* 1 (3): 99–102.

71 Zanzi, L., Albanese, A., Sciancalepore, V., and Costa-Pérez, X. (2020). NSBchain: A secure blockchain framework for network slicing brokerage. In *ICC 2020 - 2020 IEEE International Conference on Communications (ICC).*

72 Chendeb, N., Khaled, N., and Agoulmine, N. (2020). Integrating blockchain with IoT for a secure healthcare digital system. In: *8th International Workshop on ADVANCEs in ICT Infrastructures and Services (ADVANCE 2020)*, 1–8.

73 Gong, Y., Sun, S., Wei, Y., and Song, M. (2021). Deep reinforcement learning for edge computing resource allocation in blockchain network slicing broker framework. 2021 IEEE 93rd Vehicular Technology Conference (VTC2021-Spring).

74 Boussard, M., Papillon, S., Peloso, P. et al. (2019). STewARD:SDN and blockchain-based trust evaluation for automated risk management on IoT devices. IEEE INFOCOM 2019 - IEEE Conference on Computer Communications Workshops (INFOCOM WKSHPS).

75 Rebello, G., Camilo, G.F., Silva, L., Guimaraes, L. et al. (2019). Providing a sliced, secure, and isolated software infrastructure of virtual functions through blockchain technology. In *2019 IEEE 20th International Conference on High Performance Switching and Routing (HPSR)*.

76 Saad, S.B., Ksentini, A., and Brik, B. (2021). A trust architecture for the SLA management in 5G networks. ICC 2021 - IEEE International Conference on Communications.

77 Theodorou, V., Lekidis, A., Bozios, T. et al. (2021). Blockchain-based zero touch service assurance in cross-domain network slicing. In: *2021 Joint European Conference on Networks and Communications 6G Summit (EuCNC/6G Summit)*, 395–400. https://doi.org/10.1109/EuCNC/6GSummit51104.2021.9482602.

78 Scheid, E.J., Rodrigues, B.B., Granville, L.Z. and Stiller, B. (2019). Enabling dynamic SLA compensation using blockchain-based smart contracts. In: *2019 IFIP/IEEE Symposium on Integrated Network and Service Management (IM)*, 53–61. IEEE.

79 Ylianttila, M., Kantola, R., Gurtov, A. et al. (2020). 6G white paper: research challenges for trust, security and privacy. *arXiv e-prints*.

80 Jiang, W., Han, B., Habibi, M.A., and Schotten, H.D. (2021). The road towards 6G: a comprehensive survey. *IEEE Open Journal of the Communications Society* 2: 334–366. https://doi.org/10.1109/OJCOMS.2021.3057679.

81 Porambage, P. and Liyanage, M. (2019). Security in network slicing. In: *Wiley 5G Ref: The Essential 5G Reference* Online, 1–12.

82 Cheng, C., Lu, R., Petzoldt, A., and Takagi, T. (2017). Securing the internet of things in a quantum world. *IEEE Communications Magazine* 55 (2): 116–120. https://doi.org/10.1109/MCOM.2017.1600522CM.

83 O-RAN Alliance (2018). O-RAN: Towards an open and smart ran. https://www.the5gexchange.com/lg_landing.%3Cp%3Easp?%3Cp%3Epiddl%3Cp%3E_lgid_docid=765514 (accessed 01 September 2021).

84 Plantin, J.C. (2021). The political hijacking of open networking. the case of open radio access network. *LSE Research Online Documents on Economics* 36 (4): https://doi.org/10.1177/02673231211028375.

85 Singh, S.K., Singh, R., and Kumbhani, B. (2020). The evolution of radio access network towards open-RAN: challenges and opportunities. In: *2020 IEEE Wireless Communications and Networking Conference Workshops (WCNCW)*, 1–6. https://doi.org/10.1109/WCNCW48565.2020.9124820.

6

Physical Layer Security for RF-Based Massive IoT

Marcus de Ree[1], Seda Dogan-Tusha[2,3], Elmehdi Illi[2], Marwa Qaraqe[2], Saud Althunibat[4], Georgios Mantas[1,5], and Jonathan Rodriguez[6]

[1] *Instituto de Telecomunicações, Universidade de Aveiro, Aveiro, Portugal*
[2] *Division of Information and Computing Technology, College of Science and Engineering, Hamad Bin Khalifa University, Qatar Foundation, Doha, Qatar*
[3] *Wireless Institute, College of Engineering, University of Notre Dame, South Bend, IN, USA*
[4] *Faculty of Engineering, Al-Hussein Bin Talal University, Ma'an, Jordan*
[5] *Faculty of Engineering and Science, University of Greenwich, Chatham Maritime, UK*
[6] *Faculty of Computing, Engineering and Science, University of South Wales, Pontypridd, UK*

6.1 Introduction

This chapter focuses on methods that exploit physical layer parameters to enforce security of data exchanged between nodes in Internet of Things (IoT) and massive IoT networks. In particular, this chapter focuses on the unique characteristics of the radio frequency (RF) signals, representing the exchanged data, and how these characteristics can be exploited to achieve physical layer security (PLS).

Conventionally, security primitives such as node authentication, data integrity, and data confidentiality are achieved through the implementation of cryptographic methods (i.e. through symmetric key cryptography and/or public key cryptography) at the higher layers of the protocol stack. These cryptographic methods are an effective way of securing our modern computer and communication networks given that these possess sufficient computing capabilities. Public key cryptography relies on more complex mathematical operations, whereas symmetric key cryptography relies on more simplistic operations such as exclusive-or (XOR), substitutions, and permutations. However, the resource requirement and consumption are still considered significant with respect to the capabilities of many IoT devices. Namely, IoT devices are typically small battery-powered devices with limited computational and memory storage capabilities. Furthermore, in a wide variety of applications, the IoT devices are deployed in such a manner that they must remain powered for long periods of time (e.g. months, or even years). This puts a strain on the amount of energy that can be consumed by computation alone. Even though different IoT networks will have different requirements in terms of security, there will be networks where cryptography-based methods are too costly, and thus more energy-efficient security solutions are needed [1].

Security and Privacy for 6G Massive IoT, First Edition.
Edited by Georgios Mantas, Firooz B. Saghezchi, Jonathan Rodriguez, and Victor Sucasas.

In contrast to the security measures that have been proposed at the higher layers of the protocol stack, PLS takes advantage of existing protocols at the physical layer to minimize its impact on resource consumption. Namely, a wide range of PLS schemes suggest that only minor alterations or additions are necessary to the generation and processing of RF signals to achieve certain security goals. Each of these schemes exploits the inherent randomness that is characterized by the wireless communication link shared between a pair of nodes. Namely, RF signals are subject to randomness caused by variations in a node's hardware due to imperfections introduced during the manufacturing process. The randomness caused by hardware imperfections is unique per node and can therefore be exploited to achieve certain security goals. For example, hardware imperfections induce a unique variation in the RF signal generated by a particular node which allows a receiver to determine the source of the signal. This source of randomness can therefore provide security goals including, but not limited to, node authentication and the detection of malicious impersonation behavior. RF signals are also subject to randomness caused by objects within the direct environment of communicating nodes since these affect the signal propagation paths due to reflection, absorption, and diffraction. The randomness caused by the signal propagation is unique to the nodes at the ends of the wireless communication link, which can also be exploited to achieve certain security goals. For example, the nodes at the end of the wireless communication link can utilize this shared source of randomness to extract a shared secret key which would only be known to them. This shared secret key can subsequently be used to provide security goals including, but not limited to, data confidentiality and integrity. PLS is therefore an excellent solution to secure communication between resource-constrained devices since it limits the computational workload and transmission cost which leads to minimal energy consumption [2].

This chapter will provide an overview of the various physical layer techniques and approaches that have been proposed to provide node authentication, data confidentiality, and malicious node detection in RF-based IoT networks. Physical layer techniques and approaches that are not aligned with the characteristics and resource constraints of IoT devices, such as those based on the multiple-input multiple-output (MIMO) technology, are out of our scope.

This chapter is organized as follows and illustrated in Figure 6.1. Section 6.2 focuses on physical layer-based key establishment (PLKE). In particular, two approaches are covered which enable any pair of nodes to establish a shared secret key that is a prerequisite to key-based node authentication and data confidentiality schemes. Sections 6.3 and 6.4 focus on physical layer-based node authentication and data confidentiality, respectively, both including key-based and keyless approaches. Section 6.5 focuses on physical layer-based detection of malicious nodes and Section 6.6 concludes the chapter.

6.2 Physical Layer-Based Key Establishment

6.2.1 Introduction

This section covers the concept of PLKE, a physical layer approach where two arbitrary nodes exploit characteristics of their shared wireless channel to generate a key that will only be known to them [3]. PLKE therefore follows a similar philosophy to that of the cryptographic community where a key, a shared secret, is used to achieve security goals such as node authentication, data integrity, and data confidentiality. The major benefit of PLKE lies in its capacity to establish and update keys on demand and in an energy-efficient manner, whereas cryptographic key establishment either requires the pre-distribution of symmetric keys or suffers from the computational burden associated with public key cryptography (e.g. Diffie–Hellman key agreement).

It is important to make a clear distinction between PLKE schemes and physical layer-based data confidentiality schemes since both families of schemes enable the secure exchange of information. In this chapter, we distinguish between PLKE schemes and physical layer-based data confidentiality schemes based on whether the nodes are capable of transmitting a sequence of bits of their choosing. If not, the nodes exchange a random sequence of bits which can then be utilized as a shared key. If so, the nodes can securely exchange chosen messages. Notice that the secure exchange of a chosen message could potentially be used to securely distribute a key. This means that keyless physical layer-based data confidentiality schemes can, by default, also function as a PLKE scheme.

This section covers two approaches for PLKE, as illustrated in Figure 6.2. The first approach, covered in Section 6.2.2, takes advantage of the channel reciprocity principle to extract a secret key from common randomness in the wireless channel. The second approach, covered in Section 6.2.3, takes advantage of the inherent challenge of linking an anonymous signal to its transceiver. Two nodes, exchanging anonymous signals, can exploit the awareness of their roles to establish a secret key.

6.2.2 Channel Reciprocity-Based Key Establishment

6.2.2.1 Principles and Assumptions

The channel reciprocity-based key establishment approach relies on three fundamental principles – the channel reciprocity principle, the temporal variation principle, and the spatial decorrelation principle [4].

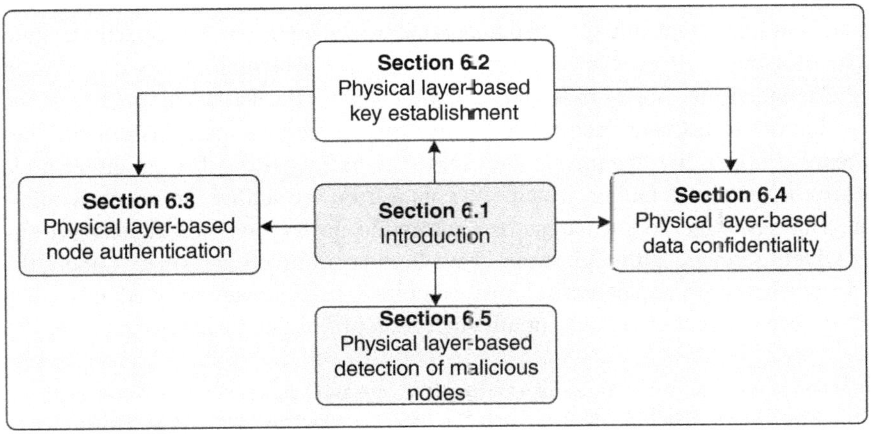

Figure 6.1 Organization of this chapter.

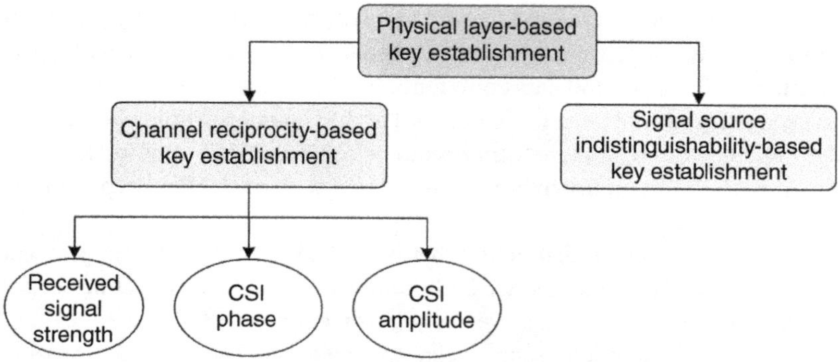

Figure 6.2 The main categorization of PLKE schemes.

- **Channel Reciprocity** implies that the wireless channel between two nodes is reciprocal since simultaneously exchanged RF signals would propagate along the same (multi-)path(s) to reach the other node. This means that exchanged RF signals are subject to the same statistical features (e.g. channel gain, phase shift, and time delay) which impact channel characteristics such as the received signal strength (RSS) and channel state information (CSI) [5]. The estimation of these channel characteristics on both sides of the wireless link forms the basis for two nodes to generate an identical key. This principle has been verified for indoor environments in [6–8].
- **Temporal Variation** implies that the statistical features of the wireless channel change with time due to changes within the environment. Namely, movement from people or objects causes changes in the reflection, refraction, and scattering of the signal propagation path(s). This principle has been verified through experiments in both indoor and outdoor environments [9–11]. The rate of change of the statistical features of a wireless channel is influenced by the dynamicity of its environment; a mobile environment induces rapid changes (i.e. fast fading channel), whereas a static environment induces no or slow changes (i.e. slow fading channel).
- **Spatial Decorrelation** implies that the statistical features of a wireless channel are decorrelated in space, typically specified to one-half wavelength.[1] This means that, when either the transmitter or receiver node moves one-half wavelength in any direction in space, the change in the signal propagation path(s) causes the statistical features of the wireless channel to become decorrelated from their previous state. This also implies that an eavesdropper at a distance greater than one-half wavelength from either node experiences uncorrelated fading, which prevents it from estimating the node's wireless channel. This principle, in particular the assumption on the one-half wavelength distance, is widely accepted, although studies have found contradicting results. Namely, the spatial decorrelation principles were confirmed in [6–8, 13], but were rejected in [14, 15]. The importance of a scatter-rich multipath environment was demonstrated in [2, 16, 17],

1 The assumption that the measured statistical features of the wireless channel at one-half wavelength are decorrelated is based on a model that considers a rich multipath environment with uniform scattering and without a line-of-sight (LoS) connection. When the number of scatterers grows to infinity, the signal decorrelates at a distance $d = 0.4\lambda$ (approximately one-half wavelength) [12].

showing the severe effects of eavesdropping when nodes have a (strong) line of sight connection.

6.2.2.2 Evaluation Metrics

The main goal of a channel reciprocity-based key establishment scheme is to ensure the establishment of a shared key between legitimate nodes while preventing any information about this shared key from being leaked to an adversary (i.e. eavesdropper). A secondary goal is to maximize the energy efficiency by minimizing the computational and communication costs. The tangible evaluation metrics include key entropy, bit generation rate (BGR), and bit mismatch rate (BMR).

- **Key Entropy** is an indicator of a key's inherent randomness. The aim is to generate a key with maximum entropy in the eyes of the adversary, meaning that the adversary considers the value of each key bit to be equally likely between 0 and 1. This also means that the key space (the set of all possible keys) follows a uniform probability distribution. An important aspect for achieving maximum entropy is to prevent correlation between consecutive (sequences of) key bits [18]. The test suite by NIST [19] is often utilized to verify the entropy requirement of generated keys, inspecting aspects such as the frequency of a symbol's occurrence, the presence of long runs, and other periodic features [20].
- **Bit Generation Rate (BGR)** is often defined as the number of secret bits that are generated per unit of time. However, the BGR becomes bounded when we impose that consecutively generated bits must be uncorrelated from each other. Namely, the generated bits are based on the estimated channel characteristics. These channel characteristics, in turn, decorrelate based on movement within the environment of the legitimate nodes (i.e. temporal variation) and based on movement from the legitimate nodes themselves (i.e. spatial decorrelation). The BGR is therefore intricately linked to the deployment scenario of the channel reciprocity-based key establishment scheme [9].
- **Bit Mismatch Rate (BMR)** is defined as the ratio of the number of mismatching bits with respect to the total number of generated bits [9]. These mismatches are caused by minor deviations in each node's estimates of the channel characteristics. These deviations can be the result of implementation in time-division duplexing (TDD) systems [4, 18] (i.e. channel estimation signals are exchanged non-simultaneously and therefore subject to minor changes in the channel conditions) or the quality of the device's hardware. Full-duplex radio systems can resolve the issue caused by TDD systems, although very few IoT devices have full-duplex communication capabilities.

6.2.2.3 Key Establishment Model

The channel reciprocity-based key establishment process is generally described in four steps, consisting of advantage distillation, quantization, information reconciliation, and privacy amplification [21]. It allows two legitimate nodes, generally referred to by the names Alice and Bob, to establish a shared key while remaining secret to any eavesdropper, generally referred to by the name Eve. This process is illustrated in Figure 6.3.

1. **Advantage Distillation** consists of the exchange of pilot signals between Alice and Bob with the aim of estimating certain characteristics of their shared wireless channel

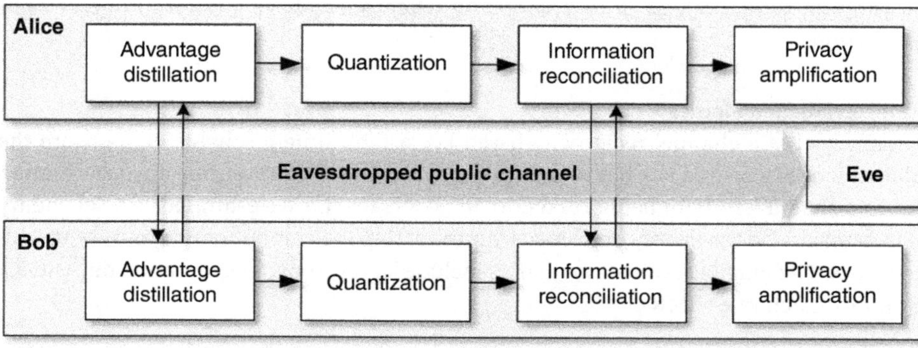

Figure 6.3 The process for channel reciprocity-based key establishment. Source: Adapted from Huth et al. [22].

(also known as channel probing). Popular channel characteristics include the RSS and CSI where the latter contains rich information from the channel impulse, frequency, and phase response [23]. RSS is a popular choice as the source of common randomness since many commercially available off-the-shelf (COTS) devices have integrated functionalities to estimate the average received signal power of each received data frame [24, 25] which simplifies the implementation and experimentation process [20, 26]. Unfortunately, the security guarantees of RSS-based schemes are often in doubt due to their vulnerability against predictable channel attacks [9, 20, 27]. CSI is a popular choice for the source of common randomness since it allows for a higher BGR [28] while it is simultaneously robust against predictable channel attacks [27, 29, 30]. Unfortunately, obtaining CSI may not always be readily available on COTS devices (including IoT devices). To ensure correlation between the channel estimates measured by Alice and Bob, they should transmit their pilot signals within their channel's coherence time (a measure for the minimum amount of time required for the statistical features of the wireless channel to become uncorrelated from their previous state). The correlation between Alice and Bob's channel estimates is maximized when both transmit their pilot signals simultaneously [4].

The important metrics related to advantage distillation are key entropy and BGR. It is possible to maximize the BGR while ensuring uncorrelated channel estimates by adapting the channel probing rate to correspond with the channel's coherence time. It was shown in [31] that 50% of the coherence time is not sufficient to guarantee uncorrelated channel estimates. This strategy also prevents resource consumption from redundant channel re-estimation [4].

2. **Quantization** takes as input the channel estimates and converts these into binary values. The aim of each quantization scheme is to generate a random-looking bit sequence at a maximum BGR while keeping the BMR below a certain threshold [20]. These are contrasting metrics and, to achieve the optimal performance, requires finding the right balance.

   As a baseline, assume that Alice and Bob utilize a threshold value that they evaluate against their estimate of a chosen channel characteristic; quantizing the channel estimate as the bit 0 if the channel estimate is below the threshold, and 1 if the channel estimate is above the threshold. Researchers have proposed a variety of techniques to increase the BGR or reduce the BMR. The BGR can be increased by estimating and quantizing a multitude of channel characteristics [32] and by increasing the number of quantization intervals [4, 33]. The latter of these techniques can have a negative effect on the BMR due to deviations in Alice and Bob's estimates [31, 34]. The BMR can be reduced by reducing the time delay between the exchange of probing signals (e.g. having probing signals prioritized at the medium access control layer [18]), by utilizing Gray codes in multi-bit quantization schemes [4], or by rejecting channel estimates that fall within some close range of a quantization threshold [4]. The latter of these techniques negatively affects the BGR.

3. **Information Reconciliation** deals with the mismatching bits in Alice and Bob's bit sequences after quantization. To resolve the issue of mismatching bits, Alice and Bob exchange information related to their bit sequences over an authenticated public channel. This exchange of information is accessible to Eve which will be able to learn some information about their bit sequences [18, 20]. Information reconciliation methods include the use of error correcting codes (e.g. BCH codes, Golay codes, low-density parity check (LDPC) codes, or Turbo codes) or fuzzy techniques (e.g. CASCADE [31, 35, 36]). If the number of mismatched bits is limited to the correcting capabilities of the used method, then Alice and Bob can determine an error-free bit sequence while only leaking a limited amount of information about this bit sequence to Eve. There is no consensus on which information reconciliation method is better since their performance is highly dependent on the number of mismatched bits between Alice and Bob's bit sequence [18, 37].
   An evaluation metric related to information reconciliation, though not often used, is the key disagreement rate (KDR) which is defined as the ratio between the number of keys that will disagree after information reconciliation with respect to the total number of generated keys. This value equals the probability that the BMR is beyond the capabilities of the information reconciliation scheme.
4. **Privacy Amplification** intends to negate the information leakage that could have occurred during information reconciliation. Privacy amplification methods include the use of fuzzy extractors [38, 39], universal hash functions [9, 40], and cryptographic hash functions [31, 41, 42]. These methods take a bit sequence as input (where the sequence contains more bits than the amount of provided entropy) and then output a shorter bit sequence, used as the shared key, such that the length of this key matches the entropy.

6.2.3 Signal Source Indistinguishability-Based Key Establishment

6.2.3.1 Principles and Assumptions

The signal source indistinguishability-based key establishment approach relies on the intrinsic challenge of mapping an anonymous transmission (i.e. no true identifiers were used in the transmission) to the transmitter. The main idea is that the two legitimate nodes are aware of their role (i.e. transmitter or receiver) during protocol execution and utilize this knowledge to establish a secret key [43]. These nodes can coordinate over a public channel, making each other aware of any potential parameters, before initiating the key establishment protocol. An unfortunate consequence of the objective of signal source indistinguishability is that it is orthogonal to the objective of physical layer-based node authentication, meaning that only one can be achieved.

6.2.3.2 Evaluation Metrics

The signal source indistinguishability-based key establishment schemes can be evaluated based on the energy efficiency of the proposed scheme or the robustness against relevant attacks:

- **Energy Efficiency:** The energy efficiency of a proposed signal source indistinguishability-based key establishment scheme can largely be determined by the expected number of exchanges required by Alice and Bob to establish their shared secret key.
- **Temporal Indistinguishability:** This means that a passive eavesdropper Eve must not be able to determine the source of a transmission based on the transmit time (i.e. timing analysis) [44, 45].
- **Spatial Indistinguishability:** This means that a passive eavesdropper Eve must not be able to determine the source of a transmission based on signal characteristics such as the RSS (i.e. power analysis) [44], angle-of-arrival estimations, or signal frequency and constellation symbol offsets caused by hardware imperfections (i.e. RF analysis) [45, 46]. In the case of power analysis and assuming that Eve is aware of the location of the legitimate nodes, she can estimate the path loss for free space [47], rural [48], urban [49], or indoor [50] environments and determine the supposed transmit power at Alice or Bob. This gives Eve a probabilistic advantage and potentially a deterministic advantage through triangulation with observations from multiple eavesdropping devices.
- **Adversarial Interference:** An active adversary Malory can exploit the same signal source indistinguishability problem to interfere in an active key establishment protocol with its own anonymous transmissions. The legitimate nodes, Alice and Bob, would then perceive this anonymous transmission as having been transmitted by the other and could prevent them from establishing a shared secret key (also known as a key poisoning denial of service (DoS) attack [44]).

6.2.3.3 Key Establishment Model

Most protocol designs for signal source indistinguishability-based key establishment consist of k rounds. In each round, either Alice or Bob transmits a signal where the transmitter is determined by a random process. Both nodes can generate a random time within the round's time slot such that the node that generated the earlier time gets to transmit during that round [44, 51] or the transmitter is determined by a channel contention algorithm like the one implemented by the 802.11 family protocols [52]. An example of the exchange is illustrated in Figure 6.4.

There are numerous methods to encode the transmission. Even though the encoding process often allows Alice and Bob to choose the bit value they wish to convey, this bit value should be determined by a pseudorandom process. For example, the transmitter (i.e. Alice or Bob) could broadcast a data packet containing only a source and destination. The transmitter could fill the source and destination fields correctly to convey the 1-bit, or incorrectly (i.e. the source field contains the address of the destination node, and the destination field

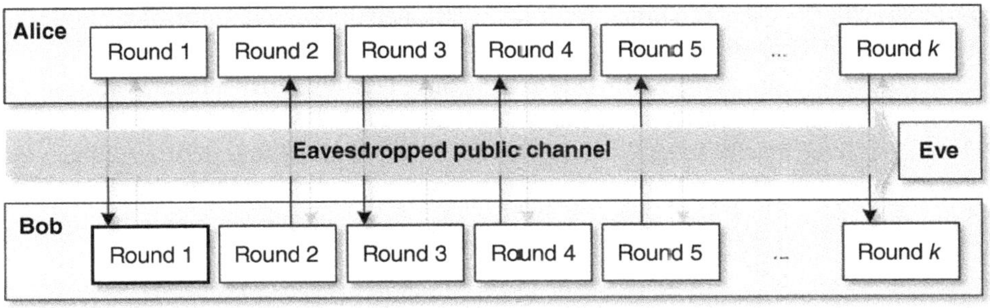

Figure 6.4 Example of the exchange for signal source indistinguishability-based key establishment where the faded arrow indicates a potential secondary transmission per round.

contains the address of the source node) to convey the 0-bit, to append to their shared key [44]. Alice and Bob could also broadcast anonymous data packets (i.e. with an empty or random entry in the source and destination field) where Alice's packet contains the complement of the bit value intended for their shared key whereas Bob's packet contains the true bit value intended for their shared key [51, 52]. A drawback of these encoding methods is that adversarial interference would cause Alice and Bob to append their key with mismatching bits, leading to disagreeing keys. Even though error correcting codes can deal with certain amounts of mismatching bits (caused by either adversarial interference or packet loss) [51], it is still possible that Alice and Bob will end up with disagreeing keys. Having to rerun the entire protocol would severely affect the energy efficiency.

Another encoding method is for Alice and Bob to both broadcast an empty anonymous data packet per round and determine the secret bit based on their transmission order [45]. Namely, they could append their shared key with the 1-bit if Alice was the first to transmit and the 0-bit if Bob was the first to transmit during that round. Even though this encoding method doubles the number of exchanges compared to the previous encoding methods, it is more robust against adversarial interference. Namely, Alice and Bob can detect malicious behavior when they receive multiple anonymous data packets during a single round and manage the situation by discarding that round of exchanges.

Finally, suppose that Alice and Bob are both equipped with a full-duplex transceiver such that they can broadcast anonymous data packets simultaneously over one of two frequencies, f_0 or f_1. They experience either a collision (i.e. both transmit on the same frequency and listen to the other) or a successful exchange. In the case of a successful exchange, they append their secret key with the 0-bit when Alice and Bob transmit at f_0 and f_1, respectively, and with the 1-bit when they transmit at f_1 and f_0, respectively [53]. Unfortunately, IoT devices generally do not have full-duplex capabilities.

6.2.3.4 Security and Performance Evaluation

Temporal indistinguishability can be achieved when implemented in carrier sense multiple access (CSMA)-based systems, preferably with collision detection (CSMA/CD) and avoidance (CSMA/CA), since this allows Alice and Bob to transmit at random times [44]. Spatial indistinguishability against power analysis can be achieved by moving and turning the devices during protocol execution when the devices are within close proximity [44, 45] or partially achieved using random transmit powers when the devices are located elsewhere [51, 52, 54]. Spatial indistinguishability against RF analysis can be achieved when Alice and Bob are capable of mimicking each other's RF signals. Unfortunately, this may require the transceivers to have software-defined radio (SDR) capabilities [45, 46, 51].

The capabilities of the signal source indistinguishability-based key establishment approach were tested through implementation in the OpenMote-CC2538, representing a resource-constrained IoT device. The results showed that a 128-bit key can be established in a matter of seconds [52].

6.2.4 Final Remarks

PLKE is an efficient and simplistic alternative compared to the conventional cryptographic key distribution or key agreement and is thus suitable for resource-constrained devices in IoT networks. It is a preliminary method for achieving security goals such as authentication, data integrity, and confidentiality. The channel reciprocity-based key establishment approach relies on dynamic multipath fading conditions to prevent an adversary from making predictions about certain channel characteristics and to ensure that these channel characteristics are continuously decorrelating for rapid key generation. This approach is therefore suitable in crowded indoor and outdoor environments (e.g. daytime urban areas, office spaces, and factories) although solutions to deal with the low BGR in static networks have been proposed [55, 56]. The signal source indistinguishability-based key establishment approach is independent of multipath fading conditions and unaffected by slow fading channels in terms of BGR. This approach is therefore more suited for static environments and scenarios where devices have a strong line of sight (LoS) connection.

6.3 Physical Layer-Based Node Authentication

6.3.1 Introduction

This section covers the concept of physical layer-based authentication (PLA), a physical layer approach where the receiver of a wireless signal attempts to verify the source based on signal characteristics that could only have been induced by one particular transceiver.

Conventionally, authentication is tasked to the upper layer security protocol stack using cryptography-based techniques [57, 58]. Such techniques include symmetric key cryptographic methods such as message authentication codes (MACs) and asymmetric (i.e. public) key cryptographic methods such as digital signatures. For example, a simple method to achieve message authentication is by having the source sign the message with its own private key, enabling the receiver to verify the integrity and the source of the message with the corresponding public key [59–61]. Conventional methods have been effective in securing various traditional wireless networks; however, IoT networks possess unique properties that hinder the adoption of conventional authentication methods. Despite the enhanced performance of IoT devices throughout the years, their power and computational processing resources are still limited [62]. Therefore, conventional authentication methods that require high computational complexity cannot be used in such networks. In addition, IoT networks often contain hundreds or thousands of distributed nodes as part of the network (i.e. massive IoT network). Therefore, the use of secret keys in the conventional method for authentication requires lengthy keys which magnifies processing delay, decreases the quality of service (QoS), particularly in delay-sensitive applications, and is inefficient in terms of bandwidth, time, and power resources since IoT devices usually communicate short (i.e. few bits) and periodic reports. Additionally, key distribution techniques used in conventional cryptography usually require a third party, which makes it impractical for IoT applications [63] as large-scale wireless networks require keys to be frequently updated, and that will induce a high overhead in the network. Moreover, if the adversary gains knowledge of the generating pattern in conventional security schemes that rely on confidential keys, the whole security system is destroyed, and the network is compromised [64–66].

The aforementioned limitations and operational complexity of the conventional cryptography-based authentication techniques have been critical drawbacks for their implementation in IoT networks, particularly in massive IoT networks. To this end, this motivated academia and industry to research lightweight and energy-efficient PLA schemes suited for resource-constrained IoT devices to secure emerging IoT services and use cases of future massive IoT. PLS has been gaining attention as a suitable security approach for IoT environments. In particular, the physical layer can be exploited to develop authentication protocols that are both lightweight and can achieve the desired security level.

This section evaluates three main approaches for achieving PLA, as illustrated in Figure 6.5. The first approach, covered in Section 6.3.2, takes advantage of a pre-established shared secret key to provide authentication. The second approach, covered in Section 6.3.3, is a keyless authentication approach that takes advantage of channel-based characteristics. The third approach, covered in Section 6.3.4, is another keyless authentication approach but takes advantage of device-based characteristics.

6.3.2 Key-Based Physical Layer Authentication

One effective method of authentication is through the use of a shared secret key that is generated based on a pre-defined protocol (such as those covered in Section 6.2). The key-based

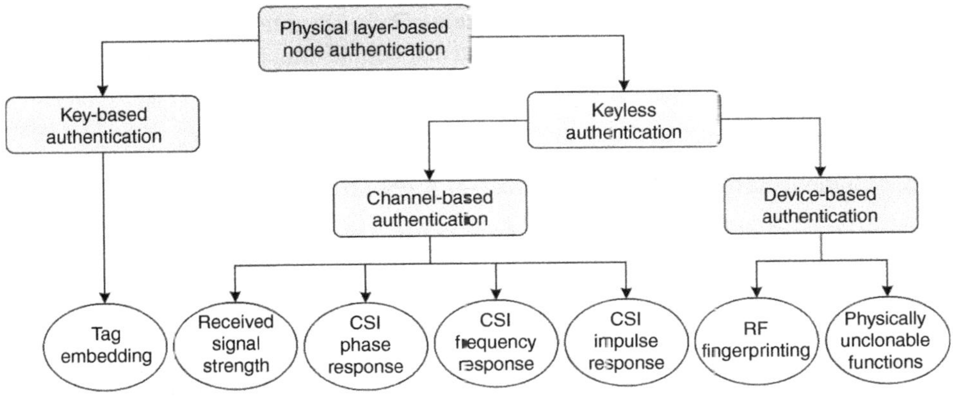

Figure 6.5 The main categorization of PLA schemes.

PLA approach is generally based on an algorithm that generates an authentication tag and subsequently embeds the data signal with the tag signal in the same frame.

6.3.2.1 Tag-Embedding Authentication

The tag-embedding fingerprint approach is based on embedding an authentication tag with the message and then transmitting them as one RF signal. Consequently, fingerprint embedding does not require additional bandwidth and is safe against brute force attacks, as in conventional authentication schemes.

In [67], a message authentication scheme based on tag-embedding is proposed. In this scheme, an authentication tag is generated from a shared secret key after which the tag signal (at a relatively low power) is superimposed with the message signal (at a relatively high power). Thus, the proposed scheme has the advantage of saving spectral resources as both the tag and the message are integrated and emitted using the same bandwidth and time resources. Furthermore, the system's secrecy is improved as the low power assigned to the tag makes it challenging for the eavesdropper to detect it. The secrecy of these tags becomes particularly important when the messages and the corresponding tags are short since an adversary would be able to search the entire key space. As mentioned previously, many IoT devices periodically transmit sensing reports which may only be a few bits long.

The authors of [68] proposed a novel PLA scheme for a non-coherent massive single-input multiple-output (SIMO) system to meet the low latency requirements of Industrial IoT (IIoT) communication systems. Since increasing the power allocated to the embedded authentication tag could increase the error rate of the message, the authors proposed finding an optimal 1-bit tag-embedding scheme.

In [69], the authors demonstrated a superimposition-based tag-embedding scheme for both single-input single-output (SISO) and MIMO systems using a SDR. Interestingly, the intended receiver exhibits the ability to authenticate the received transmission even when the embedded authentication tag has been allocated with very low power while only exhibiting a small effect on the bit error rate (BER) of the message.

The performance of tag-embedding PLA where the message signal is superimposed together with the tag signal was analyzed in [70]. In particular, analytical expressions of the authentication fail probability for joint and successive detection techniques were provided. This analytical work was extended in [71], proposing a novel tag-embedding scheme where the tag symbol is punctured by a random number of bits. The punctured number of bits is based on the reciprocal channel phase, estimated at both legitimate nodes. The reciprocal channel phase therefore acts as a shared secret key. An illustration of the proposed scheme is shown in Figure 6.6.

6.3.3 Channel-Based Keyless Physical Layer Authentication

A unique attribute of PLS is that it allows for authentication without the need for a shared secret key, also known as keyless authentication. These techniques can achieve authentication through either channel-based or device-based attributes. In this subsection, we examine the former. Channel-based keyless PLA schemes exploit characteristics of a wireless RF signal that has been transmitted through the reciprocal channel such as RSS, CSI (i.e. channel impulse response (CIR) or channel frequency response (CFR)), and pilot utilization.

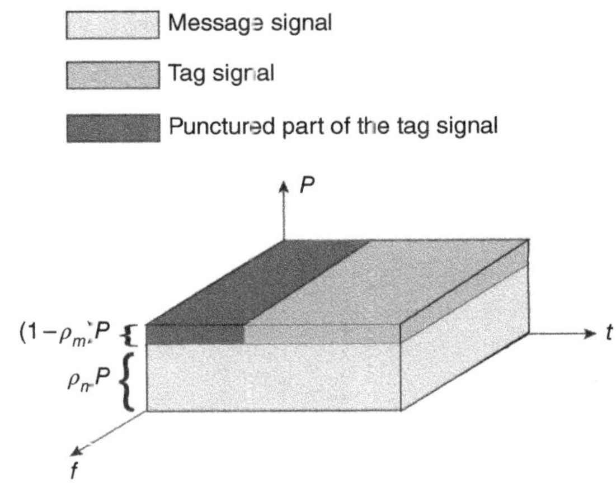

Figure 6.6 Illustration of the proposed tag-embedding PLA scheme of [71]. Source: Qaisi et al. [71]/With permission of IEEE.

6.3.3.1 RSS-Based Authentication

The PLA scheme proposed in [72] utilizes the RSS. A receiving node can verify whether two incoming signals originated from the same transmitter based on the similarity between the RSS measurements of each received signal. More specifically, if the difference in the RSS measurements of two consecutively received signals is within a certain limit, the receiver will assume that the two signals originated at the same transmitter.

Alternatively, in [73], the authors presented an RSS-based PLA scheme which makes it possible to detect spoofing attacks in wireless networks in both static and dynamic environments. In this scheme, the interactions between the intended receiver and spoofers are formulated as a zero-sum authentication game.

6.3.3.2 Frequency-Based Authentication

Additional to phase-based PLA schemes, the authors in [74] adopted the CFR of the wireless channel to determine if two consecutive RF signals are transmitted from the same node or not. Specifically, it suggests measuring the CFR sampled from the received signal. Unfortunately, the evaluation results are based on the assumption that the two communicating nodes are stationary and may therefore not apply to more dynamic scenarios. This work was generalized and presented in [75]. It is assumed that a burst includes multiple frames, where intra-burst authentication is accomplished. On the other hand, inter-burst authentication is performed using the first frame at each burst.

For determining the optimal test threshold and spoofing frequency for dynamic environments, reinforcement learning algorithms (Q-learning and Dyna-Q-learning) are utilized in a game theory scheme proposed in [73] to optimize the strategy of each agent. The explored results showed that the performance of the authentication has been significantly improved.

6.3.3.3 CIR-Based Authentication

In [76], the authors propose a PLA scheme for wireless communication which is based on the CIR in a time-varying multipath environment. The main idea behind this scheme is that a receiver monitors the variation in CIRs between consecutively received signals from a transmitter of interest and thereby distinguishes its identity. According to this test, the noise-mitigated CIR is compared with a specific threshold to determine whether the transmitter is a legitimate node or not.

6.3.3.4 Pilot-Based Authentication

Since pilot positions are changed according to a probability distribution in each authentication phase, a normal receiver can access the information about the pilot subcarriers' positions within every orthogonal frequency division multiplexing (OFDM) symbol, while attackers cannot do so. In this regard, in [77–79], the authors benefited from the concept of using channel pilots to estimate the wireless channel and hence enhance PLA in the IoT devices. The proposed scheme relies on pilot subcarriers built by OFDM systems to estimate the channel.

6.3.3.5 Machine Learning-Based Authentication

The growing popularity of machine learning led to the development of machine learning-based PLA [80]. These approaches follow the philosophy that the authentication performance is optimized when taking multiple (if not all) physical layer attributes, such as RSS, carrier frequency offset (CFO), and CIR, into consideration [81]. The main drawback of this approach is the computational complexity and associated resource consumption, especially when applied to IoT networks. It was proposed in [80] that compressed random samples are utilized as input into the machine learning algorithm to reduce the complexity while maintaining a solid authentication performance.

6.3.4 Device-Based Keyless Physical Layer Authentication

As mentioned previously, PLA can be achieved through either channel-based or device-based attributes. In this subsection, we examine the latter. Device-based keyless PLA schemes exploit the unique and hard-to-clone hardware imperfections, also referred to as RF fingerprints.

6.3.4.1 Radio Frequency Fingerprinting

Radio frequency fingerprinting (RFF) is a device-based PLA method that enables legitimate IoT devices to be recognized based on the unique signature of their hardware which is manifested in the transmitted signal. Such an approach can provide a universal, distinctive, permanent, and unique fingerprint for PLA without incurring any additional overhead on the transmitted signal. In RFF, the obtained fingerprint is unique to a device due to manufacturing imperfections, and thus cannot be cloned. Moreover, it is independent of the carried information on the transmitted signal [65].

RFF extraction requires the understanding of the transmitted signal analysis in different domains including time, frequency, and space. The system architecture of RFF consists of four main steps as shown in the block diagram of Figure 6.7.

There has been a dedicated effort in the literature for investigating RFF for the authentication of IoT devices. The authors of [65] proposed to segment the received signal into several portions to facilitate extracting different feature sets. The authors of [82] adopted time-varying CFO characteristics as a fingerprint for the purpose of detecting the spoofing attack. The proposed framework procedure is initiated by the estimation of the variable CFO

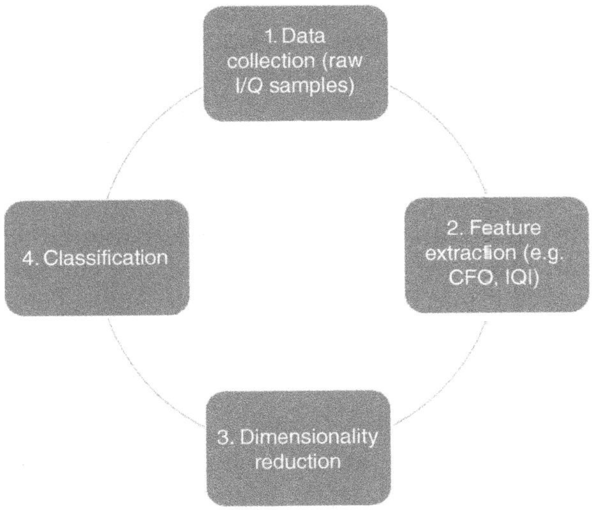

Figure 6.7 RFF methodology.

characteristics of the received signal at different times, followed by a comparison between the estimated and stored values to confirm the identity of the transmitting device.

Differently, the authors of [83] utilized the imperfection of the transmitted signal caused by noises during communication. More specifically, the authors employed phase noise features to detect the illegal nodes by comparing their noise features with the stored features of the legal nodes.

6.3.4.2 Physically Unclonable Function

Physically unclonable functions (PUFs) offer an alternative solution for lightweight hardware security suitable for IoT networks [84]. The concept of a PUF is based on the fact that two identical chips, having the same characterization and going through the same production line, will never share the same physical properties due to inevitable manufacturing variation during the fabrication of the chips [85]. PUF circuits exploit these features and utilize such variations to generate secret keys and authenticate IoT devices. To this end, a "good" PUF requires three characteristics to be met, namely, uniqueness (i.e. differences between responses of different PUFs should be large), reliability (i.e. same response achieved from same PUF given the same challenge irrespective of time), and unpredictability (i.e. ensure difficulty to predict the response of a PUF to a challenge).

Several types of PUFs have been proposed. The authors in [86] categorized two different types of PUFs, memory-based and delay-based PUFs. Memory-based PUFs exploit the initial binary sequences of memory when it is powered on whereas delay-based PUFs exploit delay variations between propagation signals of the circuit. However, more commonly, other authors classified PUFs into three types: weak, strong, and controlled PUFs [87], each with its own security properties and applications.

There are several PUF architectures suitable for IoT devices, including the arbiter PUF, ring oscillator (RO) PUF, static random access memory (SRAM) PUF, and the lightweight hybrid PUF. During the manufacturing variations of the multiplexers, different delay paths are formed, where one path is usually faster than the others. Depending on the input challenge bits, each multiplexer will select the next path to be switched to the arbiter, which provides multiple combinations of bit path selections. The arbiter PUF [88] operates by comparing two-path delays and generates a response bit “0” or “1” depending on the faster path being selected by the latch at the output. Specifically, the PUF scheme proposed in [89] is suitable for low-cost platforms, such as radio frequency identification (RFID) devices. In [90], the proposed scheme performs Wi-Fi authentication for IoT devices. Moreover, the authors in [91] perform PUF-based authentication and privacy preservation in IoT devices with a data provenance protocol (in terms of the data source). The proposed scheme is also robust against physical and cloning attacks with low energy consumption.

6.3.5 Final Remarks

This section is dedicated to the potential of key-based and keyless PLA schemes for IoT networks. In this regard, the categorization of key-based and keyless-based PLA techniques has been provided considering the utilized signal feature for the purpose of authentication.

Key-based tag-embedding authentication is very suitable in dynamic environments with high nodes’ mobility, whereas channel-based keyless authentication generates excessive false alarms due to the rapid fading decorrelation with respect to time. Also, the tag-embedding mechanism can be performed to ensure either (i) high robustness by providing the tag with the same signal-to-noise ratio (SNR) as the message (e.g. replacing some frame bits with tag ones) or (ii) high security by superimposing the message and tag signals together. Nonetheless, such schemes trade off their robustness with security. The higher the power allocated to the tag, the better the tag detection probability. However, this decreases correct message detection by Bob and/or increases tag detection by Eve. Therefore, selecting an optimal power splitting coefficient for target reliability or security level is crucial for such schemes.

For the channel-based keyless authentication schemes proposed in the literature, authentication based on successive analysis of subsequent transmission may lead to security issues, particularly if an attacker manages to get a message authenticated since this would cause subsequent messages to be authenticated as well. Additionally, analysis of the work in the literature has shown this method to be inefficient in highly mobile scenarios due to the highly dynamic channel characteristics. Furthermore, despite its high robustness compared to the RSS-based authentication, the implementation of CSI-based authentication is more complex due to the fine-grained CSI estimation loops. To this end, few commercial IoT devices can provide instantaneous CSI.

It has been shown that RFF attributes extracted from wireless signals are impacted by the wireless channel fading along with the inherent device noise. In addition to this, thorough mathematical modeling for the device-based features is needed by considering nonlinearities in the hardware. Furthermore, research on RFF using PUFs has shown that it is challenging to satisfy the unpredictability, uniqueness, and reliability properties simultaneously. Therefore, research should look into algorithms that aim at jointly optimizing all three conditions to be met when building the PUF.

Finally, based on the above per-category limitations, we can obviously conclude that major efforts are needed in order to drive PLA forward to the applicability level in realistic

IoT networks such as IoT mesh networks and military IoT. For instance, cooperative and/or adaptive PLA schemes can be designed to leverage multiple device-based and channel-based attributes together, or through combining keyless and key-based PLA to design robust multifactor PLA schemes.

6.4 Physical Layer-Based Data Confidentiality

6.4.1 Introduction

This section covers the concept of physical layer-based data confidentiality, a physical layer approach where two arbitrary nodes can exchange confidential messages by exploiting certain characteristics of their shared wireless medium. The conventional method for providing data confidentiality is through cryptographic encryption executed at the upper layers of the protocol stack [92]. The need for lightweight solutions, tailored to resource-constrained devices, led to a series of international competitions including the ECRYPT Stream Cipher project (eSTREAM) [93], the Competition for Authenticated Encryption: Security, Applicability, and Robustness (CAESAR) [94], and the Lightweight Cryptography project [95]. The effort necessary to develop cryptographic solutions suitable for resource-constrained devices is a clear indicator of its complexity. A major advantage of physical layer solutions is that these can be inherently efficient.

As previously mentioned in Section 6.2.1, physical layer-based data confidentiality schemes enable the secure exchange of data from one node to another and can therefore also function as a PLKE scheme. In this chapter, we distinguish between physical layer-based data confidentiality schemes and PLKE schemes based on whether the nodes are capable of transmitting a sequence of bits of their choosing. If not, we consider the scheme for PLKE. If so, we consider the scheme for physical layer-based data confidentiality.

This section covers two approaches for physical layer-based data confidentiality, as illustrated in Figure 6.8. The first approach, covered in Section 6.4.2, takes advantage of a pre-established shared secret key to encrypt the data during a physical layer process (e.g. channel coding, signal modulation) such that only the intended receiver can decrypt and recover the data. The second approach, covered in Section 6.4.3, employs secrecy codes, based on channel codes, to exploit an advantage in the quality of the main channel with respect to the eavesdropper channel and leverages physical layer techniques such as artificial noise and directional modulation to guarantee the existence of such an advantage.

6.4.2 Key-Based Data Confidentiality – Physical Layer Encryption

This section provides an overview of the physical layer-based data confidentiality schemes that rely on a secret key shared between two (or more) legitimate nodes. We refer to such schemes as physical layer encryption (PLE) schemes. We consider channel coding and signal modulation as the subsequent physical layer processes which can be modified to include an encryption procedure to achieve data confidentiality. There exist generic encryption methods that can be applied before either process, including the bitwise XOR of data bits

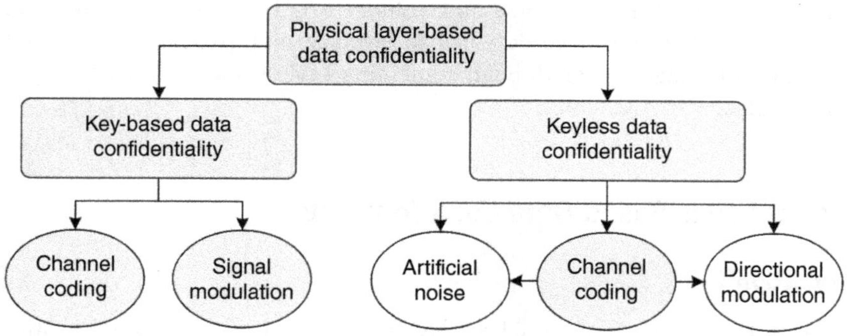

Figure 6.8 The main categorization of PL-based data confidentiality schemes.

with key bits [96–99] and the insertion of dummy bits where their positions are defined by a secret key [96]. The exact nature of the key establishment process is, in general, not relevant to the encryption technique since any secret key can be transformed into the required format.

6.4.2.1 Channel Coding-Based Physical Layer Encryption

The channel coding process encodes binary data such that a receiver is capable of correcting certain amounts of channel-induced errors. These encoding schemes are also known as error correcting codes and their error correcting capabilities are highly dependent on the amount of added redundancy. Channel coding therefore makes a tradeoff between data rate and reliability. No significant research has proposed PLE schemes based on channel coding.

6.4.2.2 Signal Modulation-Based Physical Layer Encryption

The signal modulation process takes the channel coded binary data as input and encodes it onto an analog carrier signal. This process is also referred to as bit-to-symbol mapping where the symbol refers to the constellation symbol on a constellation diagram. Popular modulation types include amplitude shift keying (ASK), frequency shift keying (FSK), and phase shift keying (PSK), where the value of the bits are encoded by varying the amplitude, frequency, and phase, respectively, of the carrier signal [100, 101].

- **Generic Signal Modulation-Based PLE:** A generic encryption method is to permute blocks of symbols where the permutation pattern is defined by a shared key [100–102]. Decryption is synonymous to restoring the order of the received symbols such that the channel code is decoded correctly. Even though the referenced schemes are proposed for OFDM systems, where the permutation pattern determines which symbols are encoded onto which subcarrier signal, the encryption method can be applied more generically.
- **Phase Shift Keying (PSK)-Based PLE:** Signal modulation-based on PSK encodes binary data onto a carrier signal by varying the phase of the signal. The carrier signal can be encrypted by inducing a constellation rotation (i.e. an additional phase-offset) [103–107] or by constellation swapping [108, 109] determined by a shared key. Constellation swapping is essentially the same as performing XOR on the binary data with key bits. Both of these encryption methods can also be applied to quadrature amplitude

modulation (QAM) [104]. Decryption is synonymous to reversing the constellation rotation or swapping such that the correct symbol is decoded.

- **Quadrature Amplitude Modulation (QAM)-Based PLE:** Signal modulation-based on QAM can be considered as a combination of ASK and PSK where binary data is encoded onto a carrier signal by varying the amplitude and phase of the signal. The carrier signal can be encrypted by inducing an additional offset in both the amplitude and phase determined by a shared key [99]. Decryption is synonymous to removing the amplitude and phase offsets such that the correct symbol is decoded.
- **Orthogonal Frequency Division Multiplexing (OFDM)-Based PLE:** The signal modulation process of OFDM systems consist of mapping blocks of symbols onto orthogonal subcarriers. An encryption procedure that either modifies [97, 98, 102, 110, 111] or scrambles [112] the in-phase and/or quadrature components of time-domain OFDM symbols causes the orthogonality of OFDM symbols to be violated and creates inter-carrier interference (ICI). Decryption is synonymous to restoring the orthogonality such that the correct symbol is decoded.
- **Chirp Spread Spectrum (CSS)-Based PLE:** Signal modulation based on CSS is a modification of FSK modulation where binary data is encoded onto a carrier signal whose frequency increases linearly over time. The binary data determines the starting frequency f_s within the range of frequencies $[f_{min}, f_{max}]$, increases linearly to the maximum frequency f_{max}, instantly drops to the minimum frequency f_{min}, and increases again until reaching the starting frequency f_s. The receiver can decode each signal by estimating the starting frequency f_s to determine the corresponding binary data. CSS modulation can be encrypted by utilizing a shared key which induces a shift in the starting frequency [113]. Decryption is synonymous to removing the frequency shift such that the correct symbol is decoded.
- **Frequency Hopping Spread Spectrum (FHSS)-Based PLE:** FHSS is a spread spectrum technology in which a wide frequency band is divided into smaller sub-bands. The modulated data is transmitted onto a carrier signal whose frequency changes after a pre-determined time interval, thus "hopping" from one frequency to the next. The secret frequency hopping pattern can be defined by a shared key to reduce the impact of passive eavesdroppers and active jammers [114–116].

6.4.3 Keyless Data Confidentiality

This section provides an overview of physical layer-based data confidentiality schemes without the need for a shared key. This approach is based on the seminal works of Wyner [117] and Csiszár and Körner [118] who determined that any pair of nodes can achieve information-theoretically secure communication exclusively from channel coding while in the presence of an eavesdropper although only under the condition that the eavesdropper channel (i.e. the channel between a transmitter Alice and eavesdropper Eve) is a degraded version of the main channel (i.e. the channel between a transmitter Alice and intended receiver Bob) [118]. Namely, this condition indicates that Bob is affected by fewer channel-induced errors compared to Eve. Alice can therefore encode the binary data in such a way that Bob is capable of correcting the channel-induced errors and recovering the confidential data whereas Eve cannot. These channel coding-based schemes are often referred to as secrecy schemes.

6.4.3.1 Security Metrics

There have been numerous security metrics proposed to evaluate the secrecy properties or performance under various channel conditions. We define the most important metrics below.

- **Channel Capacity:** The channel capacity C_C is defined as the maximum data rate R_C (i.e. the ratio between data and redundancy bits) at which a transmitter–receiver pair can communicate reliably. The probability that the receiver decodes the noisy observation in error is therefore arbitrarily small.
- **Secrecy Capacity:** The secrecy capacity C_S is defined as the maximum data rate R_S at which a transmitter–receiver pair can achieve both reliability and secrecy against eavesdroppers [119]. This means that the probability that the receiver decodes its noisy observation in error is arbitrarily small while the eavesdropper is affected by uncertainty during the decoding process since its noisier observation is expected to be beyond the error correcting capabilities of the channel code, causing the eavesdropper to learn nothing about the data. It was shown that the secrecy capacity (C_S) is the difference between the capacity of the main channel $\left(C_C^M\right)$ and the capacity of the eavesdropper channel $\left(C_C^E\right)$ for the discrete memoryless channel (DMC) [117] and Gaussian channel model [120] (i.e. $C_S = C_C^M - C_C^E$). This indicates a strictly positive secrecy capacity when the eavesdropper channel is noisier than the main channel [118]. The secrecy rate R_S is a data rate at which data can be confidentially transmitted and is upper bounded by the secrecy capacity C_S.

6.4.3.2 Channel Coding-Based Data Confidentiality

As mentioned previously, the seminal work from Wyner [117] demonstrated that it is possible to achieve data confidentiality exclusively through channel coding. The corresponding network model consists of a transmitter Alice, intended receiver Bob, and eavesdropper Eve, as illustrated in Figure 6.9.

Wyner illustrated the idea for channel coding-based data confidentiality with a special case where the channel is modeled as a binary symmetric channel (BSC) with crossover probability $p_B = 0$ for Bob (i.e. noiseless) and $0 < p_E \leq 0.5$ for Eve. The encoding set $C = \{0, 1\}^N$, consisting of 2^N binary vectors of length N, is divided into two non-overlapping

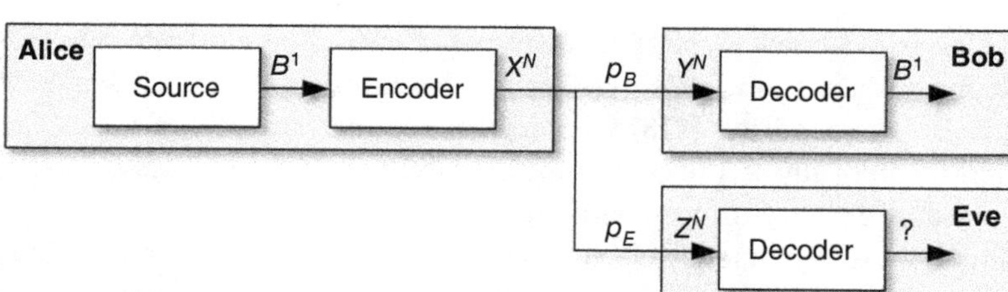

Figure 6.9 The model for channel coding-based data confidentiality.

subsets, C_0 and C_1. The subset C_0 consists of the binary vectors of even parity (i.e. with an even number of 1's) and C_1 consists of the binary vectors of odd parity (i.e. with an odd number of 1's). The encoding scheme takes as input a bit $B \in \{0, 1\}$ and outputs a binary vector X^N, chosen at random, from the subset C_B. The binary vector X^N is then transmitted through the channel and received by Bob and Eve as binary vectors Y^N and Z^N, respectively. In this special case, the probability of erroneous decoding at Bob is always zero whereas the probability for erroneous decoding at Eve depends on the value of N. The interesting result is that, as $N \to \infty$, the erroneous decoding probability for Eve approaches 0.5. This particular encoding scheme therefore achieves the goals of reliable communication with the intended receiver and secrecy from any eavesdropping. A drawback of this encoding scheme is that, as $N \to \infty$, the data rate $R \to 0$.

The general idea of exploiting an advantage in the quality of the main channel with respect to the eavesdropper channel (i.e. a positive secrecy capacity) by cleverly encoding the binary data is equally applicable in other channel models. Unfortunately, designing channel codes with these secrecy properties is very complex. Thus far, channel coding schemes based on LDPC codes [121–123] or polar codes [124–126] have been proposed as a practical solution to achieve keyless data confidentiality. These proposed schemes evaluated their performance in terms of the secrecy rate and whether they achieve secrecy capacity under a variety of channel models.

6.4.3.3 Artificial Noise-Based Extensions

The reliance on a strictly positive secrecy capacity is a major drawback for establishing confidential communication based on secrecy codes. The main problem lies in the inability to measure the capacity of the eavesdropper channel due to their passive and noncooperative behavior. Without knowledge of the eavesdropper's channel capacity, a transmitter is unable to estimate the secrecy capacity. The introduction of artificial noise, also referred to as cooperative jamming, has been proposed to resolve this challenge [127]. Artificial noise can be interpreted as a dummy signal that is transmitted simultaneously, either by the transmitter itself or by a cooperative third party [128–130], along with the legitimate data signal although only in the direction of any potential eavesdroppers. It has been proposed to modulate the artificial noise in the null space of the intended receiver. This prevents the main channel from degrading while degrading any other channel that is not co-located with the intended receiver. To summarize, artificial noise can degrade the quality of the eavesdropper channel irrespective of their location and without it affecting the quality of the main channel such that a strictly positive secrecy capacity can be guaranteed [129].

Challenges of artificial noise-based schemes include (i) the number of antennas committed to transmitting artificial noise must outweigh the number of antennas from the eavesdropper(s) to guarantee a positive secrecy capacity [131], (ii) imperfect CSI at the transmitter(s) of artificial noise would cause some of the artificial noise to be leaked into the main channel [131], and (iii) the close co-location of eavesdroppers with the intended receiver limits the benefits of artificial noise. With respect to the first challenge, cooperative third party nodes will be required to guarantee the positive secrecy capacity for securing uplink communication in IoT networks under the assumption of single antenna equipped IoT devices.

6.4.3.4 Directional Modulation-Based Extensions

Directional modulation is another approach proposed to guarantee a strictly positive secrecy capacity. In contrast to artificial noise, directional modulation aims to guarantee a strictly positive secrecy capacity by improving the capacity of the main channel. Directional modulation is a beamforming technique which can modulate signals in the direction of the intended receiver [132]. Unfortunately, the utilization of directional modulation requires an antenna array at the transmitter which limits its applicability to mainly downlink communication for IoT networks [133]. Data confidentiality can only be achieved for uplink communication when the secure downlink is utilized for key distribution.

6.4.4 Final Remarks

This section provided a high-level overview of two main approaches, the key-based and keyless approach, for providing physical layer-based data confidentiality. The key-based approach, referred to as PLE, consists of a wide variety of methods for achieving data confidentiality. This includes methods with the potential to achieve perfect secrecy under the condition that the key generation rate (KGR) is greater than or equal to the desired data rate. However, such demanding security guarantees come with a significant increase in energy consumption which is far from ideal for resource-constrained devices in IoT networks. Fortunately, as demonstrated by traditional cryptography-based solutions, acceptable levels of security can also be achieved without the guarantees of perfect secrecy and sufficient PLE schemes have been proposed, which can achieve acceptable levels of secrecy while having a small impact on the resource consumption. Furthermore, PLE schemes do not require knowledge about the location or channel conditions of eavesdroppers to achieve data confidentiality. The main drawback of PLE schemes is their reliance on a suitable key establishment scheme.

The keyless approach, referred to as secrecy schemes, is a promising solution to enable data confidentiality between resource-constrained devices. This approach has been extensively studied with respect to its theoretical limits and capabilities and under a wide variety of channel models and assumptions on eavesdroppers. Even though it is difficult to make assumptions about the eavesdropper (e.g. its channel capacity) which are generally applicable to a wide range of deployment scenarios, extensions such as artificial noise and directional modulation offer interesting solutions to guaranteeing a strictly positive secrecy capacity. However, these extensions also come with their own challenges. The utilization of artificial noise to secure uplink communication in IoT networks requires cooperative jammers. Due to the resource consumption associated with jamming, it is not ideal to have other nearby IoT devices act as these jammers. Directional modulation, though an interesting solution, may also not be suited to secure uplink communication due to the multi-antenna requirement. Fortunately, when it is possible to make assumptions about the eavesdropper based on a particular deployment scenario, the keyless approach can be a viable solution. For example, IoT networks deployed in indoor environments that are inaccessible by the eavesdropper (e.g. offices, factories, and homes) can benefit from assumptions on the eavesdropper's channel capacity and exploit the advantage in channel quality to secure communication.

6.5 Physical Layer-Based Detection of Malicious Nodes

6.5.1 Introduction

This section covers the concept of physical layer-based detection of malicious nodes, a physical layer approach that exploits characteristics of the wireless medium to detect malicious activity within the network and identifies the perpetrator. The detection and identification of compromised and malicious nodes is necessary to remove them from the network. Awareness of malicious nodes can also be exploited to achieve security goals (e.g. data confidentiality) without informing them that they were caught.

The scope of this section is to cover physical layer solutions against active adversaries that launch malicious attacks and have not yet been covered in the previous sections. Therefore, spoofing (i.e. identity impersonation) attacks and eavesdropping attacks are not considered in this section since solutions mitigating these malicious attacks have already been covered in Sections 6.3 and 6.4, respectively.

A high-level overview is given of the proposed physical layer approaches for the detection of three kinds of malicious behavior, the detection of data injection attacks, Sybil attacks, and sleep deprivation attacks. The data injection attack is characterized by a malicious node that transmits false data instead of benign data and could severely affect the QoS provided by an IoT system [134]. The approaches for detecting data injection attacks are covered in Section 6.5.2. The Sybil attack is characterized by a malicious node that identifies itself with a multitude of pseudonyms and could severely affect network functions such as routing, resource allocation, and voting [135]. The approaches for detecting Sybil attacks are covered in Section 6.5.3. The sleep deprivation attack is characterized by a compromised node whose behavior and functionalities are altered to drain its battery, which subsequently degrades the capabilities of an IoT system [134]. The approaches for detecting sleep deprivation attacks are covered in Section 6.5.4 (Figure 6.10).

6.5.2 Data Injection Attack Detection

The malicious data injection attack is characterized by a malicious node, also referred to as Byzantines [136], which transmits false data instead of benign data on one or more

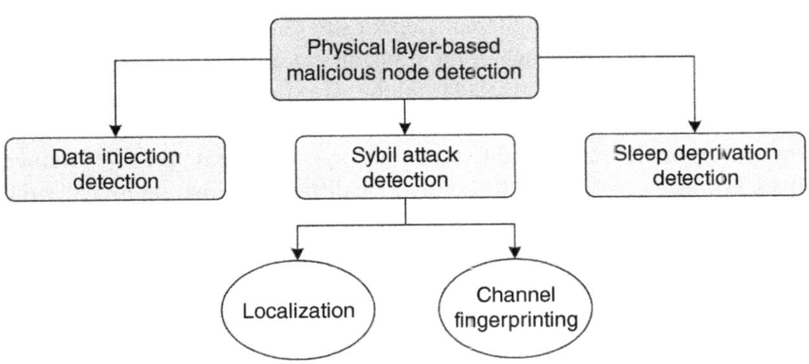

Figure 6.10 The main categorization of PL-based malicious node detection schemes.

attributes (e.g. location data, temperature, atmospheric pressure, moisture levels, and gas concentration levels). This issue also applies to honest but malfunctioning nodes. In the context of an IoT system, the false data may cause the system to provide incorrect services which further negatively impacts the effectiveness of the system itself as well as puts the other nodes at risk [134]. Many detection schemes against malicious data injection attacks are based on determining correlation coefficients or anomalies in the received data, often through a comparison of data received from clusters of nodes that report on the same attribute. Such data analysis-based detection schemes do not exploit any physical layer parameters of the wireless medium for detection purposes and are therefore out of scope for physical layer-based malicious node detection.

The injection of malicious location data, however, can be detected from physical layer parameters. Location spoofing and wormhole attacks could be the result of malicious beacon nodes providing sensing nodes with false location data, subsequently affecting routing protocols and preventing data from reaching its destination. One method to detect these malicious beacon nodes is for them to monitor each other by periodically impersonating themselves as an ordinary resource-constrained device and requesting location information. The beacon nodes can verify the honesty of its nearby beacon nodes by verifying the received location data using distance estimations between them from physical layer parameters such as RSS [136], time of arrival (ToA) or time difference of arrival (TDoA) [137].

6.5.3 Sybil Attack Detection

The Sybil attack [138] is characterized by a malicious node that transmits data under a multitude of identities, therefore impersonating itself as a group of nodes and increasing its leverage within the network. These identities can be fabricated or stolen from a legitimate node and used in a simultaneous or nonsimultaneous manner to prevent detection [135]. The Sybil attack therefore poses a significant threat to IoT networks on important functions such as routing, resource allocation, misbehavior detection, and voting [135]. An identifying characteristic of a Sybil node is that all the transmitted signals from the supposedly different nodes can be traced back to a single transceiver.

Under the assumption of a static network with time-invariant channels, it is possible to detect a Sybil node through localization or channel fingerprinting.

- **Localization Schemes** utilize channel characteristics to determine the location of a transmitter and detect the Sybil node when the transmission from supposedly different nodes (i.e. with different identities) originated from the same location. These localization schemes take advantage of channel fading and signal propagation characteristics to detect Sybil nodes when, over time, supposedly different nodes appear to be at the same distance from the receiver. Localization schemes based on node distance estimation includes the utilization of RSS measurements [139–141] and pulse-based signals in ultra-wideband (UWB) systems [142]. The distance estimation in UWB systems are highly accurate and therefore achieves a better Sybil node detection performance compared to RSS-based distance estimations. The localization accuracy can be improved further through triangulation although this requires the cooperation of multiple honest nodes. A drawback of these methods is the need to store location estimations per node. Another localization method is through neighborhood information sharing [143] and it is based on the idea that the probability of two honest nodes, in a dense network, that have the exact same set of neighbors within their transmission range is extremely low. A protocol where neighboring nodes broadcast information about their local neighborhood provides the necessary information which allows the detection of Sybil nodes. A drawback of this method is that its performance is limited when Sybil nodes only exhibit a few Sybil identities [144].

- **Channel Fingerprinting Schemes** utilize similarities in channel characteristics and detect the Sybil node when the channel characteristics from supposedly different nodes are deemed identical, meaning that the signals are propagated through the exact same wireless channel. Similarities in channel characteristics such as the CIR [145], general CSI [146], power gain, and delay spread [147], potentially in combination with further data processing and analysis [148], have been proposed for Sybil node detection. The attentive reader may recognize that channel fingerprinting schemes share many similarities with keyless PLA-based solutions, as covered in Sections 6.3.3 and 6.3.4, and that these solutions would also be effective for Sybil node detection.

Alternative methods for Sybil node detection are required when considering a mobile network with time-variant channels. Based on the premise that the identities of a Sybil node share the same location over time, it is possible to utilize the localization schemes as discussed above and extend these to track the movements and locations of each node over time and identify the supposedly different nodes associated with Sybil node [144]. Another method is through estimation of the movement-induced Doppler shift of each node and comparing their mobility patterns to detect a Sybil node attack [149].

6.5.4 Sleep Deprivation Attack Detection

The sleep deprivation attack, also referred to as denial of sleep attack, aims to drain the battery of resource-constrained IoT devices. These devices are usually programmed to follow a sleep routine during inactivity to reduce power consumption and extend their battery life. However, an adversary can alter its program to break the sleep routines and drain the device's battery. Examples include the execution of a malicious code that contains an infinite loop or by artificially increasing its power consumption [134].

The detection of sleep deprivation attacks may be impossible depending on the strategy of the attacker. However, some strategies can be detected using physical layer techniques. One method for detecting sleep deprivation attacks is through collective neighborhood monitoring, where nodes within each other's transmission range monitor their transmission behavior (e.g. the frequency of broadcasting) [150, 151]. The monitoring nodes may notice that one (or more) node(s) stop broadcasting sensing reports which could indicate that its central processing unit (CPU) is occupied executing malicious code or that its battery has already been depleted. The monitoring nodes may also notice a significant increase in the number of broadcasts [152] or a significant increase in the RSS [153] of the broadcast transmissions to artificially increase the compromised node's power consumption.

6.5.5 Final Remarks

This section provided a high-level overview of physical layer-based approaches for detecting malicious nodes launching attacks such as the data injection attack, Sybil attack, and sleep deprivation attack.

The detection of data injection attacks seems inherently complicated to solve through physical layer techniques. This is largely based on the fact that variations in the transmitted data is independent of any physical layer parameters. The detection of most data injection attacks therefore propose solutions based on data analysis, executed by a more powerful device such as a gateway or fusion center, which is a solution that falls out of our scope. However, when the false data injection is related to attributes such as location data, there are a few physical layer approaches that offer viable solutions.

In contrast, the detection of Sybil attacks seems inherently easier to solve through physical layer techniques than through techniques at the upper layers of the protocol stack, a statement that can be backed by the amount of literature proposing physical layer-based solutions. Since every device is unique, physical layer techniques can exploit channel characteristics to construct a so-called fingerprint such that incoming transmissions can be traced back to its transceiver. This prevents Sybil attacks since a particular device would no longer be able to identify itself under varying network identities. However, few solutions have been proposed for dynamic networks with time-varying channels. Fortunately, device-based PLA schemes seem to be an excellent candidate to offer additional suitable solutions for this scenario.

The detection of sleep deprivation attacks may be solved through physical layer techniques, although it is highly dependent on the strategy of the attacker. The detection becomes especially problematic under the assumption that the compromised IoT device does not periodically report its battery status to other network devices and that it does not show any signs of altered behavior. However, if either of these assumptions is false, detection through physical layer techniques seems possible and even likely.

6.6 Conclusion

This chapter covered a wide spectrum of PLS methods for achieving node authentication, data confidentiality, and malicious node detection. In addition, this chapter discussed two approaches of PLKE that enable any pair of nodes to establish a shared secret key which is a prerequisite to key-based node authentication and data confidentiality schemes. The covered PLS approaches are aligned with the characteristics of IoT devices which makes them suitable solutions for massive IoT networks.

The section on PLKE, Section 6.2, covered the channel reciprocity and signal source indistinguishability-based approaches. Both approaches are suitable for massive IoT since they enable energy-efficient and on-demand key distribution. However, both approaches rely on assumptions that could restrict the amount of deployment use cases. The channel reciprocity-based approach relies on the temporal and spatial decorrelation principle which may limit this approach for key establishment to dynamic scatter-rich environments, whereas the signal source indistinguishability approach relies on the anonymity of RF signals and stands orthogonal to the objective of physical layer authentication.

The section on PLA, Section 6.3, covered the key-based and keyless approaches. The key-based approach provides node authentication and data integrity, even when short authentication tags are generated from relatively small keys (<128 bits). These tags also enable the authentication of the data's source when transmitted over multiple hops. A drawback of the key-based approach is its reliance on an underlying key establishment scheme. The keyless approach also provides node authentication and data integrity by identifying unique channel- or device-induced features of the received RF signal. A drawback of the keyless approach is that a receiver is unable to authenticate the data's origin if this was forwarded by relay nodes (i.e. limited to single-hop authentication and data integrity).

The section on physical layer-based data confidentiality, Section 6.4, also covered a key-based and keyless approach. The key-based approach, also referred to as PLE, consists of a diverse portfolio of PLE methods such that any massive IoT network can benefit from data confidentiality. A drawback of the PLE approach is its reliance on an underlying key establishment scheme. The keyless approach achieves data confidentiality by exploiting an advantage in the main channel with respect to the eavesdropper channel. To guarantee this advantage, extensions based on artificial noise and directional modulation have been proposed, although they require the collaboration of nearby third party nodes.

The section on physical layer-based detection of malicious nodes, Section 6.5, covered detection approaches for malicious data injection attacks, Sybil attacks, and sleep deprivation attacks. Depending on the attack strategy of the malicious nodes, it may not be possible to detect data injection attacks or sleep deprivation attacks. However, Sybil attacks seem inherently easy to solve using physical layer techniques since channel- or device-based fingerprints allow the mapping of any RF signal to its unique transceiver.

List of Abbreviations

ASK	amplitude shift keying
BER	bit error rate
BGR	bit generation rate
BMR	bit mismatch rate
BSC	binary symmetric channel
CSS	chirp spread spectrum
CFO	carrier frequency offset
CFR	channel frequency response
CIR	channel impulse response
CPU	central processing unit
CSI	channel state information
CSMA	carrier sense multiple access
CSMA/CA	CSMA with collision avoidance
CSMA/CD	CSMA with collision detection
DoS	denial of service
FSK	frequency shift keying
ICI	inter-carrier interference

IIoT	Industrial IoT
IoT	Internet of Things
KGR	key generation rate
KDR	key disagreement rate
LDPC	low-density parity check
LoS	line of sight
MIMO	multiple-input multiple-output
OFDM	orthogonal frequency division multiplexing
PSK	phase shift keying
PLS	physical layer security
PLA	physical layer-based authentication
PLE	physical layer encryption
PLKE	physical layer-based key establishment
PUF	physically unclonable function
QoS	quality of service
RF	radio frequency
RFF	radio frequency fingerprinting
RFID	radio frequency identification
RO	ring oscillator
SDR	software-defined radio
SIMO	single-input multiple-output
SISO	single-input single-output
SNR	signal-to-noise ratio
SRAM	static random access memory
TDoA	time difference of arrival
ToA	time of arrival
XOR	exclusive-or

References

1 Trappe, W., Howard, R., and Moore, R.S. (2015). Low-energy security: limits and opportunities in the internet of things. *IEEE Security & Privacy* 13 (1): 14–21.

2 Trappe, W. (2015). The challenges facing physical layer security. *IEEE Communications Magazine* 53 (6): 16–20.

3 Capkun, S. (2021). Physical layer & telecommunications security knowledge area. The Cyber Security Body of Knowledge (CyBOK).

4 Zhang, J., Duong, T.Q., Marshall, A., and Woods, R. (2016). Key generation from wireless channels: a review. *IEEE Access* 4: 614–626.

5 Xu, W., Zhang, J., Huang, S. et al. (2021). Key generation for internet of things: a contemporary survey. *ACM Computing Surveys* 54 (1): 1–37.

6 Hamida, S.T.-B., Pierrot, J.-B., and Castelluccia, C. (2010). Empirical analysis of UWB channel characteristics for secret key generation in indoor environments. IEEE International Symposium on Personal, Indoor and Mobile Radio Communications (PIMRC), Istanbul, Turkey.

7 Madiseh, M.G., He, S., McGuire, M.L. et al. (2009). Verification of secret key generation from UWB channel observations. IEEE International Conference on Communications (ICC).

8 Zenger, C.T., Zimmer, J., and Paar, C. (2015). Security Analysis of quantization schemes for channel-based key extraction EAI International Conference on Mobile and Ubiquitous Systems (MobiQuitous), Coimbra, Portugal.

9 Jana, S., Premnath, S.N., Clark, M. et al. (2009). On the effectiveness of secret key extraction from wireless signal strength in real environments. ACM International Conference on Mobile Computing and Networking (MobiCom), Beijing, China.

10 Liu, H., Yang, J., Wang, Y., and Chen, Y. (2012). Collaborative secret key extraction leveraging received signal strength in mobile wireless networks. IEEE International Conference on Computer Communications (INFOCOM), Orlando, FL, USA.

11 Liu, H., Yang, J., Wang, Y. et al. (2014). Group secret key generation via received signal strength: protocols, achievable rates, and implementation. *IEEE Transactions on Mobile Computing* 13 (12): 2820–2835.

12 Goldsmith, A. (2005). *Wireless Communications*. Cambridge University Press.

13 Marino, F., Paolini, E., and Chiani, M. (2014). Secret key extraction from a UWB channel: analysis in a real environment. IEEE International Conference on Ultra-WideBand (ICUWB), Paris, France.

14 Edman, M., Kiayias, A., and Yener, B. (2011). On passive inference attacks against physical-layer key extraction. European Workshop on System Security (EuroSec), Salzburg, Austria.

15 He, X., Dai, H., Shen, W., and Ning, P. (2013). Is link signature dependable for wireless security? IEEE International Conference on Computer Communications (INFOCOM), Turin, Italy.

16 Zhang, J., Woods, R., Duong, T.Q. et al. (2016). Experimental study on key generation for physical layer security in wireless communications. *IEEE Access* 4: 4464–4477.

17 Zenger, C.T., Vogt, H., Zimmer, J. et al. (2016). The passive eavesdropper affects my channel: secret-key rates under real-world conditions. IEEE Global Communications Conference (GLOBECOM) Workshops, Washington, DC, USA.

18 Zeng, K. (2015). Physical layer key generation in wireless networks: challenges and opportunities. *IEEE Communications Magazine* 53 (6): 33–39.

19 Bassham, L.E., Rukhin, A.L., Soto, J. et al. (2010). *A Statistical Test Suite for Random and Pseudorandom Number Generators for Cryptographic Applications*, Special Publication 800-22 Rev 1a. Gaithersburg, MD: National Institute of Standards and Technology (NIST).

20 Bottarelli, M., Epiphaniou, G., Kbaier Ben Ismael, D. et al. (2018). Physical characteristics of wireless communication channels for secret key establishment: a survey of the research. *Computers & Security* 78: 454–476.

21 Zhang, J., Li, G., Marshall, A. et al. (2020). A new frontier for IoT security emerging from three decades of key generation relying on wireless channels. *IEEE Access* 8: 138406–138446.

22 Huth, C., Guillaume, R., Strohm, T. et al. (2016). Information reconciliation schemes in physical-layer security: a survey. *Computer Networks* 109 (1): 84–104.

23 Alhasanat, M., Althunibat, S., Darabkh, K.A. et al. (2020). A physical-layer key distribution mechanism for IoT networks. *Mobile Networks and Applications* 25: 173–178.

24 Gollakota, S. and Katabi, D. (2011). Physical layer wireless security made fast and channel independent. IEEE International Conference on Computer Communications (INFOCOM), Shanghai, China.

25 Li, Z., Xu, W., Miller, R., and Trappe, W. (2006). Securing wireless systems via lower layer enforcements. ACM Workshop on Wireless Security (WiSe), Los Angeles, CA, USA.

26 Li, Z., Pei, Q., Markwood, I. et al. (2018). Secret key establishment via RSS trajectory matching between wearable devices. *IEEE Transactions on Information Forensics and Security* 13 (3): 802–817.

27 Mathur, S., Miller, R., Varshavsky, A. et al. (2011). ProxiMate: proximity-based secure pairing using ambient wireless signals. ACM International Conference on Mobile Systems, Applications, and Services (MobiSys), Bethesda, MD, USA.

28 Liu, Y., Draper, S.C., and Sayeed, A.M. (2012). Exploiting channel diversity in secret key generation from multipath fading randomness. *IEEE Transactions on Information Forensics and Security* 7 (5): 1484–1497.

29 Liu, H., Wang, Y., Yang, J., and Chen, Y. (2013). Fast and practical secret key extraction by exploiting channel response. IEEE International Conference on Computer Communications (INFOCOM), Turin, Italy.

30 Alshamaseen, T., Althunibat, S., and Qaraqe, M. (2021). Secure key distribution for IoT networks based on the physical layer security. IEEE International Workshop on Computer Aided Modeling and Design of Communication Links and Networks (CAMAD).

31 Zhang, J., Kasera, S.K., and Patwari, N. (2010). Mobility assisted secret key generation using wireless link signatures. IEEE International Conference on Computer Communications (INFOCOM), San Diego, CA, USA.

32 Usman, M., Althunibat, S., and Qaraqe, M. (2022). A channel state information-based key generation scheme for internet of things. *Security and Communication Networks* 2022 (7976319): 1–15.

33 Usman, M., Althunibat, S., and Qaraqe, M. (2021). A channel magnitude based key generation scheme for static and dynamic environments. IEEE International Workshop on Computer Aided Modeling and Design of Communication Links and Networks (CAMAD).

34 Zeng, K., Wu, D., Chan, A., and Mohapatra, P. (2010). Exploiting multiple-antenna diversity for shared secret key generation in wireless networks. IEEE International Conference on Computer Communications (INFOCOM), San Diego, CA, USA.

35 Brassard, G. and Salvail, L. (1993). Secret-key reconciliation by public discussion. International Conference on the Theory and Applications of Cryptographic Techniques (EUROCRYPT), Lofthus, Norway.

36 Aono, T., Higuchi, K., Ohira, T. et al. (2005). Wireless secret key generation exploiting reactance-domain scalar response of multipath fading channels. *IEEE Transactions on Antennas and Propagation* 53 (11): 3776–3784.

37 Nguyen, K.-C., van Assche, G., and Cerf, N.J. (2004). Side-information coding with turbo codes and its application to quantum key distribution. arXiv ePrint Archive cs/0406001.

38 Dodis, Y., Ostrovsky, R., Reyzin, L., and Smith, A. (2008). Fuzzy extractors: how to generate strong keys from biometrics and other noisy data. *SIAM Journal on Computing* 38 (1): 97–139.

39 Wang, Q., Su, H., Ren, K., and Kim, K. (2011). Fast and scalable secret key generation exploiting channel phase randomness in wireless networks. IEEE International Conference on Computer Communications (INFOCOM), Shanghai, China.

40 Premnath, S.N., Jana, S., Croft, J. et al. (2013). Secret key extraction from wireless signal strength in real environments. *IEEE Transactions on Mobile Computing* 12 (5): 917–930.

41 Wei, Y., Zeng, K., and Mohapatra, P. (2013). Adaptive wireless channel probing for shared key generation based on PID controller. *IEEE Transactions on Mobile Computing* 12 (9): 1842–1852.

42 Ambekar, A., Hassan, M., and Schotten, H.D. (2012). Improving channel reciprocity for effective key management systems. International Symposium on Signals, Systems, and Electronics (ISSSE), Potsdam, Germany.

43 Alpern, B. and Schneider, F.B. (1983). Key exchange using 'keyless cryptography'. *Information Processing Letters* 16 (2): 79–81.

44 Castelluccia, C. and Mutaf, P. (2005). Shake them up! A movement-based pairing protocol for CPU-constrained devices. ACM International Conference on Mobile Systems, Applications, and Services (MobiSys), Seattle, WA, USA.

45 Zhang, Y., Xiang, Y., Wang, T. et al. (2018). An over-the-air key establishment protocol using keyless cryptography. *Future Generation Computer Systems* 79 (1): 284–294.

46 Danev, B., Luecken, H., Capkun, S., and El Defrawy, K.M. (2010). Attacks on physical-layer identification. ACM International Conference on Wireless Network Security (WiSec), Hoboken, NJ, USA.

47 Rappaport, T.S. (2002). *Wireless Communications: Principles and Practice*. Upper Saddle River, NJ: Prentice Hall.

48 Barsocchi, P., Oligeri, G., and Potorti, F. (2009). Measurement-based frame error model for simulating outdoor Wi-Fi networks. *IEEE Transactions on Wireless Communications* 8 (3): 1154–1158.

49 Hata, M. (1980). Empirical formula for propagation loss in land mobile radio services. *IEEE Transactions on Vehicular Technology* 29 (3): 317–325.

50 Bandemer, B., Oestges, C., Czink, N., and Paulraj, A. (2009). Physically motivated fast-fading model for indoor peer-to-peer channels. *Electronics Letters* 45 (10): 515–517.

51 di Pietro, R. and Oligeri, G. (2013). COKE: crypto-less over-the-air key establishment. *IEEE Transactions on Information Forensics and Security* 8 (1): 163–173.

52 Sciancalepore, S., Oligeri, G., Piro, G. et al. (2019). EXCHANGe: securing IoT via channel anonymity. *Computer Communications* 134: 14–29.

53 Usman, M., Raponi, S., Qaraqe, M., and Oligeri, G. (2021). KaFHCa: key-establishment via frequency hopping collisions. IEEE International Conference on Communications (ICC), Montreal, QC, Canada.

54 di Pietro, R. and Oligeri, G. (2015). ESC: an efficient, scalable, and crypto-less solution to secure wireless networks. *Computer Networks* 84: 46–63.

55 Aldaghri, N. and Mahdavifar, H. (2020). Physical layer secret key generation in static environments. *IEEE Transactions on Information Forensics and Security* 15: 2692–2705.

56 Staat, P., Elders-Boll, H., Heinrichs, M. et al. (2021). Intelligent reflecting surface-assisted wireless key generation for low-entropy environments. IEEE International Symposium on Personal, Indoor and Mobile Radio Communications (PIMRC), Helsinki, Finland.

57 International Organization for Standardization (ISO) (2011). Information Security - Security Techniques - Message Authentication Codes (MACs) - Part 1: Mechanisms using a Block Cipher. *Technical report, ISO/IEC 9797-1:2011*, International Organization for Standardization (ISO), Geneva, Switzerland.

58 Hankerson, D.R., Hoffman, D.G., Leonard, D.A. et al. (2000). *Coding Theory and Cryptography: The Essentials*, 2e. CRC Press.

59 Bernstein, D.J. and Tanja, L. Disasters, 27 January 2014 [Online]. Available: https://competitions.cr.yp.to/disasters.html (accessed 28 October 2021).

60 Rogaway, P. and Wagner, D. (2003). A critique of CCM. Cryptology ePrint Archive 2003/070.

61 Kohno, T., Viega, J., and Whiting, D. (2003). The CWC authenticated encryption (associated data) mode. Cryptology ePrint Archive 2003/106.

62 Stinson, D.R. (2006). *Cryptography: Theory and Practice*, 3e. Boca Raton, FL: CRC Press.

63 Xie, N., Li, Z., and Tan, H. (2021). A survey of physical-layer authentication in wireless communications. *IEEE Communications Surveys & Tutorials* 23 (1): 282–310.

64 Shor, P. (1994). Algorithms for quantum computation: discrete logarithms and factoring. IEEE Symposium on Foundations of Computer Science (SFCS), Santa Fe, NM, USA.

65 Zhang, J., Rajendran, S., Sun, Z. et al. (2019). Physical layer security for the internet of things: authentication and key generation. *IEEE Wireless Communications* 26 (5): 92–98.

66 Shakiba-Herfeh, M., Chorti, A., and Poor, H.V. (2021). Physical layer security: authentication, integrity, and confidentiality. In: *Physical Layer Security* (ed. K.N. Le), 129–150. Cham: Springer.

67 Yu, P.L., Verma, G., and Sadler, B.M. (2015). Wireless physical layer authentication via fingerprint embedding. *IEEE Communications Magazine* 53 (6): 48–53.

68 Gu, Z., Chen, H., Xu, P. et al. (2020). Physical layer authentication for non-coherent massive SIMO-based industrial IoT communications. *IEEE Transactions on Information Forensics and Security* 15: 3722–3733.

69 Verma, G., Yu, P., and Sadler, B.M. (2015). Physical layer authentication via finterprint embedding using software-defined radios. *IEEE Access* 3: 81–88.

70 Qaisi, M., Althunibat, S., and Qaraqe, M. (2021). Performance analysis of tag embedded based message authentication scheme. IEEE Global Communications Conference (GLOBECOM), Madrid, Spain.

71 Qaisi, M., Althunibat, S., and Qaraqe, M. (2022). Phase-assisted dynamic tag-embedding message authentication for IoT networks. *IEEE Internet of Things Journal* 9 (20): 20620–20629.

72 Zeng, K., Govindan, K., and Mohapatra, P. (2010). Non-cryptographic authentication and identification in wireless networks. *IEEE Wireless Communications* 17 (5): 56–62.

73 Xiao, L., Li, Y., Han, G. et al. (2016). PHY-layer spoofing detection with reinforcement learning in wireless networks. *IEEE Transactions on Vehicular Technology* 65 (12): 10037–10047.

74 Xiao, L., Greenstein, L., Mandayam, N., and Trappe, W. (2007). Fingerprints in the ether: using the physical layer for wireless authentication. IEEE International Conference on Communications (ICC), Glasgow, UK.

75 Xiao, L., Greenstein, L., Mandayam, N., and Trappe, W. (2008). A physical-layer technique to enhance authentication for mobile terminals. IEEE International Conference on Communications (ICC), Beijing, China.

76 Liu, F.J., Wang, X., and Tang, H. (2011). Robust physical layer authentication using inherent properties of channel impulse response. Military Communications Conference (MILCOM), Baltimore, MD, USA.

77 Efstathiou, D. (2018). IoT physical layer security enhancement. Global Information Infrastructure and Networking Symposium (GIIS), Thessaloniki, Greece.

78 Aygül, M.A., Büyükçorak, S., da Costa, D.B. et al. (2020). Signal relation-based physical layer authentication. IEEE International Conference on Communications (ICC), Dublin, Ireland.

79 Gao, N., Ni, Q., Feng, D. et al. (2020). Physical layer authentication under intelligent spoofing in wireless sensor networks. *Signal Processing* 166 (107272): 1–13.

80 Senigagliesi, L., Baldi, M., and Gambi, E. (2020). Physical layer authentication techniques based on machine learning with data compression. IEEE Conference on Communications and Network Security (CNS), Avignon, France.

81 Fang, H., Wang, X., and Hanzo, L. (2019). Learning-aided physical layer authentication as an intelligent process. *IEEE Transactions on Communications* 67 (3): 2260–2273.

82 Hou, W., Wang, X., Chouinard, J.-Y., and Refaey, A. (2014). Physical layer authentication for mobile systems with time-varying carrier frequency offsets. *IEEE Transactions on Communications* 62 (5): 1658–1667.

83 Zhao, C., Huang, M., Huang, L. et al. (2017). A robust authentication scheme based on physical-layer phase noise fingerprint for emerging wireless networks. *Computer Networks* 128: 164–171.

84 Wang, D., Bai, B., Zhao, W., and Han, Z. (2018). A survey of optimization approaches for wireless physical layer security. *IEEE Communications Surveys & Tutorials* 21 (2): 1878–1911.

85 Schinianakis, D. (2019). Lightweight security for the internet of things: a soft introduction to physical unclonable functions. *IEEE Potentials* 38 (2): 21–28.

86 Maes, R. and Verbauwhede, I. (2010). Physically unclonable functions: a study on the state of the art and future research directions. In: *Towards Hardware-Intrinsic Security* (ed. A.R. Sadeghi and D. Naccache), 3–37. Berlin: Heidelberg, Springer.

87 Rührmair, U., Sehnke, F., Sölter, J. et al. (2010). Modeling attacks on physical unclonable functions. ACM Conference on Computer and Communications Security (CCS), Chicago, IL, USA.

88 Gassend, B., Clarke, D., van Dijk, M., and Devadas, S. (2002). Silicon physical random functions. ACM Conference on Computer and Communications Security (CCS), Washington, DC, USA.

89 Gao, Y., Li, G., Ma, H. (2016). Obfuscated challenge-response: a secure lightweight authentication mechanism for PUF-based pervasive devices. IEEE International Conference on Pervasive Computing and Communication (PerCom) Workshops, Sydney, NSW, Australia.

90 Mahalat, M.H., Saha, S., Mondal, A., and Sen, B. (2018). A PUF based light weight protocol for secure WiFi authentication of IoT devices. International Symposium on Embedded Computing and System Design (ISED), Cochin, India.

91 Aman, M.N., Basheer, M.H., and Sikdar, B. (2019). Data provenance for IoT with light weight authentication and privacy preservation. *IEEE Internet of Things Journal* 6 (6): 10441–10457.

92 de Ree, M., Mantas, G., Rodriguez, J. et al. (2021). Data confidentiality for IoT networks: cryptographic gaps and physical-layer opportunities. IEEE International Workshop on Computer Aided Modeling and Design of Communication Links and Networks (CAMAD).

93 Bernstein, D.J. and Tanja, L. eSTREAM: the ECRYPT Stream Cipher Project, 27 January 2014 [Online]. Available: https://competitions.cr.yp.to/estream.html (accessed 24 January 2023).

94 Bernstein, D.J. and Lange, T. CAESAR submissions, 20 February 2019 [Online]. Available: https://competitions.cr.yp.to/caesar-submissions.html (accessed 24 January 2023).

95 Bassham, L., Chang, D., Hong, D. et al. Lightweight cryptography. National Institute of Standards and Technology (NIST), 9 November 2022 [Online]. Available: https://csrc.nist.gov/projects/lightweight-cryptography ().

96 Zhang, J., Marshall, A., Woods, R., and Duong, T.Q. (2017). Design of an OFDM physical layer encryption scheme. *IEEE Transactions on Vehicular Technology* 66 (3): 2114–2127.

97 Huo, F. and Gong, G. (2014). A new efficient physical layer OFDM encryption scheme. IEEE International Conference on Computer Communications (INFOCOM), Toronto, ON, Canada.

98 Huo, F. and Gong, G. (2015). XOR encryption versus phase encryption, an in-depth analysis. *IEEE Transactions on Electromagnetic Compatibility* 57 (4): 903–911.

99 Hua, J., Jiang, S., Lu, W. et al. (2018). A novel physical layer encryption algorithm based on statistical characteristics of time-selective channels. *IEEE Access* 6: 38225–38233.

100 Khan, M.A., Asim, M., Jeoti, V., and Manzoor, R.S. (2007). On secure OFDM system: chaos based constellation scrambling. International Conference on Intelligent and Advanced Systems (ICIAS), Kuala Lumpur, Malaysia.

101 Li, H., Wang, X., and Chouinard, J.-Y. (2015). Eavesdropping-resilient OFDM system using sorted subcarrier interleaving. *IEEE Transactions on Wireless Communications* 14 (2): 1155–1165.

102 Liu, J., Ren, A., Sun, R. et al. (2019). A novel chaos-based physical layer security transmission scheme for internet of things. IEEE Global Communications Conference (GLOBECOM), Waikoloa, HI, USA.

103 Reilly, D. and Kanter, G.S. (2009). Noise-enhanced encryption for physical layer security in an OFDM radio. IEEE Radio and Wireless Symposium (RWS), San Diego, CA, USA.

104 Ma, R., Dai, L., Wang, Z., and Wang, J. (2010). Secure communications in TDS-OFDM system using constellation rotation and noise insertion. *IEEE Transactions on Consumer Electronics* 56 (3): 1328–1332.

105 Tahir, M., Jarot, S.P.W., and Siddiqi, M.U. (2010). Wireless physical layer security using encryption and channel pre-compensation. International Conference on Computer Applications and Industrial Electronics (ICCAIE), Kuala Lumpur, Malaysia.

106 Hou, Y., Li, G., Dang, S. et al. (2022). Physical layer encryption based on dynamic constellation rotation. IEEE Vehicular Technology Conference (VTC Fall), London, UK.

107 Abushattal, A., Althunibat, S., Qaraqe, M.K., and Arslan, H. (2021). A secure downlink NOMA scheme against unknown internal eavesdroppers. *IEEE Wireless Communications Letters* 10 (6): 1281–1285.

108 Pechetti, S.V., Jindal, A., and Bose, R. (2017). Channel-based mapping diversity for enhancing the physical layer security in the internet of things. IEEE International Symposium on Personal, Indoor and Mobile Radio Communications (PIMRC), Montreal, QC, Canada.

109 Pechetti, S.V., Jindal, A., and Bose, R. (2019). Exploiting mapping diversity for enhancing security at physical layer in the internet of things. *IEEE Internet of Things Journal* 6 (1): 532–544.

110 Li, H., Wang, X., and Zou, Y. (2014). Dynamic subcarrier coordinate interleaving for eavesdropping prevention in OFDM systems. *IEEE Communications Letters* 18 (6): 1059–1062.

111 Pan, F., Song, H., and Wen, H. (2015). An OFDM cross-layer encryption secure communication scheme. IEEE/CIC International Conference on Communications in China (ICCC) Workshops, Shenzhen, China.

112 Li, H., Wang, X., and Hou, W. (2013). Secure transmission in OFDM systems by using time domain scrambling. IEEE Vehicular Technology Conference (VTC Spring), Dresden, Germany.

113 Taha, F.A. and Althunibat, S. (2021). Improving data confidentiality in chirp spread spectrum modulation. IEEE International Workshop on Computer Aided Modeling and Design of Communication Links and Networks (CAMAD).

114 Alsadi, A. and Mohan, S. (2020). A new frequency hopping scheme to secure the physical layer in the internet of things (IoT). Wireless Telecommunications Symposium (WTS), Washington, DC, USA.

115 Letafati, M., Kuhestani, A., Ng, D.W.K., and Behroozi, H. (2020). A new frequency hopping-aided secure communication in the presence of an adversary jammer and an untrusted relay. IEEE International Conference on Communications (ICC) Workshops, Dublin, Ireland.

116 Letafati, M., Kuhestani, A., Behroozi, H., and Ng, D.W.K. (2020). Jamming-resilient frequency hopping-aided secure communication for internet-of-things in the presence of an untrusted relay. *IEEE Transactions on Wireless Communications* 19 (10): 6771–6785.

117 Wyner, A.D. (1975). The wire-tap channel. *Bell System Techical Journal* 54 (8): 1355–1387.

118 Csiszár, I. and Körner, J. (1978). Broadcast channels with confidential messages. *IEEE Transactions on Information Theory* 24 (3): 339–348.

119 Bloch, M. and Barros, J. (2011). *Physical-Layer Security: From Information Theory to Security Engineering*. Cambridge University Press.

120 Leung-Yan-Cheong, S.K. and Hellman, M.E. (1978). The Gaussian wire-tap channel. *IEEE Transactions on Information Theory* 24 (4): 451–456.

121 Thangaraj, A., Dihidar, S., Calderbank, A.R. et al. (2007). Applications of LDPC codes to the wiretap channel. *IEEE Transactions on Information Theory* 53 (8): 2933–2945.

122 Liu, R., Liang, Y., Poor, H.V., and Spasojevic, P. (2007). Secure nested codes for Type II wiretap channels. IEEE Information Theory Workshop (ITW), Tahoe City, CA, USA.

123 Klinc, D., Ha, J., McLaughlin, S.W. et al. (2011). LDPC codes for the Guassian wiretap channel. *IEEE Transactions on Information Forensics and Security* 6 (3): 532–540.

124 Mahdavifar, H. and Vardy, A. (2011). Achieving the secrecy capacity of wiretap channels using polar codes. *IEEE Transactions on Information Theory* 57 (10): 6428–6443.

125 Koyluoglu, O.O. and El Gamal, H. (2012). Polar coding for secure transmission and key agreement. *IEEE Transactions on Information Forensics and Security* 7 (5): 1472–1483.

126 Sasoglu, E. and Vardy, A. (2013). A new polar coding scheme for strong security on wiretap channels. IEEE International Symposium on Information Theory (ISIT), Istanbul, Turkey.

127 Illi, E., Qaraqe, M.K., El Bouanani, F., and Al-Kuwari, S.M. (2022). Can artificial noise boost further the secrecy of dual-hop RIS-aided networks?2022.

128 Negi, R. and Goel, S. (2005). Secret communication using artificial noise. IEEE Vehicular Technology Conference (VTC Fall), Dallas, TX, USA.

129 Goel, S. and Negi, R. (2008). Guaranteeing secrecy using artificial noise. *IEEE Transactions on Wireless Communications* 7 (6): 2180–2189.

130 Lai, L. and El Gamal, H. (2008). The relay-eavesdropper channel: cooperation for secrecy. *IEEE Transactions on Information Theory* 54 (9): 4005–4019.

131 Zhou, X. and McKay, M.R. (2010). Secure transmission with artificial noise over fading channels: achievable rate and optimal power allocation. *IEEE Transactions on Vehicular Technology* 59 (8): 3831–3842.

132 Zhang, B., Liu, W., Li, Q. et al. (2020). Directional modulation design under a given symbol-independent magnitude constraint for secure IoT networks. *IEEE Internet of Things Journal* 8 (20): 15140–15147.

133 Nusenu, S.Y., Huaizong, S., Pan, Y., and Basit, A. (2021). Directional modulation with precise legitimate location using time-modulation retrodirective frequency diversity array for secure IoT communications. *IEEE Systems Journal* 15 (1): 1109–1119.

134 Hassija, V., Chamola, V., Saxena, V. et al. (2019). A survey on IoT security: application areas, security threats, and solution architectures. *IEEE Access* 7: 82721–82743.

135 Newsome, J., Shi, E., Song, D., and Perrig, A. (2004). The sybil attack in sensor networks: analysis & defenses. Proceedings of the 3rd International Symposium on Information Processing in Sensor Networks (IPSN), Berkeley, CA, USA.

136 Vempaty, A., Ozdemir, O., Agrawal, K. et al. (2013). Localization in wireless sensor networks: Byzantines and mitigation techniques. *IEEE Transactions on Signal Processing* 61 (6): 1495–1508.

137 Liu, D., Ning, P., and Du, W. (2005). Detecting malicious beacon nodes for secure location discovery in wireless sensor networks. IEEE International Conference on Distributed Computing Systems (ICDCS), Columbus, OH, USA.

138 Douceur, J.R. (2002). The sybil attack. International Workshop on Peer-to-Peer Systems (IPTPS), Cambridge, MA, USA.

139 Demirbas, M. and Song, Y. (2006). An RSSI-based scheme for sybil attack detection in wireless sensor networks. IEEE International Symposium on a World of Wireless, Mobile and Multimedia Networks (WoWMoM), Buffalo-Niagara Falls, NY, USA.

140 Wang, J., Yang, G., Sun, Y., and Chen, S. (2007). Sybil attack detection based on RSSI for wireless sensor network. International Conference on Wireless Communications, Networking and Mobile Computing (WiCom), Shanghai, China.

141 Jan, M.A., Nanda, P., He, X., and Liu, R.P. (2015). A sybil attack detection scheme for a centralized clustering-based hierarchical network. IEEE International Joint Conference on Trust, Security and Privacy in Computing and Communications (TrustCom), on Big Data Science and Engineering (BigDataSE), and on Parallel and Distributed Processing with Applications (ISPA), Helsinki, Finland.

142 Sarigiannidis, P., Karapistoli, E., and Economides, A.A. (2015). Detecting sybil attacks in wireless sensor networks using UWB ranging-based information. *Expert Systems with Applications* 42 (21): 7560–7572.

143 Ssu, K.-F., Wang, W.-T., and Chang, W.-C. (2009). Detecting sybil attacks in wireless sensor networks using neighboring information. *Computer Networks* 53 (18): 3042–3056.

144 Jamshidi, M., Zangeneh, E., Esnaashari, M., and Meybodi, M.R. (2017). A lightweight algorithm for detecting mobile sybil nodes in mobile wireless sensor networks. *Computers and Electrical Engineering* 64: 220–232.

145 Xiao, L., Greenstein, L.J., Mandayam, N.B., and Trappe, W. (2009). Channel-based detection of sybil attacks in wireless networks. *IEEE Transactions on Information Forensics and Security* 4 (3): 492–503.

146 Pan, F., Pang, Z., Xiao, M. et al. (2018). Clone detection based on physical layer reputation for proximity service. *IEEE Access* 7: 3948–3957.

147 Li, Q., Zhang, K., Cheffena, M., and Shen, X.S. (2016). Exploiting dispersive power gain and delay spread for sybil detection in industrial WSNs. IEEE/CIC International Conference on Communications in China (ICCC), Chengdu, China.

148 Li, Q. and Cheffena, M. (2019). Exploiting dispersive power gain and delay spread for sybil detection in industrial WSNs: a multi-Kernel approach. *IEEE Transactions on Wireless Communications* 18 (3): 1805–1818.

149 Dogan-Tusha, S., Althunibat, S., and Qaraqe, M. (2022). A novel sybil attack detection mechanism for mobile IoT networks. IEEE Global Communications Conference (GLOBECOM), Rio de Janeiro, Brazil.

150 Liu, F., Cheng, X., and Chen, D. (2007). Insider attacker detection in wireless sensor networks. IEEE International Conference on Computer Communications (INFOCOM), Anchorage, AK, USA.

151 Stetsko, A., Folkman, L., and Matyas, V. (2010). Neighbor-based intrusion detection for wireless sensor networks. International Conference on Wireless and Mobile Communications (ICWMC), Valencia, Spain.

152 Ghugar, U., Pradhan, J., Bhoi, S.K., and Sahoo, R.R. (2019). LB-IDS: securing wireless sensor network using protocol layer trust-based intrusion detection system. *Journal of Computer Networks and Communications* 2019 (2054298): 1–13.

153 Bhuse, V. and Gupta, A. (2006). Anomaly intrusion detection in wireless sensor networks. *Journal of High Speed Networks* 15 (1): 33–51.

7

Quantum Security for the Tactile Internet

Shima Hassanpour[1], Riccardo Bassoli[2,3,4], Janis Nötzel[5], Frank H. P. Fitzek[2,3], Holger Boche[5], and Thorsten Strufe[1,3,6]

[1]Chair of Privacy and Network Security, Technische Universität Dresden, Dresden, Germany
[2]Deutsche Telekom Chair of Communication Networks, Technische Universität Dresden, Dresden, Germany
[3]Centre for Tactile Internet with Human-in-the-Loop (CeTI), Cluster of Excellence, Dresden, Germany
[4]Quantum Communication Networks Research Group, Fakultät Elektrotechnik und Informationstechnik, Institut für Nachrichtentechnik, Technische Universität Dresden, Dresden, Germany
[5]Lehrstuhl für Theoretische Informationstechnik, TUM School of Computation, Information and Technology, Technische Universität München, Karlsruher Institut für Technologie, München, Germany
[6]Chair of IT-Security, Karlsruher Institut für Technologie, Karlsruhe, Germany

7.1 Introduction

The idea of communicating audiovisual and tactile information for remote control and interaction has a long tradition in research. The first works mentioning this date back to the end of the 1960s. Nevertheless, at that time, communication networks and technologies were not ready for this application. With the new millennium, the evolution and improvement of the cellular networks and the Internet has provided a fertile ground for so-called *haptic communications* [1]. As a result, some of the first real experiments became possible over the Internet, such as remote surgery [2] in 2007.

Next, in the last decade, the vision of 5G and beyond networks provided the full inclusion and support of haptic verticals in the communication ecosystem (e.g. massive IoT networks). In particular, this facilitated a new and popular field of research, called the Tactile Internet (TI). This encouraged the International Telecommunication Union Telecommunication Standardization Sector (ITU-T) for providing a technology watch report about TI in 2014. This initial work inspired the subsequent creation of various standardization working groups like the IEEE 1918.1 Tactile Internet, the Society of Motion Picture and Television Engineers (SMPTE), and a working group within the European Telecommunications Standards Institute (ETSI) [3].

Today, the research on and the design of future 6G networks extends the vision of the TI to the more general concept of telepresence that addresses the entire human sensory experience. As can be seen from the current vision of 6G (e.g. that of the European Union [EU] flagships Hexa-X and Hexa-X II), the convergence of physical, human, and digital world will provide a mixed reality in which machines, humans, and virtual representations will interact with each other [4]. However, this vision poses significant challenges in terms of

Security and Privacy for 6G Massive IoT, First Edition.
Edited by Georgios Mantas, Firooz B. Saghezchi, Jonathan Rodriguez, and Victor Sucasas.

ensuring the necessary security at the required low latency and reliability. In fact, one of the key aspects is the trustworthiness of the end-to-end system, which negatively affects the communication and computing load. New approaches, for instance relying on quantum physics are hence raising interest. They seem to promise more efficient and effective solutions to security and, more generally, to trustworthiness [5].

The TI is a 5G-and-beyond vertical that mainly relies on ultra-reliable low-latency communication (URLLC) infrastructure. Furthermore, the standard IEEE 1918.1 complements the definition and specification of URLLCs, by also including Key Performance Indicators (KPIs), particularly focused on haptic data communications.

To realize the TI, requires a network infrastructure that can ensure the URLLC Quality of Service (QoS) at anytime and anywhere among TI users. This has become possible via the realizations of network abstractions.

Figure 7.1 depicts a network architecture that helps realize the connection between two end users in TI, which we call *Tactile Edges*. Normally, the communication happens between the *Master Domain* – usually humans in the TI with Human-in-the-Loop (TaHiL) – and the *Controlled Domain*, which consists of machines. The peculiar characteristic of TI communication is the fact of being a two-way communication (i.e. with feedback), which multiplexes different kinds of data streams (audio, video, and haptic), with different KPIs and hence QoS requirements. In fact, not only the input from the Master should be transmitted to the Controlled device, but also the input from the Controlled entity should be rapidly and reliably sent back to the Master. This allows the Master for the successful and correct perception of the remote action, which has been performed. In fact, the scope behind the design is to make the communication between Tactile Edges seamless, without sensing the negative contributions of the *Network domain* (latency, losses, etc.).

The Network domain mainly consists of six parts. First, the Radio Access Network (RAN) is the *ecosystem* of heterogeneous wireless access technologies. It includes the generations of cellular networks, wireless local area networks (WLANs), and the 5G New Radio (NR) with mmWave antennas. Second, the fronthaul is the wired/wireless network, which contains aggregation points to get traffic from the RAN and forward it to the edge. The edge network is generally a wired network including edge small data centers for edge and cloud RAN (C-RAN) computing. The RAN and edge can usually have the size of either a corporate (campus) area network (CAN) or a metropolitan area network (MAN). Next, the backhaul is a wired network. It connects the edge and the core of the operators' network. The core network is generally a wide area network (WAN) and it also has small data centers for network computing. Finally, the Internet connects different networks globally, and it provides access to big data centers for highly-demanding cloud computing services. The network can also involve satellite communications in any of its areas.

Figure 7.2 displays the proposed logical structure of a tactile device at the Tactile Edge according to the standard IEEE 1918.1. The tactile device is normally equipped with at least a sensor (S) or an actuator (A), without networking functions. Instead, sensor nodes (SNs) or actuator nodes (ANs) not only perform sensing and actuation functions but also networking. Side by side, sensor gateways (SGs) or actuator gateways (AGs) are needed as interfaces between S (A) and SN (AN). Next, a controller node (CN) is responsible for

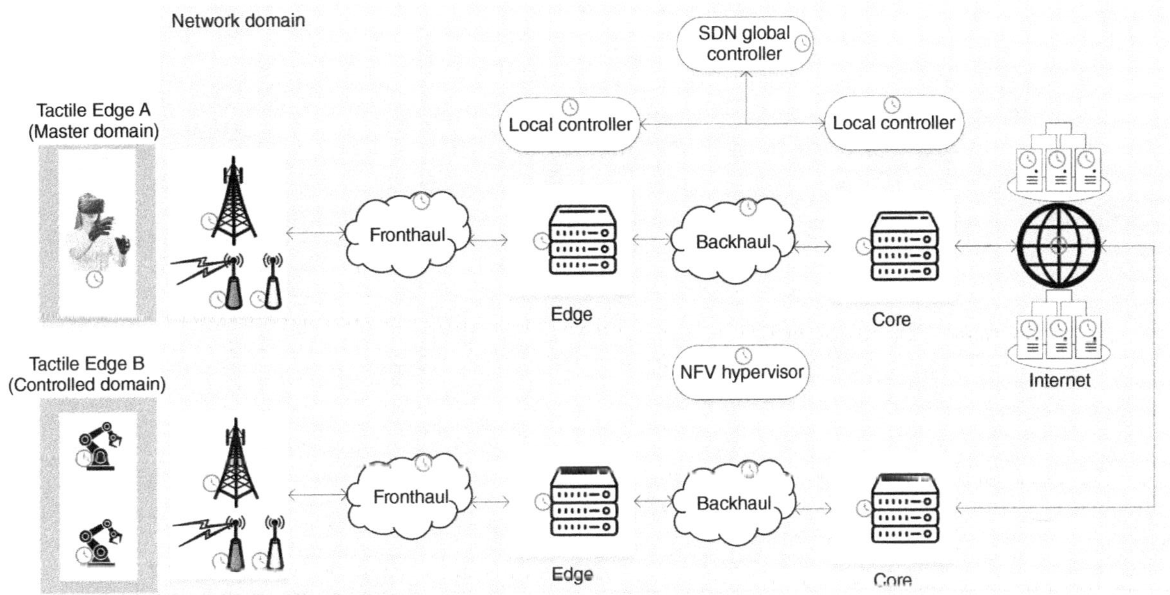

Figure 7.1 The Tactile Internet network architecture in the 5G and beyond ecosystem.

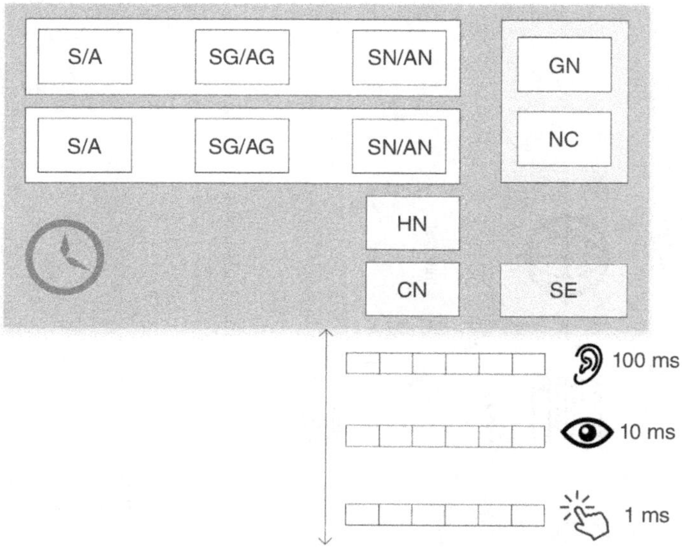

Figure 7.2 Architecture of a tactile device at the Tactile Edge according to standardization work of IEEE 1918.1.

the control operations of SNs and ANs. On the other hand, the haptic node (HN) processes and compresses tactile information (i.e. kinaesthetic and audiovisual) by using tactile codecs [6].

In particular, these codecs permit the realization of remote sensory perception by avoiding the negative effect of propagation latency while also minimizing the amount of information to be sent. From a networking perspective, the gateway node (GN) is the interface between the Tactile Edge and the Network domain, carrying the data-plane information. In parallel, the network controller (NC) deals with control-plane messages (e.g. congestion control, resource management, etc.). Finally, the storage entity (SE) is a logic entity that can host the high-demanding computing tasks that cannot run on the limited resources available at the tactile device.

As previously mentioned, the stream of data that is transmitted from the Tactile Edge to the Network domain and vice versa, is heterogeneous, and so are the respective KPIs. In fact, a human user requires an auditory reaction time of 100 ms, a typical visual reaction time of 10 ms, and a latency for the manual control of a visual scene (performing a command) of 1 ms. The *network* (or control plane) and the *haptic synchronization* (or data plane) are two-dimensional synchronisation structures of the TI data streams. The former is the pillar for supporting the end-to-end connection between Tactile Edge A and Tactile Edge B. The latter maintains the synchronisation among the different haptic data, referred to as the multisensory information.

As the above discussion indicates, the entire architecture of TI is designed for the extremely low delay requirements. Convincing solutions to guarantee security in this context are rare. In this sense, the recent advancements in quantum technologies for security could provide an opportunity to achieve low latency security in the TI. Sections 7.4–7.7 introduce and explain some of the major solutions that the research community is envisioning for the integration of quantum technologies in future communication networks. Moreover, the discussion highlights some pros and cons in respect of their application and integration in the TI. Sections 7.2 and 7.3 present the preliminaries of quantum mechanics and security, respectively. Then, in Section 7.4, the quantum key distribution (QKD) is discussed. Physical layer security and known attacks on quantum security are discussed in Sections 7.5 and 7.6, respectively. In Section 7.7 another cryptographic primitive, oblivious transfer, is explained in more detail, and finally, a conclusion is given in Section 7.8.

7.2 Preliminaries – Quantum Mechanics

To provide foundations, we begin by introducing the necessary background on quantum mechanics.

7.2.1 Finite Dimensional Quantum Systems

A finite dimensional (FD) quantum system is described via a finite dimensional Hilbert space $\mathcal{H}$, where we use $\langle u, v\rangle$ to denote the scalar product between two elements u and v of $\mathcal{H}$. In the rest of the chapter, d denotes the dimension of $\mathcal{H}$. The Hilbert space is a vector space over the complex number $\mathbb{C}$. In case that $d = 2$, the quantum system is called qubit. The qubit is considered the quantum analog to classical bit, and it is one of the basic resources in the theoretical analysis of quantum information processing tasks. By using the Dirac's notation [5], a qubit is typically written as

$$|\psi\rangle = \alpha|0\rangle + \beta|1\rangle \tag{7.1}$$

where α and β are complex numbers satisfying $|\alpha|^2 + |\beta|^2 = 1$ and $\{|0\rangle, |1\rangle\}$ form an orthonormal basis of $\mathbb{C}$. Each quantum state can be represented by a vector in the Bloch sphere (see Figure 7.3), where $\alpha = \cos\frac{\theta}{2}$ and $\beta = e^{i\phi}\sin\frac{\theta}{2}$.

Another possible representation describes the state of a quantum system as a $d \times d$ matrix with complex entries, which is a linear operator $\rho : \mathcal{H} \to \mathcal{H}$ such that $\rho \geq 0$ and $tr(\rho) = 1$, where $\rho \geq 0$ is equivalent to $\langle v, \rho v\rangle \geq 0$ for all $v \in \mathcal{H}$. The expression tr denotes the trace, which can be defined as $tr(\rho) = \sum_{i=1}^{d} \rho_{i,i}$ where $\rho_{i,j}$ are the entries of the matrix ρ. For a pure state $|\psi\rangle = \sum_{i=1}^{d} \psi_i|i\rangle$, its density matrix has entries $\rho_{i,j} = \psi_i\psi_j^*$ and it can be written as $\rho = |\psi\rangle\langle\psi|$.

7.2.2 Quantum Gates

Unitary operations are operations on quantum systems that do not *reduce* (or *collapse*) their quantum state. Normally, unitary operations are represented mathematically via operators that act on a qubit, and graphically by their respective quantum gates. Pauli gates are an

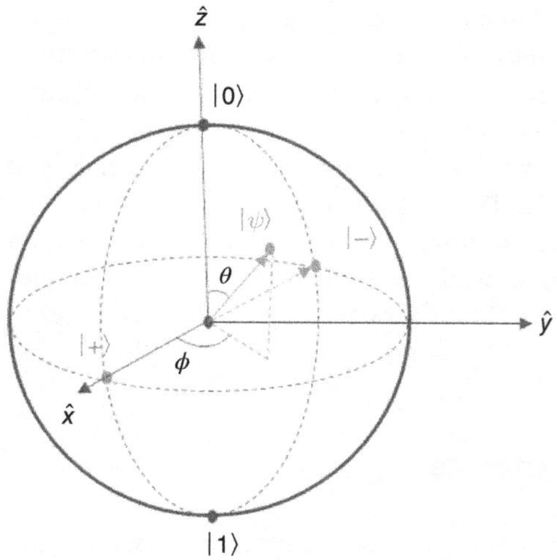

Figure 7.3 Geometrical representation of a two-dimensional quantum state (Bloch sphere).

example of single-qubit gates that can perform unitary operations on single quantum systems. Their properties are described by the so-called Pauli matrices

$$\begin{aligned} I &= |0\rangle\langle 0| + |1\rangle\langle 1|, \quad \sigma_x = |0\rangle\langle 1| + |1\rangle\langle 0|, \\ i\sigma_y &= |0\rangle\langle 1| - |1\rangle\langle 0|, \quad \sigma_z = |0\rangle\langle 0| - |1\rangle\langle 1| \end{aligned} \tag{7.2}$$

Another important single-qubit quantum gate is the so-called Hadamard gate, defined by

$$H = |+\rangle\langle 0| + |-\rangle\langle 1| \tag{7.3}$$

where $|+\rangle = \frac{1}{\sqrt{2}}(|0\rangle + |1\rangle), |-\rangle = \frac{1}{\sqrt{2}}(|0\rangle - |1\rangle)$ and $|0\rangle, |1\rangle$ are defined as Hadamard and computational basis, respectively.

Additionally, there can also be quantum gates performing operations on two qubits.

7.2.3 Quantum Measurement

A property of a quantum state cannot be determined until it is measured. This process *reduces* the quantum state to one of the possible values (probabilistic outputs). Mathematically, the measurement in the computational basis is a projection from Hilbert space to the $\{|0\rangle, |1\rangle\}$ basis. The measurement collapses the coherence of the input quantum state so that the original quantum state is lost and cannot be reconstructed. Clearly, $|\alpha|^2$ and $|\beta|^2$ determine the probabilities of obtaining $|0\rangle$ and $|1\rangle$, respectively, by measuring the quantum state $|\psi\rangle$ on the computational basis, $\{|0\rangle, |1\rangle\}$ (Figure 7.4).

Any quantum measurement can be described by a set of measurement operators, M_c. The operators must satisfy $\sum_c M_c^\dagger M_c = \mathbb{1}$. If the state of the quantum system is $|\psi\rangle$ at the time of the measurement, then the probability of measuring outcome c is $p_c = \langle\psi|M_c^\dagger M_c|\psi\rangle$,

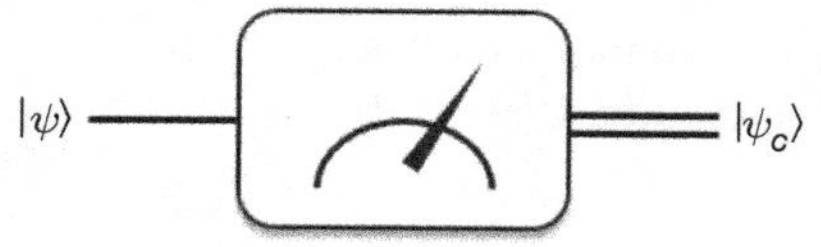

Figure 7.4 Diagram of a quantum measurement. The result of the measurement is a classical variable c according to a probability distribution.

where c is the result of the measurement. After the measurement, the quantum state collapses into the state:

$$|\psi_c\rangle = \frac{M_c|\psi\rangle}{\sqrt{\langle\psi|M_c^\dagger M_c|\psi\rangle}} \tag{7.4}$$

The mixed state is represented by density matrix $\rho = \sum_i p_i|\psi_i\rangle\langle\psi_i|$ (if $p_i = 1$, the state is called a pure state). For mixed states, the probability of obtaining outcome c is $p_M(c) := tr(\rho M_c^\dagger M_c)$ and the post-measurement corresponding to outcome c is

$$\rho_c = \frac{M_c \rho M_c^\dagger}{p_M(c)} \tag{7.5}$$

7.2.4 Fock Space

The Fock space model systems, which are made up of a variable number of particles, obey certain symmetry rules. Mathematically, it is a tensor algebra over a Hilbert space $\mathcal{H}$.

$$F_x[\mathcal{H}] = \bigoplus_{n=0}^{\infty} S_x \mathcal{H}^{\otimes n} = F^b \otimes F^f \tag{7.6}$$

where the vectors in the bosonic and fermionic subspaces can be identified with the symmetric and antisymmetric tensor products, respectively. Depending on whether the Hilbert space describes particles, obeying bosonic ($x = +$) or fermionic ($x = -$) statistics, the bosonic (symmetric) Fock space or the fermionic (antisymmetric) Fock space then arise. The case $x = -$ is relevant for systems, which are made up of photons. Each vector v in $F\_[\mathcal{H}]$ can be written as

$$v = \sum_{n=0}^{\infty} v_n |n\rangle \tag{7.7}$$

where $\{|n\rangle\}_{n=0}^{\infty}$ is an orthonormal basis, which is also called the number-state basis. This representation becomes helpful for the description of the so-called continuous variable quantum states.

7.2.5 Continuous Variable Quantum States

A continuous variable (CV) quantum system is a quantum system represented in an infinite-dimensional Hilbert space [7–9]. A bosonic CV system is a quantized field, such as the electromagnetic field. This system can be modeled as a collection of non-interacting quantum harmonic oscillators of different frequencies, where each oscillator is called a mode of the system. Mathematically, a CV system of M bosonic modes of the electromagnetic field is described by a Hilbert space $\mathcal{H} = \otimes_{k=1}^{M} \mathcal{H}_k$, where $\mathcal{H}_k$ represents a single-mode Hilbert space. In the case of the electromagnetic field, the Hamiltonian of the whole system can be defined as:

$$\hat{H} = \sum_{k=1}^{N} \hat{H}_k, \quad \hat{H}_k = \hbar\omega_k \left(\hat{a}_k^\dagger \hat{a}_k + \frac{1}{2}\right) \tag{7.8}$$

where $\hat{H}_k$ refers to the reduced Hamiltonian of the kth mode. $\hat{a}_k^\dagger$ and $\hat{a}_k$ are the creation and annihilation operators in mode k. For the analysis of these systems, it turns out to be useful to study elementary operations, like the creation and the destruction of a photon. Given a number state of M modes, they act as follows:

$$\hat{a}_k^\dagger |n_1\rangle \otimes \cdots \otimes |n_M\rangle = \sqrt{n_k+1}|n_1\rangle \otimes \cdots \otimes |n_k+1\rangle \otimes \cdots \otimes |n_M\rangle \tag{7.9}$$

$$\hat{a}_k |n_1\rangle \otimes \cdots \otimes |n_M\rangle = \sqrt{n_k}|n_1\rangle \otimes \cdots \otimes |n_k-1\rangle \otimes \cdots \otimes |n_M\rangle \tag{7.10}$$

$$[\hat{a}_k, \hat{a}_{k'}] = [\hat{a}_{k'}^\dagger, \hat{a}_{k'}^\dagger] = 0, \quad [\hat{a}_k, \hat{a}_{k'}^\dagger] = \delta_{kk'} \tag{7.11}$$

where $\delta_{kk'}$ is the Kronecker delta function.[1] For each mode k, the quadrature operators are expressed as

$$\hat{q}_k = \hat{a}_k + \hat{a}_k^\dagger, \quad \hat{p}_k = i(\hat{a}_k^\dagger - \hat{a}_k) \tag{7.12}$$

The vector of quadrature operators can be arranged as $\hat{R} = (\hat{q}_1, \hat{p}_1, \ldots, \hat{q}_M, \hat{p}_M)$, and the commutation relation between the quadrature operators can be defined as

$$[\hat{R}_n, \hat{R}_m] = 2i\Omega_{nm} \tag{7.13}$$

where $\hat{R}_n(\hat{R}_m)$ is the nth (mth) element of the vector $\hat{R}$, and Ω_{nm} is the element of the nth row and mth column of the matrix Ω

$$\Omega = \bigotimes_{k=1}^{M} \omega, \quad \omega = \begin{pmatrix} 0 & 1 \\ -1 & 0 \end{pmatrix} \tag{7.14}$$

The infinite-dimensional single-mode Hilbert space $\mathcal{H}_k$ is spanned by the Fock or number-state basis $|n_k\rangle_{n=0}^{\infty}$ satisfying

$$\hat{n}_k|n_k\rangle = n_k|n_k\rangle \tag{7.15}$$

where the Fock state $|n_k\rangle$ is the eigenstate of the number operator $\hat{n}_k = \hat{a}_k^\dagger \hat{a}_k$. The action of the bosonic field operators on the Fock number states is given by

$$\hat{a}_k|n_k\rangle = \sqrt{n_k}|n_k - 1\rangle, \quad \hat{a}_k^\dagger|n_k\rangle = \sqrt{n_k+1}|n_k+1\rangle \tag{7.16}$$

with $\hat{a}_k|0\rangle = 0$, where $|0\rangle$ is the vacuum state.

7.2.6 Gaussian States

Gaussian states are at the heart of quantum information theory. A Gaussian state is defined as any state whose characteristic functions are Gauss function, known as the normal distribution. The Gaussian quantum state can be completely described by the first moment, displacement (d), or mean value, and second moment, covariance matrix (V), as follows [10, 11]:

$$\begin{aligned} d &= (\langle \hat{R}_1 \rangle, \langle \hat{R}_2 \rangle, \ldots, \langle \hat{R}_M \rangle) \\ V_{nm} &= \frac{1}{2}\langle \hat{R}_n\hat{R}_m + \hat{R}_m\hat{R}_n \rangle - \langle \hat{R}_n \rangle \langle \hat{R}_m \rangle \end{aligned} \tag{7.17}$$

1 The Kronecker function is 1 if the variables are equal and 0 otherwise.

A Gaussian state with M modes is a CV state whose Wigner function (a way of representing a quantum state in phase space), $W(R)$, follows a Gaussian distribution.

$$W(R) = \frac{1}{(2\pi)^{2M}\sqrt{det(V)}} exp\left(-\frac{1}{2}(R-\langle R\rangle)V^{-1}(R-\langle R\rangle)^T\right) \tag{7.18}$$

Important examples of Gaussian states are vacuum states, thermal states, and coherent states.

7.2.7 Conjugate Coding

Conjugate coding (CC) is a famous idea by Wiesner [12]. It can be described as follows: There are two bases available, which are denoted as $+ = \{e_0, e_1\}$ and $\times = \{d_0, d_1\}$, where the basis vectors of the $\times$ basis are defined as $d_0 = \frac{1}{\sqrt{2}}(e_0 + e_1)$ and $d_1 = \frac{1}{\sqrt{2}}(e_0 - e_1)$. These are used to generate signal states $q_{s,i}$ where $s \in \{+, \times\}$ and $i \in \{0, 1\}$.

A sender can thus encode two bits (s, i) into one qubit $q_{s,i}$. However, if a person who is unaware of the choice (s, i) makes a measurement on the qubit $q_{s,i}$ either in basis $+$ or $\times$, then the probability of that person choosing the wrong basis is $1/2$, and in case the wrong basis is chosen, the measurement result m of that person will be independent of i. However, if the person was to know s, then i would be revealed.

Wiesner suggested CC to introduce the concept of quantum money, which is treated as well in the more recent work [13]. His quantum money scheme works as follows:

Protocol 1:

1. Each banknote b gets a serial number s.
2. The bank that issues b stores s along with a random string $\theta^n \in \{-, \times\}^n$ of angles and a binary random string $x^n \in \{0, 1\}$.
3. The banknote is made to store the string $\otimes_{i=1}^{n} q_{\theta_i, x_i}$ of qubits.
4. A counterfeiter has a probability of 2^{-n} of guessing θ^n correctly.
5. Any index i where he guessed incorrectly will change the original banknote and produce a damaged copy.
6. At any index i where the basis got changed, the bank has a probability of $1/2$ of detecting this.
7. In summary, the probability of being able to create a fake banknote is $(\frac{1}{2} + \frac{1}{2}\frac{1}{2})^n$.

7.2.8 Quantum Entanglement

Entanglement is one of the most distinctive quantum phenomena with no classical counterpart, and it is recognized as a fundamental resource for quantum communication [14]. Let us consider a composite quantum system consisting of two subsystems A and B in the Hilbert spaces $\mathcal{H}_A$ and $\mathcal{H}_B$, respectively. A composite quantum system is described by a quantum state ρ_{AB} acting on the Hilbert space $\mathcal{H}_A \otimes \mathcal{H}_B$. This is defined as separable if it can be written as

$$\rho_{AB} = \sum_i p_i \rho_i^A \otimes \rho_i^B \tag{7.19}$$

where ρ_i^A and ρ_i^B are states in $\mathcal{H}_A$ and $\mathcal{H}_B$ respectively and $\{p_i\}_i$ form a probability distribution.

According to the defined criteria, a quantum system is *entangled* if it is *not separable*. An important two qubit maximally entangled state is the *Bell state*, also called Einstein–Podolsky–Rosen (*EPR*) *pairs* [15]. One of the key factors in achieving highly reliable quantum communication is the generation of a pair of maximally entangled states between two parties. The more the shared pair is entangled, the less the qubit at the receiver will differ from the original qubit at the transmitter.

$$
\begin{aligned}
|\phi^+\rangle &= \frac{1}{\sqrt{2}}(|00\rangle + |11\rangle) \\
|\phi^-\rangle &= \frac{1}{\sqrt{2}}(|00\rangle - |11\rangle) \\
|\psi^+\rangle &= \frac{1}{\sqrt{2}}(|01\rangle + |10\rangle) \\
|\psi^-\rangle &= \frac{1}{\sqrt{2}}(|01\rangle - |10\rangle)
\end{aligned}
\tag{7.20}
$$

7.2.9 Entanglement Swapping

Entanglement swapping is a technique that can allow for entangling two particles that have never interacted with each other [16]. Let us consider a scenario of two EPR pairs as shown in Figure 7.5. Node 1 has entangled particles one and two, and Node 2 has entangled particles three and four. Node 1 and Node 2 send their respective first particles to an intermediate

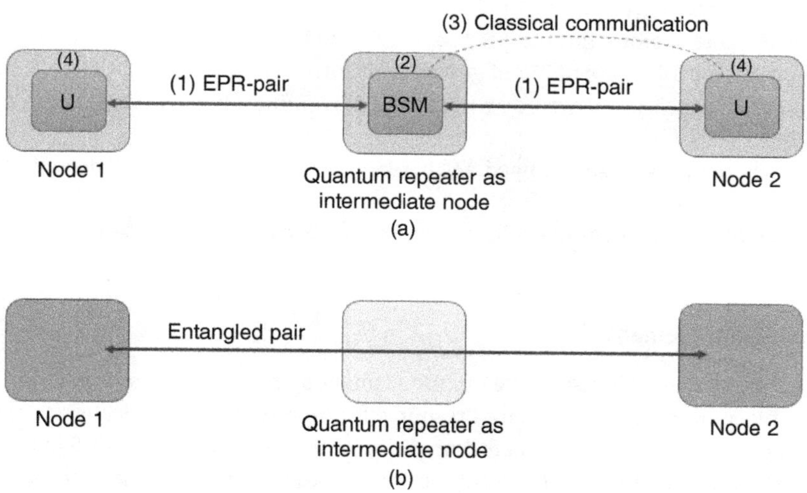

Figure 7.5 Entanglement swapping (a) Each node shares an EPR pair with quantum repeater as an intermediate node, a bell state measurement (BSM), classical bits obtained from the measurement operation and a Pauli operation (U) are performed to complete the swapping process. (b) The result is the entanglement between two nodes over a longer distance.

node. If the intermediate node now performs a Bell state measurement on these particles, their state collapses, and the second particles remaining at Nodes 1 and 2 become entangled with each other, as it appears from Eq. (7.21). Thus, through the process of entanglement swapping, it is possible to achieve entanglement between distant particles.

$$\begin{aligned} |\phi^+\rangle_{12} \otimes |\phi^+\rangle_{34} &= \frac{1}{\sqrt{2}}(|00\rangle + |11\rangle)_{12} \otimes \frac{1}{\sqrt{2}}(|00\rangle + |11\rangle)_{34} \\ &= \frac{1}{2}(|00\rangle_{12}|00\rangle_{34} + |00\rangle_{12}|11\rangle_{34} + |11\rangle_{12}|00\rangle_{34} + |11\rangle_{12}|11\rangle_{34}) \\ &= \frac{1}{2}(|00\rangle_{14}|00\rangle_{23} + |01\rangle_{14}|01\rangle_{23} + |10\rangle_{14}|10\rangle_{23} + |11\rangle_{14}|11\rangle_{23}) \\ &= \frac{1}{2}(|\phi^+\rangle_{14}|\phi^+\rangle_{23} + |\phi^-\rangle_{14}|\phi^-\rangle_{23} + |\psi^+\rangle_{14}|\psi^+\rangle_{23} + |\psi^-\rangle_{14}|\psi^-\rangle_{23}) \end{aligned} \tag{7.21}$$

7.3 Preliminaries – Security

Data security implies achieving confidentiality (i.e. disclosure of information to only authorized parties), integrity (i.e. the absence of unauthorized tampering with transmitted data), and availability of data. Cryptography provides ciphers and integrity checks that are frequently leveraged to achieve confidentiality and integrity. A cryptosystem is a set of algorithms: a randomized key generation algorithm that outputs a secret key, an encryption algorithm that takes as input a plain text message, and outputs a cipher text, and a decryption algorithm that takes as input the cipher text, a corresponding key, and outputs the original plain text message.

The long history of cryptography has been characterized by the struggle between finding ways to better achieve confidentiality and integrity, and its counterpart of cryptanalytic attacks that aim to break them. In 1917, the so-called one-time pad encryption, which uses a symmetric random secret key, was proposed, sometimes also called the Vernam cipher after its inventor [17]. It prevents all possible attacks where the adversary gets access to the ciphertext only. Shannon famously proved that these requirements on the key (i.e. has to be as long as the message and entirely random) are indeed necessary to achieve perfect secrecy, sometimes also called information-theoretic security [18]. It implies that, given these two properties of the key, no attack can be better than randomly guessing each symbol of the plain text.

The shared secret between the two parties that share a legitimate communication, commonly called "Alice" (A) and "Bob" (B), has to be generated, distributed, and managed. Requiring the key string to be entirely random and as long as the message poses severe burdens on such key management. Specifying the distribution of such a long secret key causes a main challenge in all types of conventional encryption schemes, which is called the distribution problem.

Asymmetric or public-key cryptography implies one solution for the key distribution problem. The security is based on the computational complexity, or expected time to solve certain mathematical problems. Public-key cryptography uses a public key is available to everyone, and a private key is only known to the receiver. Past asymmetric schemes are commonly based on the assumption, that it takes a long time to solve either discrete logarithms or factor large integers. Both are challenged under the assumption that quantum computers with several thousand qubits width may, at some stage, exist. Two examples of challenged cryptosystems are the public-key cipher Rivest–Shamir–Adleman (RSA) [19], and the Diffie-Hellman key exchange [20]. The corresponding attack is due to an algorithm proposed by Shor in 1994 [21], which solves the factorization problem in efficient time. With symmetric encryption, and due to certain properties of quantum mechanics, such as superposition, entanglement, and the quantum no-cloning theorem, secure communication in a quantum world is still possible. The most prominent example of the latter field of quantum cryptography is QKD [22].

7.3.1 Formal Cryptographic Frameworks

In modern cryptography, security proofs are usually given in a theoretical framework that allows for careful analysis and easy understanding. The most common framework is the game-based framework [23], which formalizes a game between the adversary and a challenger. Security is proved by probabilistic analysis of the adversary winning the game. The universal composability (UC) [24] extends on this model. It guarantees the security of the protocol in an arbitrarily adversarial environment. The adversary is considered as a distinguisher who has access to an unknown system and has to guess which of the two systems (the real world and the ideal world) she is interacting with.

The intuition for direct application to quantum security approaches is as follows: for $\rho_{KE} := \sum_{k\in K} P_K(k)|k\rangle\langle k| \otimes \rho_E^k$, where a key K distributed according to P_K, $\{|k\rangle\}_{k\in K}$ is an orthonormal basis of Hilbert space $\mathcal{H}_K$, and the state ρ_E^k is the adversary's system. The quantum part (ρ_E^k) is independent of K, implying that no information on K can be learned by observing ρ_E^k. Thus, if ρ_{KE} and $\rho_U \otimes \rho_E$, where ρ_U is a fully mixed state on $\mathcal{H}_K$, are ε-close in terms of their trace distance $\delta(\rho, \sigma)$, where ρ and σ are two density operators, then the real system ρ_{KE} behaves as the ideal system $\rho_K \otimes \rho_E$. The main idea of this approach is that the ideal system is indistinguishable from the real one. The trace distance has the operational meaning that the $\frac{1}{2} + \frac{1}{2}\delta(\rho, \sigma)$ is the probability of success in distinguishing two density operators by a measurement [25].

7.3.2 Practical Source and Security

The generation of single photons with brightness, purity, and indistinguishability is essential to achieve secure, high-speed quantum cryptography. Brightness represents the maximum rate at which single photons can be emitted, and purity characterizes the probability of emitting more than one photon per pulse. In an ideal single-photon source, the probability of emitting a single photon is 100%, while the probability of emitting multiple photons is 0%. However, the source is deterministic, indistinguishable, and the repetition rate is fast, where a single photon can be emitted at any arbitrary time defined by the user. Any deviation from these ideal properties that occurs in the real world must be taken into account when assessing the security of a practical implementation [26].

7.4 Quantum Key Distribution

Quantum key distribution (QKD) aims at exploiting quantum mechanic properties to agree on secret, shared randomness between a sender and a receiver for the purpose of encrypting some plaintext. There are many ways to implement QKD protocols. They can be categorized by the way the information is prepared: prepare-and-measure or entanglement-based (see Figure 7.6), or by the type of variables: discrete variables or continuous variables [27].

7.4.1 Discrete Variable Protocols

Currently, the majority of QKD implementations are for discrete variable (DV) protocols. DV-QKD protocols are good candidates for long distance applications. Based on the dimensionality, discrete variable QKD can be divided into two groups; qubit-based QKD and qudit-based QKD. In qubit-based QKD, quantum information can be encoded in two modes, polarization or different time bins. BB84 is the first and simplest qubit-based protocol, which is introduced by Bennet and Brassard [22]. Using the idea of an entangled photon source, the E91 protocol was proposed by Ekert [28] as an entanglement-based protocol in 1991. The main point of using an entangled photon source is that the protocol can be secure even if the source (a third party other than Alice and Bob) is untrusted. The BBM92 protocol was proposed by Bennett et al. [29] shortly after the E91 protocol and is also classified as an entanglement-based protocol. B92 is another important protocol, introduced by Bennett in 1992 [28]. It differs from BB84 by using only two states, $|0\rangle$ and $|+\rangle$.

With qudit-based QKD, the transmitter can encode high-dimensional key information, thus improving the performance of the system. Differential phase shift (DPS) [30] and coherent one-way (COW) [31] protocols are two DV protocols that do not use independent

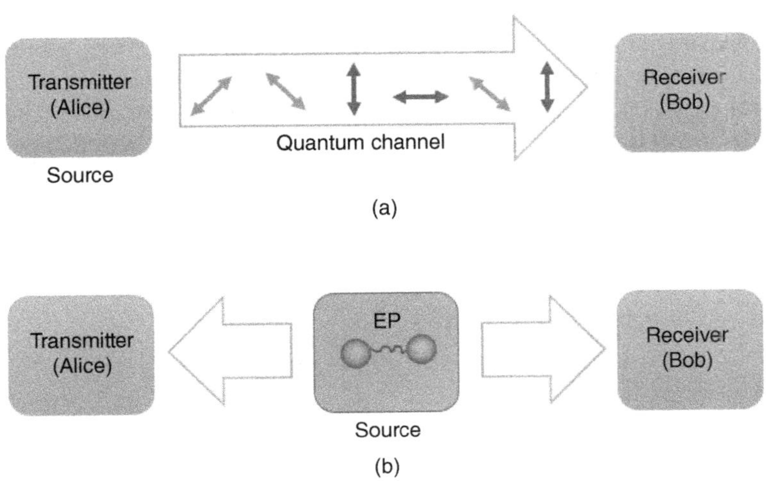

Figure 7.6 Various types of QKD. (a) Prepare-and-measure QKD, where Alice transmits quantum states to Bob through an insecure quantum channel. (b) Entanglement-based QKD, where entangled pairs are generated and distributed to Alice and Bob.

quantum states. The information is encoded in the relative phase or time-bin between light pulses, namely distributed phase protocols. The first DPS protocol was proposed in 2002 [30]. In the DPS protocol, the transmitter generates coherent states that are randomly phase modulated, and the receiver extracts the information by measuring the differential phase using an interferometer. In the COW protocol, the transmitter prepares one of three states, a coherent state followed by the vacuum, the vacuum followed by a coherent state, or a decoy state, which is two coherent states, one after another. The transmitter modulates the relative phase between the coherent states by using its phase modulator. The security proof of DPS and COW is more difficult due to the use of coherent states [32, 33]. Discrete variables can be implemented using a single-photon source and a single-photon detector. To overcome the imperfection of the single-photon source, photon-counting techniques have been applied [34].

In practice, due to noise and imperfections, the so-called real device differs from the ideal one considered in the security proof. A simple idea to remove imperfections was originally proposed by Mayers and Yao [35], called the device independent (DI) protocol. In the DI-QKD protocol [36–38], no assumptions should be made about the quantum devices used in the protocol (see Figure 7.7). If the transmitter and receiver share a quantum state with high correlation, an eavesdropper Eve cannot have a strong correlation with their quantum states. Therefore, measuring the maximally entangled state results in a uniformly distributed bit that is random from Eve's point of view. In general, the implementation of DI-QKD is more challenging due to the generation of high quality entangled states with high detection efficiency. All DI-QKD experiments [39–41] require two distant parties to perform a Bell test, typically, the Clauser–Horne–Shimony–Holt (CHSH) test [42].

A protocol for QKD to remove all detector side channels [43], called measurement device independence (MDI), was proposed in 2012 [43]. In this approach, the transmitter and receiver send quantum states to an untrusted third party that performs Bell state measurements during the protocol (see Figure 7.8). There are two types of discrete variables for MDI-QKD, the prepare-and-measure and the entanglement-based. The prepare-and-measure protocol starts with the transmitter and receiver, which randomly prepare a quantum state from $\{|0\rangle, |1\rangle, |+\rangle, |-\rangle\}$ and send it to an untrusted third party. The entanglement-based protocol is the same as prepare-and-measure protocol, except that the transmitter and receiver prepare an EPR pair and send half of this state to an untrusted party. In recent years, many discrete variable MDI-QKD [44, 45] and continuous variable MDI-QKD [46, 47] have been proposed.

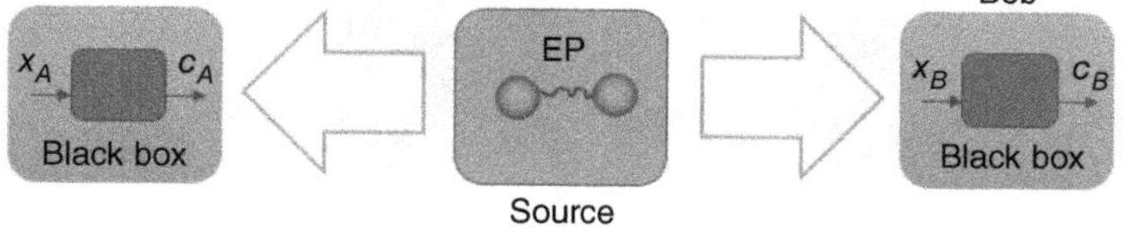

Figure 7.7 An illustration of the DI-QKD. Entangled photon pairs are distributed to Alice and Bob, who perform measurements. Alice and Bob see their quantum devices as black boxes producing classical outputs, c_A and c_B as a function of classical inputs x_A and x_B. From the observed statistics and without making any assumptions about the internal workings of the quantum devices, Alice and Bob should be able to find out whether or not they establish a secret key.

Figure 7.8 Schematic of the MDI-QKD. Alice and Bob send their quantum states to an untrusted third party, namely Charlie, for measurement.

7.4.1.1 The Example of BB84

A well-known QKD protocol is the Bennett and Brassard (BB84) protocol [22], which allows two users to generate a secure key in the presence of an eavesdropper Eve. The security proofs of QKD systems have been developed by several authors during the last decade. A quantum channel between users can be an optical fiber or free space. Single photons and weak laser pulses can be considered as quantum states. Quantum states of light can be transmitted basically without decoherence through the quantum channel, but the problem with light is losses. Losses constrain bounds on the secret key rate and the achievable distance of QKD [48, 49]. As the attenuation coefficient of air is much smaller than that of fiber, free space QKD can reach higher distances than QKD on optical fibers. More recently, satellite-based QKD between Austria and China has been demonstrated, as shown in Figure 7.9 [50].

The BB84 [22] protocol is generally considered to be the first and simplest quantum cryptography protocol. It uses quantum mechanics to share a secret key between two separate parties, namely Alice and Bob. Alice and Bob are connected via a quantum channel (e.g. optical fiber or free space) and an authenticated classical channel. It assumes that an eavesdropper, Eve, would like to obtain as much information about the secret key as possible without being detected by either party. A possible scenario for Eve could be an intercept and resend attack (IRA) strategy where she captures and measures Alice's quantum states and resends qubits to Bob according to her results. The description and the security

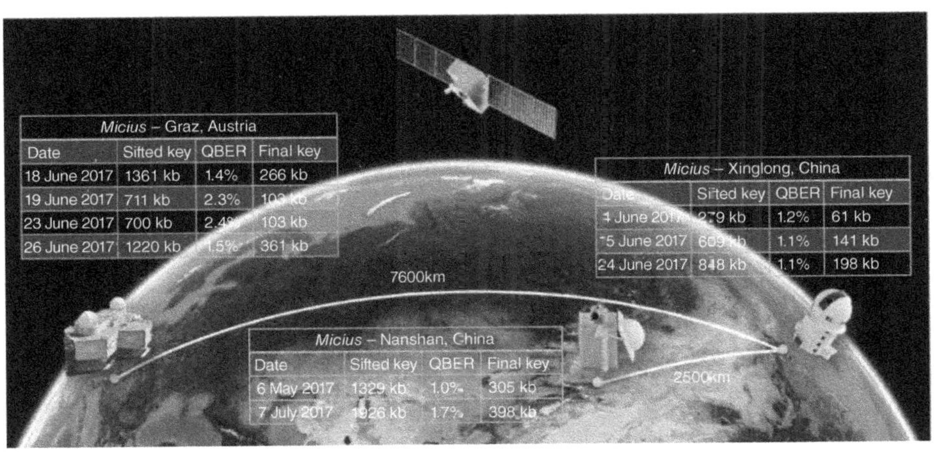

Figure 7.9 Satellite-based QKD between three cooperating ground stations between Austria and China. *Source:* Liao et al. 2018 [50]/American Physical Society.

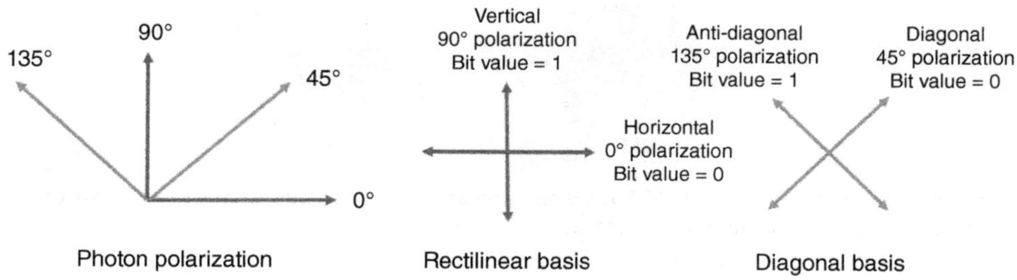

Figure 7.10 The four possible polarizations in BB84 encoding.

argument of the best-known BB84 protocol by Bennett and Brassard are explained in more detail below.

In the BB84 protocol, suppose Alice has a source of single photons whose spectral properties are well defined, and polarization is the only remaining degree of freedom. Alice and Bob align their polarizers and agree to use the polarization states called Horizontal, Vertical, Diagonal, and Anti-diagonal polarization (see Figure 7.10). The polarization state of the photon is represented by a qubit in $\mathcal{H}^2$. The Horizontal/Vertical bases, called rectilinear bases, are associated with the computational basis $\{|0\rangle, |1\rangle\}$, and the Diagonal/Anti-diagonal bases, called Diagonal basis, are associated with the Hadamard basis $\{|+\rangle, |-\rangle\}$. Horizontal and Diagonal represent bit 0, while Vertical and Anti-diagonal represent 1.

BB84 consists of the following steps (see Figure 7.11):

Alice generates a sequence of n random bits $r_1^a, \ldots, r_n^a$, called the raw key and n random bases $q_1^a, \ldots, q_n^a$. She sends a string of qubits through the quantum channel to Bob. For example, if $r_k^a = 1$ and $q_k^a = 1$ then the kth qubit that she sends is in state $|-\rangle$. Then, Bob generates random bases $q_1^b, \ldots, q_n^b$ and measures the qubits he received, $r_1^b, \ldots, r_n^b$.

Alice and Bob share their bases, q_k^a and q_k^b, respectively, via an authenticated classical channel. The classical channel must be authenticated and integrity checked, which means

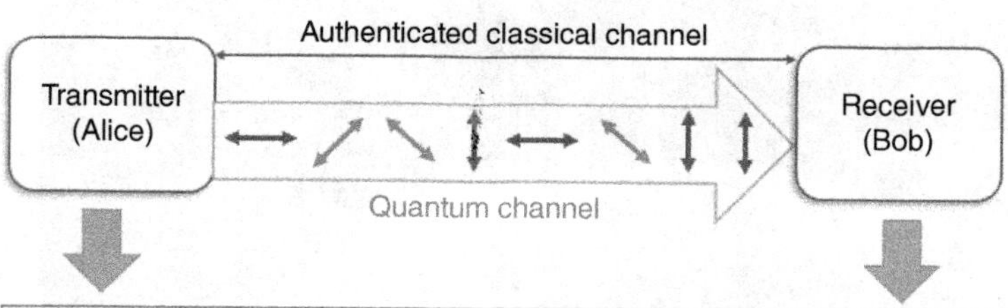

Alice's bit string	0	0	1	1	0	1	1	1
Alice's basis	+	×	×	+	+	×	+	+
Alice's photon polarization	↔	↗	↖	↕	↔	↖	↕	↕
Bob's basis	+	+	×	+	×	+	+	×
Bob's measurement	0	0	1	1	1	1	1	0
Sifted key	0	–	1	1	–	–	1	–

Figure 7.11 Diagram of the BB84 encoding.

that the identity of the sender is verified and any attempt to modify the message is detected. It does not need to have confidentiality, as the communication does not expose any useful information to the adversary. Alice and Bob only keep the bits where both have chosen the same measurement basis. This is expected to hold for around $n/2$ of the kth qubit, $q_k^a = q_k^b$. The string of bits corresponding to these qubits is called the sifted key. Alice and Bob discard the bits that were measured with incompatible bases (different q_k^a, q_k^b).

Alice randomly selects and sends $n/4$ positions with their bit values r_k^a in the sifted key. Then, Bob compares his values in those positions r_k^b with the received ones. Ideally, there should be no difference in their values. But, in practice, the errors due to the transmission are given by the quantum error rate. The quantum bit error rate (QBER) is calculated as [51]:

$$QBER = QBER_{opt} + QBER_{det} = \frac{p_{opt} + p_{noise}}{p_{photon}} \tag{7.22}$$

where p_{opt} is the probability of a photon hitting the wrong detector, and p_{noise} is the probability of dark noise (which mainly comes from the dark counts, when the detection events occur even in the darkness without any optical input) per gating pulse window.

In this way, Alice and Bob compute the QBER. If Eve intercepts the qubits Alice sends to Bob, she has only a 50% chance of selecting the right bases and resending the same prepared qubit to Bob. Thus, it lets Alice and Bob obtain 25% QBER to detect Eve. For instance, if the QBER is greater than a number p, then they detect that an eavesdropper is trying to tamper with the channel and they abort. p is a certain security threshold of the protocol.

Alice and Bob discard the $n/4$ bits that were used for computing the error rate. With the remaining $n/4$ bits, they perform post-processing steps called key reconciliation and privacy amplification. Key reconciliation uses error correction techniques [52–54] applied to ensure that an identical key is shared between Alice and Bob. After the reconciliation step, the eavesdropper Eve may still have some information about the secret key. The privacy amplification [55] is processed to ensure that this information is destroyed.

7.4.1.2 Security Argument

A QKD protocol consists of an interaction between two trusted parties, Alice and Bob. At the end of the protocol, each party outputs a secret key of length ℓ, K_A for Alice and K_B for Bob. The goal of the eavesdropper, Eve, is to obtain as much information as possible about Alice and Bob's secret keys without being detected (i.e. in the case where the protocol is not being aborted).

Correctness, secrecy, and overall security of a QKD protocol are defined as follows [56]:

Definition 7.1 ***(Correctness)*** A QKD protocol is said to be ε_{corr}-correct, if Alice and Bob's secret keys are identical with probability at least $1 - \varepsilon_{corr}$ ($Pr(K_A \neq K_B) \leq \varepsilon_{corr}$).

Definition 7.2 ***(Secrecy)*** A QKD protocol is said to be ε_{sec}-secret if, for a key of length ℓ, the key is unknown to Eve; otherwise, the protocol is aborted by detecting the existence of Eve.

$$(1 - Pr[abort])\, \|\rho_{K_AE} - \rho_{U_\ell} \otimes \rho_E\| \leq \varepsilon_{sec} \tag{7.23}$$

where $Pr[abort]$ is the probability that the protocol aborts, ρ_{K_AE} is the join state of the classical key K_A and Eve's quantum state at the end of the protocol that is called real key and describes the real situation, ρ_{U_ℓ} is the uniform key (the quantum state which is maximally mixed state on ℓ bits), and ρ_E is Eve's quantum state. The combination of ρ_{U_ℓ} and ρ_E is the ideal situation for the final key.

In order to guarantee the secrecy of a key of length ℓ, we must ensure that the eavesdropper Eve still has no information about the ℓ. The secrecy of the final key depends only on the privacy amplification step.

If the protocol is ε_{corr}-correct and ε_{sec}-secret, then it is ε^s_{QKD}-correct-and-secret for any $\varepsilon^s_{QKD} \geq \varepsilon_{corr} + \varepsilon_{sec}$.

In order to prove the security, one needs to prove both the correctness and the secrecy of the protocol.

Definition 7.3 ***(Security)*** A QKD protocol is said to be $\varepsilon^s_{QKD}, \varepsilon^c_{QKD}$-secure, if:

1. (Completeness) There exists an implementation that the protocol does not abort with probability greater than $1 - \varepsilon^c_{QKD}$.
2. (Soundness) For any implementation the protocol is ε^s_{QKD}-correct and secret.

A security proof must constrain the information available to the eavesdropper Eve. One of the important computational parameters in a QKD security proof is the key rate. This requires a description of the density operator that describes the systems of Alice, Bob, and Eve.

Based on Eve's action on the quantum channel, three types of attacks are defined as individual, collective, and coherent attacks. In all these attacks, Eve is assumed to be preparing quantum ancillas. An ancillary quantum system is attached to the particles transmitted through the quantum channel from Alice to Bob. Eve applies a fixed unitary transformation to the results, and measures the quantum state.

In the individual attack, Eve applies a fixed measurement before the classical post-processing and assumes that this measurement cannot depend on the classical information that Alice and Bob exchange for error correction and privacy amplification. The intercept and resend attack is a basic eavesdropping strategy belonging to the class of individual attacks, where Eve intercepts Alice's particles and randomly measures them in the Hadamard or Computational bases. If Eve chooses the same basis as Alice, she has obtained full information without being detected. If she chooses the wrong basis, half of the time she can only get the information if Bob chooses the same basis as Alice. So, we can only extract the final secret key only if Eve has less information about Alice's bit string than Bob, $I(A : B) < I_E$ [57].

For collective attacks, Eve collects all her probes in a quantum memory and measures them together after Alice and Bob have performed error correction and privacy amplification. This strategy gives Eve more data, which could potentially increase her information about the secret key.

In the coherent attack, Eve can perform an arbitrary operation on the quantum systems established between Alice and Bob. This operation involves all particles, which makes it difficult to analyze. Many security proofs only prove security against collective attacks [58–60]. However, the Exponential de Finetti theorem [61], the postselection technique [62], and an entropic uncertainty principle [63] can be used to prove the protocol against coherent attacks.

A critical perspective on the use of QKD protocols from a network security perspective has been given in [64].

7.4.2 Continuous Variable Protocols

In CV-QKD, the quantum information is encoded in infinite-dimensional bases, such as the amplitude and phase of light, called quadrature variables. CV protocols typically use one of two types of states, like coherent states. The processes of encoding and decoding techniques in the CV-QKD are called modulation and demodulation, respectively. Homodyne and heterodyne detectors are used as detectors to randomly measure one of the quadratures of the received signal. In contrast to DV-QKD, CV-QKD systems have low-cost implementation, easier encoding and decoding, and better performance, such as generating higher key rate in channel with low loss channel [47], thus, making them candidates for the future quantum internet. Despite all these advantages, CV-QKD faces challenges at long distances. Physical limitations include imperfections in the transmitter and receiver devices, path loss, and environmental noise, which reduce the signal-to-noise ratio (SNR). Low SNR makes reconciliation difficult and reduces its efficiency. Quantum repeaters (QRs) [65] and noiseless linear amplifiers (NLAs) [66] are good candidates for overcoming path loss and environmental noise.

The first Gaussian QKD was proposed in 2001 [67]. A year later, another Gaussian QKD based on coherent states and homodyne detection was proposed [68]. Based on the first idea, other types of CV-QKD protocols have been proposed [69, 70].

7.5 Integrated Physical Layer Security

Security in face of an omnipotent adversary is impossible, and protecting communications always implies assumptions that bound the adversaries in some sense. The accepted approach in cryptography is to bound the computational resources of an adversary, and to analyze the security of cryptographic constructions against adversaries that are assumed to be bound in probabilistic polynomial time (PPT) or, given the possible implementation of sufficiently powerful quantum computers of sufficient size, in bounded-error quantum polynomial time (BQP). If the parameters of the cryptographic construction are sufficiently large (e.g. if the key is long enough), the schemes are proven to be secure by demonstrating that a worst-case attack would take longer than the lifespan of the sun to succeed.

Physical layer security, and quantum-based security, for that matter, make an alternative assumption: they do not necessarily bound the adversary in their computation, but rather consider limitations in the communication or sensing capabilities of potential adversaries. Assumptions on the transmission and reception capabalities of adversaries are always optimistic in nature. In addition, many of the implied assumptions in physical layer security have been shown to be broken [71–76]. The domain, however, remains an interesting field for research, especially when considering quantum mechanics.

7.5.1 Conventional Approaches

The wiretap channel (see Figure 7.12), introduced by Wyner [77] in 1975, is one mathematical channel model where an eavesdropper taps a communication and is assumed to be constrained in their capacities. The model consists of three parties: a sender (Alice), a legitimate receiver (Bob), and an eavesdropper (Eve). Alice tries to transmit a message m over the channel to Bob, meanwhile Eve has a noisier channel than Bob and tries to extract as much information as possible from her channel output, having no constraints on her resources.

A communication system is considered secure if Eve's best attack strategy does not perform better than random guessing. In information theory, there are different security metrics (i.e. weak, strong, and semantic, the last translating to cryptographic semantic security against a PPT adversary, as described in Section 7.5), where semantic security, introduced in [78, 79], is the strongest security metric and, as argued in [78], it should be adopted as a standard security measure in information-theoretic security. There are many possible coding approaches to achieve information-theoretic physical layer security in the wiretap model.

An interesting way, which has received a lot of attention recently, is the approach with a seeded modular coding scheme for semantic security, which is suitable for a large class of channels [78, 80–82]. One advantage of this scheme is its modular structure: the reliability of the transmission is ensured by the coding layer, while the security is induced separately by a security layer (see Figure 7.12). All three parties know the applied functions in the security layer and also they use a seed that is a uniformly random number. Due to the separation, any existing error-correcting code can be embedded in the coding layer of the scheme. In [83], the seeded modular scheme was evaluated for the first time in an experimental environment with software defined radios. In [84], the scheme was tested with polar codes and various security metrics in software simulations.

The modular coding scheme has been integrated in [85] into 5G NR systems to measure the additional latency due to the security layers, which is negligibly small. An architecture for the integration of the scheme into future communication systems has

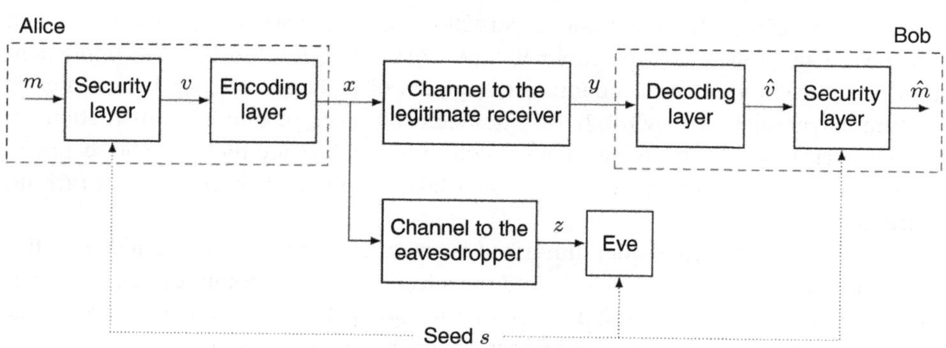

Figure 7.12 Seeded modular coding scheme for the wiretap channel, consisting of a coding layer and a security layer.

already been suggested in [85]. In [83] and [84], the best possible attack strategy within the information-theoretic model, a maximum likelihood decoder, was used for small blocklengths to evaluate the statistical security. When using the maximum likelihood decoder, increasing the blocklength and thus indirectly increasing the encoding randomness leads to prohibitively high computational complexity, so that the evaluation of the statistical security takes a very long time. In [86], a way to upper and lower bound the performance of an attack strategy, which is very close to the best attack strategy within the information-theoretic model, is shown. This approach can be used to estimate information-theoretic security within the mentioned models, especially for larger encoding randomness.

7.5.2 Quantum Approaches

Before we survey the recent results for quantum systems, we first introduce some required notation.

7.5.2.1 Notation

For a quantum system in state ρ, its (von Neumann) entropy is a measure of the amount of uncertainty associated with the quantum system. It is defined as

$$S(\rho) = -tr(\rho \log \rho) \tag{7.24}$$

The Holevo information of an ensemble of quantum states $\mathcal{N} := \{\rho_x\}_{x\in\mathcal{X}}$ and a probability distribution p on the (classical) values $x \in X$ is defined as

$$\chi(p;\mathcal{N}) := S(\overline{\rho}) - \sum_{x\in X} p(x)S(\rho_x) \tag{7.25}$$

In the context of communication, a well-known concept is that of a classical-quantum channel (cq channel). An ensemble $\mathcal{N}$ models exactly such a cq channel, and the distribution p can be understood as the distribution of code-words in a message transmission system where the transmitter uses the states of the ensemble to send data to a receiver who is allowed to perform the most general quantum measurements on the received states. The trace distance between two quantum states ρ, σ is a measure of the distance between two quantum states. It is defined as the maximum difference between the probabilities of any two outcomes when the states are measured on the same basis. It is equal to $\|\rho - \sigma\|_1/2$, where $\|\cdot\|_1$ is the one-norm.

Generally speaking, every physical medium that is termed a "channel" in the conventional information processing terminology could as well be used as a "quantum channel," the most simple example being an optical pure loss channel [[87], Equation (4.6)], and also very recent ones modeling e.g. the impact of the Kerr effect given by [88, 89]. To simplify the comparison, a conventional signal is represented as a coherent state $|\alpha\rangle$ where $\alpha \in \mathbb{C}$ models a continuous-wave laser beam of intensity proportional to $|\alpha|^2$ and phase $\arg\alpha$, where the proportionality constant depends on the frequency of the wave [90]. The pure-loss channel of transmissivity $\tau \in [0, 1]$ will act on such a state as $|\alpha\rangle \to |\tau \cdot \alpha\rangle$, but the quantum description encompasses the possibility of describing as well the action of a pure loss channel on non-classical states.

7.5.2.2 Quantum Physical Layer Security

The study of physical layer security was – from the perspective of quantum information theory – initiated in [91]. In recent works [92–94], semantic work has been studied as well. In particular, the works [92, 93] focus on robust coding schemes that are highly applicable in combination with a variety of established coding methods, thereby bridging the gap between information theory and cryptography in this context. The work [94] analyzes semantic security for general infinite-dimensional quantum systems, thereby allowing to bridge the gap between information theory and signal processing.

The works [92, 93] build upon the idea of a shared random variable between Alice and Bob, which allows them to efficiently deal with a variety of channel models. Works such as [95] show the potential of such shared random variable for the specific model of the arbitrarily varying classical-quantum channel and for the strong secrecy criterion. Both [92] and [93] employ the strong secrecy criterion as well, but take a significant step forward in the sense of applicability and generality of the results. As is visible from the literature in purely classical communication scenarios [96, 97], the existence of a shared common random variable helps Alice and Bob to establish robust communication over the transmission link between them. For the sake of achieving robust communication only, the shared random variable does not need to be kept secret from Eve. In contrast to standard methods of information theory, the so-called "derandomization" step can sometimes not be performed at all [98], proving the superiority of communication systems that can utilize pre-shared correlating resources. This observation has also been made for quantum communication systems [99–101].

The method described by [92] and [93] builds upon the large body of evidence in favor of shared, non-secret correlations. It realizes the "security layer" in Figure 7.12 via a function $f : \mathcal{X} \times \mathcal{S} \to \mathcal{A}$, which maps a pair (x, s) to a value α, which is passed into the encoder. Each such function induces a so-called "mosaic" $(D_\alpha)_{\alpha\in\mathcal{A}}$ of incidence structures, where x and s are incident in D_α if and only if $f(x, s) = \alpha$. The function f is then called the "functional form" of the mosaic $(D_\alpha)_{\alpha\in\mathcal{A}}$. In the case where every D_α is a combinatorial design, one can then specialize the analysis to the case where every point x is incident with precisely r block indices and every block index s is incident with precisely k points so that $k \cdot |\mathcal{S}| = r \cdot |\mathcal{X}|$ holds. All communicating parties own the value s. For every message x sent by Alice, Eve gets a noisy version $E \circ V(x)$ of x, where E models the encoding of an error-correcting code (E, G) consisting of encoding E and decoding G. Next, V represents the channel from Alice to Bob. The most important feature of the modular scheme is that it allows to keep a desired security level of data transmission for a variety of codes (E, G), solely by adapting f. This property can be read off [[92], Corollary 3.4], which allows adapting the parameters of f to the concatenated channel $E \circ V$. Via [[92], Corollary 3.4], it is proven that the increased number of incidence structures, that arises from increasing k, yields a higher security guarantee at the cost of a reduced communication rate.

7.6 Challenges and Known Attacks on Quantum Security

Despite the seemingly superior nature of quantum mechanics in regard to the effect of potential attacks, there are a number of challenges that arise in their context. Quantum physics implies highly beneficial properties, such as entanglement and superposition, or, for instance, the no-cloning theorem. Their effect on the security of primitives that are actually implemented, however, is not well understood. Many suggested schemes are probabilistic in nature; instead of actually preventing an adversary to break confidentiality, they rather rely on the possibility of the legitimate party to detect attacks under certain circumstances [102, 103]. However, weak measurements are another example that may require future adaptations of both design and analysis of quantum secure protocols and systems [104].

In addition, although QKD provides a string of secret bits between two separate parties, there are a number of security problems as well. In theory, QKD provides an unconditional security. The models of QKD imply, however, that all devices are ideal. This obviously does not hold true in practice, and the final secret key might be compromised because QKD implementations deviate from the theoretical models considered in the security proof. This situation is particularly serious for QKD and similar approaches that extend their trusted domain to the communication devices. Classical cryptography assumes that confidential information is kept only in memory and processed on the CPU, so any parts of the operating system or hardware beyond these components need not be trusted, as they will only handle cipher text. QKD and physical layer security, on the other hand, extend the trusted domain to the network interfaces (and to some extent even to the channel), as they are used to generate the secret for encryption, or at least detect the existence of an attack before the confidential information has been compromised. This represents a serious extension of the attack surface of the systems, which may be exploitable due to their inherent visibility to adversaries, and ease of tampering. This big gap between the theoretical model and the practical implementation is a source for various side channels that represent potential vulnerabilities to an adversary.

A variety of side-channel attacks on QKD have been proposed and experimentally demonstrated [105–111]. They can be categorized into two groups: active and passive attacks. The class of active attacks consists of all attacks in which the adversary interferes with the communication or the implementing devices. Active attacks include timing attacks [110, 111], and light injection attacks [105–109]. On the other hand, the class of passive attacks implies that adversaries are merely observing or measuring communication or devices, without active interference. One prominent example is the power attack [112].

In QKD systems, the receiver usually uses two separate detectors, namely Detector_0 and Detector_1, for detecting bits 0 and 1, respectively. The detection efficiency for bits 0 and 1 is not equal to each other. This can be due to manufacturing characteristics or to interference by an adversary Eve (see Figure 7.14). As it is shown in Figure 7.13, at time t_0 the detection efficiency for bit 0 is much higher than bit 1. Therefore, Eve can shift the arrival time of each signal to either t_0 (negative shift) or t_1 (positive shift) and steal information without being detected by Alice and Bob. As a countermeasure, the receiver uses only one

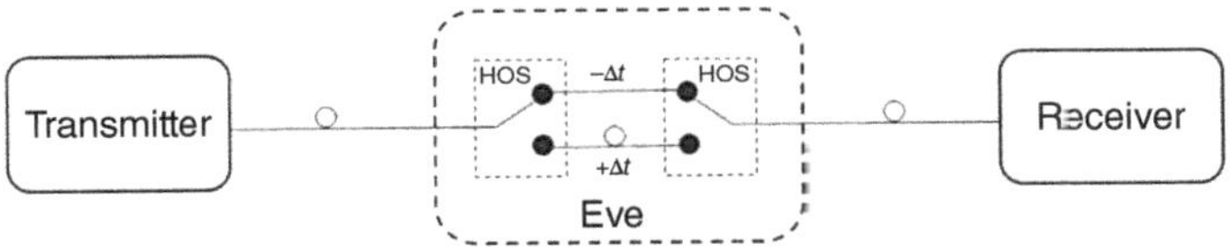

Figure 7.13 Timing attack. The scheme of Eve's attack. HOS: high-speed optical switch.

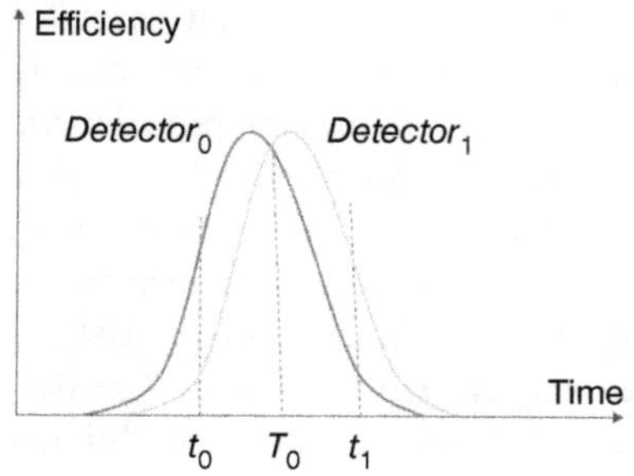

Figure 7.14 A diagram of the time-dependent efficiency of the two Detectors: at time t_0, $Detector_0$ is more sensitive than $Detector_1$ to the incoming photon.

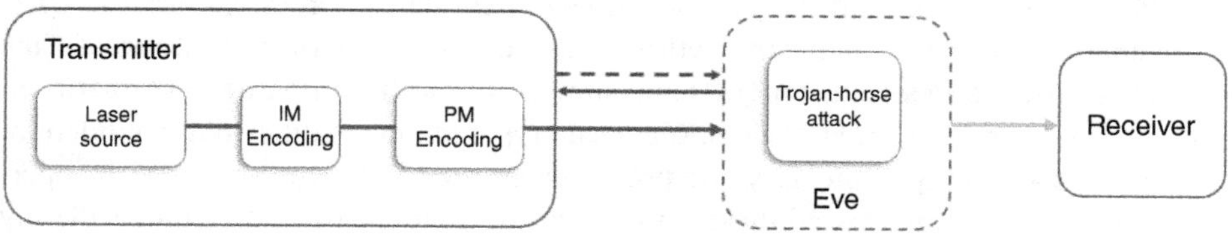

Figure 7.15 Trojan-horse attack. An attacker injects a bright light into transmitter's unit and analyzes the back-reflected light to extract information about the initial basis, phase modulator, and intensity modulator (IM) setups.

detector to detect bit 0 or bit 1 instead of using two separate detectors (Figure 7.14). Another solution might be to have strict checks for detection rate [110] and loss on the quantum channel. Phase shift setting to the receiver's phase modulator for a narrow time period [113] is another potential solution.

Unfortunately, it has been demonstrated experimentally that an active adversary, Mallory, can modify the characteristics of the components. Light injection attacks in the source, such as Trojan-horse attacks [107–109], laser-seeding attacks [106, 114], and laser-damage attacks [105] are examples. In the Trojan-horse attack (see Figure 7.15), Mallory sends a bright pulse into the transmitter's (receiver's) unit and monitors a suitable back-reflected pulse of the optical components, which might reveal information about the secret key. In this attack, Mallory tries to correct her signal with the quantum state prepared by the legitimate users. More recently, it has been shown that the injection of bright light into the cavities of some spacial lasers increases the intensity of the transmitted (received) pulses. Therefore, a countermeasure against this attack is to actively monitor the incoming light in transmitter's (receiver's) unit by putting in spectral filters. Two additional countermeasures are commonly discussed. One is adding an isolator, a passive component, whereas the other requires installing a detector, an active component. All the recent works perform Trojan-horse attacks on the standard wavelength ($\lambda = 1550$ nm), which is used by the transmitter (receiver) for their communication. It is crucial to study the threat of these attacks at different wavelengths and design suitable countermeasures.

Another light injection attack, the laser-seeding attack, is proposed in [106, 114]. In contrast to the Trojan-horse attack, it affects the operation of the laser diode in the transmitter unit. An adversary can indeed directly increase the intensity or modify the wavelength of the quantum states instead of correcting her signals with such a quantum state. To prevent this attack, the transmitter and receiver can limit the power of injected light to a suitable value [109, 115]. It has also been discussed to monitor the incoming light via the transmitter unit in order to detect injected attack. Worse, it has been demonstrated that a classical detector might be damaged by an adversary's light with high iteration rate [116]. Note that it is possible for an attacker to combine laser-seeding attack with other attacks to increase her chance to access the secret key.

Laser-damage attacks are the final type of light injection attack. They assume a high power laser that is applied to alter the characteristics of the optical component. As a high power attack, it can break the calibration of an optical attenuator, which, in many QKD systems, is employed to prepare a weak coherent state in the source. If the optical attenuation is altered, the mean photon number assumption that guarantees the security of the secret key can be broken. An optical attenuator is the last component in the transmitter unit and is closest to the quantum channel. A high power laser shining through the source can affect the performance of the attenuator. It has recently been demonstrated [105] that attenuators do not have enough tolerance to combat such high power laser-damage attacks. Adding a special passive component [117] between the vulnerable attenuator and the quantum channel [115] could be a solution to protect the vulnerable attenuator from a laser-damage attack. An exact countermeasure to this attack is currently not known and needs to be developed.

Currently, there are two common strategies to overcome such active side-channel attacks. One is the design of novel QKD protocols, such as measurement-device-independent QKD (MDI-QKD) [118], that can remove all detector side channels. MDI-QKD protocols eliminate the security assumptions about the measurement unit. However, they make additional assumptions about the quantum state preparation phase. This is a minor drawback because the quantum states need to be verified in a protected environment outside the interference of any attacker. Another approach is to monitor and analyze the parameters of a QKD system in order to detect interference [110, 119, 120]. Neither of these approaches currently provides satisfactory security guarantees, and it is unclear as to how the required thresholds for parameters can be determined.

Considering the passive power attack, an eavesdropper may discover the secret key when observing dynamic power consumption at the legitimate parties. The parity sum algorithm is a common operation in all general error-correcting methods. Low-density parity-check (LDPC) [54], Cascade [52], and Winnow [53] are frequently suggested for this purpose. The experimental results show that monitoring the power consumption of the corresponding operation makes it possible to recover the secret key with a success rate of 100% when the noise level is low. The reason is that storing a 0 bit in a register consumes less power than storing 1. Hiding techniques are a general method to mitigate side-channel attacks. An example is to randomly insert fake operations before, during, and after the algorithm's execution [112] and shuffling of operations in the algorithm [121] so as to obfuscate the actual key bits.

7.7 Oblivious Transfer (OT)

Quantum mechanics may have further applications to security in various communication tasks. QKD, for instance, has been generalized to multiple parties with quantum conference key agreement (QCKA) [122–125], where N parties provide a common secret key. Moreover, each party has a fair influence on the common secret key, and the key should not be specified fully by any party alone. QCKA plays an essential role in the development of large-scale quantum networks. Recently, experimental implementation of QCKA has been accomplished [126]. Protecting the privacy of a party among distrustful parties is a major problem in various security domains.

Oblivious transfer is a cryptographic primitive that allows a sender (Alice) to transfer a piece of information to a receiver (Bob) in such a way that Bob does not learn which piece of information was transferred, while Alice does not learn whether Bob received the information or not. Oblivious transfer can be used as a key building block to realize various cryptographic protocols such as secure multi-party computation and electronic voting systems. There are two main types of OT: 1-out-of-2 OT and 1-out-of-N OT. In 1-out-of-2 OT, Alice has two pieces of information, and Bob can choose to receive one of them without revealing which one he chose. In 1-out-of-N OT, Alice has n pieces of information, and Bob can choose to receive one of them without revealing which one he chose.

The idea of 1-out-of-2 OT goes back to Wiesner [12], who described it as "A means for transmitting two messages either but not both of which may be received." As the publication of Wiesner's work was delayed by 15 years, this idea was eventually rediscovered and formalized in 1985 by Even et al. [127], so that it can be stated in the following form:

Alice has two bits, m_0 and m_1. Bob chooses a value $c \in \{0, 1\}$. At the end of the protocol, the following holds:

1. Bob knows m_c
2. Bob gains no further information about $m_{1 \oplus c}$
3. Alice does not know c

Quantum private queries (QPQ) is a protocol that uses quantum mechanics to implement symmetric private information retrieval (SPIR), a primitive that is similar to 1-out-of-N OT [128, 129]. Specifically, the protocols ensure that a user correctly obtains an item from a database provider without leaking the retrieval address to anyone, called user privacy. Also, a user should not get any items from the database except for the one she is interested in, called database security. In this scheme, a user and a database provider share an oblivious key in such a way that the database provider knows the whole key and the user knows only few bits of the key. As a practical model of QPQ, it has attracted wide attention and has come into the focus of various investigations [130–132].

7.7.1 Probabilistic Formulation of OT

The work [133] describes the so-called Rabin OT, which is a probabilistic description of OT. In Rabin OT, Alice has a bit b. At the end of the protocol, one of two events occurs, each with probability $\frac{1}{2}$:

1. Bob learns the value of b.
2. Bob gains no further information about the value of b (other than what she already knew).

At the end of the protocol, Bob knows which of these two events actually occurred, and Alice learns nothing.

7.7.2 Relevance of OT

Oblivious Transfer is universal for secure two-party computation [133]. Using an existing OT protocol, *any* function can be securely evaluated among two parties – namely, OT enables the following:

Protocol 2: In this protocol, Alice has some secret, $i \in I$, and Bob has some secret, $j \in J$. Both have agreed on some circuit, $f : I \times J \rightarrow S$. At the end of the protocol, the following three conditions are fulfilled:

1. Both know $f(i, j)$.
2. Bob gains no information about i other than the pair $(j, f(i, j))$.
3. Alice gains no information about j other than the pair $(i, f(i, j))$.

We will now take a look at a closely related protocol, the so-called bit commitment (BC). This protocol is an important building block in classical cryptography, but has been proven to be impossible for quantum systems.

7.7.3 Bit Commitment

BC is a cryptographic protocol where one party (Alice) can commit to a chosen bit (i.e. either 0 or 1) without revealing the chosen bit to another party (Bob), until a later time when Alice chooses to reveal the bit. The protocol should ensure that Alice cannot change the committed bit after it has been committed, and that Bob cannot learn the committed bit until it is revealed. Bit commitment is used as a building block in various cryptographic protocols, such as coin flipping and secure multi-party computation. If quantum effects play no role in the modeling, then OT implies BC [133].

Given Section 7.2 it may appear that CC conceptually implements BC. However, in [134], it has been shown that BC is, in full generality, impossible when properly formulated in the quantum domain. With the work [135] which was based on an approach developed in [136] (see also a review in [134]), a previously claimed BC protocol based on a quantum-mechanical formulation [137] was rejected as false.

7.7.4 Why Bit Commitment Is Important to Oblivious Transfer

In 1991, the paper [138] claimed an implementation of OT using a quantum-mechanical system model. A stripped-down version of the essential steps in the protocol can (following the outline given in [139]) can be given as follows:

Protocol 3:

1. Alice creates a bipartite random string $(\theta^n, x^n) \in \{+, \times\}^n \times \{0, 1\}^n$.
2. Alice sends the qubit stream q_{θ_i, x_i}, $i = 1, \dots, n$ to Bob.
3. Bob chooses (and commits to) angles ϕ^n and measures all qubits, resulting in m^n.
4. Alice picks a random $M \subset \{1, \dots, n\}$. Bob reveils θ_M and m_M.
5. Alice sends θ^n to Bob if the statistics in the previous steps are correct.

6. Bob sets $\{1, \ldots, n\} = I_0 \mathring{\cup} I_1$ where $i \in I_c \Leftrightarrow \theta_i = \phi_i$.
7. Bob send I_0 and then I_1 to Alice.
8. Alice samples two independent hash functions and sends both f_i and $s_i = f_i(x_{I_i}) \oplus m_i$ to Bob.
9. Bob recovers $m_c = s_c \oplus f_c(m_{I_c})$.

As can be seen, the use of a bit commitment starting from step 3 implies that this protocol works only under the assumed availability of BC (in which case its correctness has been established in [140]). However, BC has been proven to be non-existent when unlimited quantum resources are available for Alice and Bob [135, 141]. This observation has motivated researchers to study resource constraint models that limit for example, the storage time of quantum memories. An overview of approaches and advancements covering efforts until 2016 is provided in [139]. Due to the complexity of providing and integrating quantum memory, the quantum analysis of protocols realizing either OT or BC under hardware constraints is a promising research area that combines novel hardware solutions, classical cryptography, and quantum analysis as a method to harden protocol design.

7.8 Conclusion

The Tactile Internet aims to increase the performance requirements in terms of low latency, high security, and privacy when sending audiovisual and haptic signals (i.e. massive IoT data). This chapter introduced the concepts of Tactile Internet and the challenges that are posed to future communication networks' technologies to enable it. Next, the chapter described the main paradigms in the context of quantum technologies that may mutually promise and advantage both in security and latency terms. In this context, the main highlight was on trustworthiness, and specifically on security as a fundamental challenge.

Given the recent technological and political trends, quantum security is a key potential direction for finding possible solutions. In this sense, we introduced the necessary fundamentals of quantum mechanics, as well as the general background on security and its formalization. Building on these, the idea of QKD was described, which represents one possible approach to improve security on the physical layer.

The chapter then introduced more recent extensions to QKD, and common challenges that quantum security faces. Side-channel attacks on the system that are directly accessible to potential adversaries have been mentioned, as well as the fact that the actual protection provided by quantum security has not been very well analyzed, and depends on assumptions that, in some cases, are somewhat probabilistic.

Finally, the current research questions in the field were introduced and discussed. The most interesting issues include to that extent more complex security functionalities, such as private information retrieval and oblivious transfer, can be improved with quantum mechanics.

Acknowledgment

This work has been partially funded by the German Research Foundation (DFG, Deutsche Forschungsgemeinschaft) as part of Germany's Excellence Strategy – EXC2050/1 – Project ID 390696704 – Cluster of Excellence "Centre for Tactile Internet with Human-in-the-Loop" (CeTI) of Technische Universität Dresden. R. Bassoli and F.H.P. Fitzek also acknowledge the financial support by the Federal Ministry of Education and Research of Germany in the programme of "Souverän. Digital. Vernetzt." Joint project 6G-life, project identification numbers: 16KISK001K and 16KISK002.

References

1 Steinbach, E., Hirche, S., Ernst, M. et al. (2012). Haptic communications. *Proceedings of the IEEE* 100 (4): 937–956.

2 Arata, J., Takahashi, H., Pitakwatchara, P. et al. (2007). A remote surgery experiment between Japan and Thailand over Internet using a low latency CODEC system. Proceedings 2007 IEEE International Conference on Robotics and Automation, 953–959.

3 Holland, O., Steinbach, E., Prasad, R.V. et al. (2019). The IEEE 1918.1 "tactile internet" standards working group and its standards. *Proceedings of the IEEE* 107 (2): 256–279.

4 Hexa-X (2021). D1.2 –Expanded 6G vision, use cases and societal values—including aspects of sustainability, security and spectrum.

5 Bassoli, R., Boche, H., Deppe, C. et al. (2021). *Quantum Communication Networks*, 1e. Springer.

6 Steinbach, E., Strese, M., Eid, M. et al. (2019). Haptic Codecs for the Tactile Internet. *Proceedings of the IEEE* 107 (2): 447–470.

7 Adesso, G. and Illuminati, F. (2007). Entanglement in continuous-variable systems: recent advances and current perspectives. *Journal of Physics A: Mathematical and Theoretical* 40: 7821.

8 Braunstein, S.L. and Van Loock, P. (2005). Quantum information with continuous variables. *Reviews of Modern Physics* 77: 513.

9 Eisert, J. and Plenio, M. (2003). Introduction to the basics of entanglement theory in continuous-variable systems. *International Journal of Quantum Information* 1: 479–506.

10 Wang, X.-B., Hiroshima, T., Tomita, A., and Hayashi, M. (2007). Quantum information with Gaussian states. *Physics Reports* 448: 1–111.

11 Weedbrook, C., Pirandola, S., García-Patrón, R. et al. (2012). Gaussian quantum information. *Reviews of Modern Physics* 84: 621.

12 Wiesner, S. (1983). Conjugate coding. *SIGACT News* 15 (1): 78–88.

13 Molina, A., Vidick, T., and Watrous, J. (2013). Optimal counterfeiting attacks and generalizations for Wiesner's quantum money. In: *Theory of Quantum Computation, Communication, and Cryptography* (ed. K. Iwama, Y. Kawano, and M. Murao), 45–64. Berlin, Heidelberg: Springer-Verlag.

14 Horodecki, R., Horodecki, P., Horodecki, M., and Horodecki, K. (2009). Quantum entanglement. *Reviews of Modern Physics* 81: 865.

15 Einstein, A., Podolsky, B., and Rosen, N. (1935). Can quantum-mechanical description of physical reality be considered complete? *Physical Review* 47: 777.

16 Rozpedek, F., Yehia, R., Goodenough, K. et al. (2019). Near-term quantum-repeater experiments with nitrogen-vacancy centers: overcoming the limitations of direct transmission. *Physical Review A* 99: 052330.

17 Vernam, G.S. (1926). Cipher printing telegraph systems: for secret wire and radio telegraphic communications. *Journal of the AIEE* 45: 109–115.

18 Shannon, C.E. (1949). Communication theory of secrecy systems. *The Bell System Technical Journal* 28: 656–715.

19 Rivest, R.L., Shamir, A., and Adleman, L. (1978). A method for obtaining digital signatures and public-key cryptosystems. *Communications of the ACM* 21: 120–126.

20 Diffie, W. and Hellman, M.E. (1976). New directions in cryptography. *IEEE Transactions on Information Theory* 22: 644–654.

21 Shor, P.W. (1994). *Algorithms for Quantum Computation: Discrete Logarithms and Factoring*. IEEE.

22 Bennett, C.H. and Brassard, G. (1984). Quantum cryptography: public key distribution and coin tossing. Proceedings of the International Conference on Circuits, Systems, Signal Processing, Volume 175, 175–179.

23 Boneh, D. and Zhandry, M. (2013). *Secure Signatures and Chosen Ciphertext Security in a Quantum Computing World*. Wiley-Blackwell.

24 Müller-Quade, J. and Renner, R. (2009). Composability in quantum cryptography. *New Journal of Physics* 11: 085006.

25 Dunjko, V., Fitzsimons, J.F., Portmann, C., and Renner, R. (2014). Composable security of delegated quantum computation. In: *Advances in Cryptology –ASIACRYPT 2014*. ASIACRYPT 2014, Lecture Notes in Computer Science, vol. 8874, 406–425. Berlin, Heidelberg: Springer-Verlag.

26 Uppu, R., Pedersen, F.T., Wang, Y. et al. (2020). Scalable integrated single-photon source. *Science Advances* 6 (50): eabc8268.

27 Renner, R. and Wolf, R. (2023). Quantum advantage in cryptography. *AIAA Journal* 61 (5): 1895–1910.

28 Bennett, C.H. (1992). Quantum cryptography using any two nonorthogonal states. *Physical Review Letters* 68: 3121.

29 Bennett, C.H., Brassard, G., and Mermin, N.D. (1992). Quantum cryptography without Bell's theorem. *Physical Review Letters* 68: 557.

30 Inoue, K., Waks, E., and Yamamoto, Y. (2002). Differential phase shift quantum key distribution. *Physical Review Letters* 89: 037902.

31 Stucki, D., Brunner, N., Gisin, N. et al. (2005). Fast and simple one-way quantum key distribution. *Applied Physics Letters* 87: 194108.

32 Moroder, T., Curty, M., Lim, C.C.W. et al. (2012). Security of distributed-phase-reference quantum key distribution. *Physical Review Letters* 109: 260501.

33 Wen, K., Tamaki, K., and Yamamoto, Y. (2009). Unconditional security of single-photon differential phase shift quantum key distribution. *Physical Review Letters* 103: 170503.

34 Wang, Q., Wang, X.-B., and Guo, G.-C. (2007). Practical decoy-state method in quantum key distribution with a heralded single-photon source. *Physical Review A* 75: 012312.

35 Mayers, D. and Yao, A. (1998). Quantum cryptography with imperfect apparatus. Proceedings 39th Annual Symposium on Foundations of Computer Science (Cat. No. 98CB36280), 503–509.

36 Barrett, J., Hardy, L., and Kent, A. (2005). No signaling and quantum key distribution. *Physical Review Letters* 95: 010503.

37 Curty, M. and Lo, H.-K. (2019). Foiling covert channels and malicious classical post-processing units in quantum key distribution. *npj Quantum Information* 5: 14.

38 Masanes, L., Pironio, S., and Acín, A. (2011). Secure device-independent quantum key distribution with causally independent measurement devices. *Nature Communications* 2: 238.

39 Giustina, M., Versteegh, M.A., Wengerowsky, S. et al. (2015). Significant-loophole-free test of bell's theorem with entangled photons. *Physical Review Letters* 115: 250–401.

40 Shalm, L.K., Zhang, Y., Bienfang, J.C. et al. (2021). Device-independent randomness expansion with entangled photons. *Nature Physics* 17: 452–456.

41 Zhang, W., van Leent, T., Redeker, K. et al. (2022). A device-independent quantum key distribution system for distant users. *Nature* 607: 687–691.

42 Clauser, J.F., Horne, M.A., Shimony, A., and Holt, R.A. (1970). Proposed experiment to test local hidden variable theories. *Physical Review Letters* 24: 549.

43 Xu, F., Ma, X., Zhang, Q. et al. (2020). Secure quantum key distribution with realistic devices. *Reviews of Modern Physics* 92: 025002.

44 Tamaki, K., Lo, H.-K., Fung, C.-H.F., and Qi, B. (2012). Phase encoding schemes for measurement-device-independent quantum key distribution with basis-dependent flaw. *Physical Review A* 85: 042307.

45 Xu, F., Curty, M., Qi, B., and Lo, H.-K. (2014). Measurement-device-independent quantum cryptography. *IEEE Journal of Selected Topics in Quantum Electronics* 21: 148–158.

46 Ma, H.-X., Huang, P., Bai, D.-Y. et al. (2019). Long-distance continuous-variable measurement-device-independent quantum key distribution with discrete modulation. *Physical Review A* 99: 022322.

47 Pirandola, S., Ottaviani, C., Spedalieri, G. et al. (2015). High-rate measurement-device-independent quantum cryptography. *Nature Photonics* 9: 379–402.

48 Pirandola, S., Laurenza, R., Ottaviani, C., and Banchi, L. (2017). Fundamental limits of repeaterless quantum communications. *Nature Communications* 8: 1–15.

49 Takeoka, M., Guha, S., and Wilde, M.M. (2014). Fundamental rate-loss tradeoff for optical quantum key distribution. *Nature Communications* 5: 1–7.

50 Liao, S.-K., Cai, W.-Q., Handsteiner, J. et al. (2018). Satellite-relayed intercontinental quantum network. *Physical Review Letters* 120: 030501.

51 Gisin, N., Ribordy, G., Tittel, W., and Zbinden, H. (2002). Quantum cryptography. *Reviews of Modern Physics* 74: 145.

52 Brassard, G. and Salvail, L. (1993). Secret-key reconciliation by public discussion. Workshop on the Theory and Application of Cryptographic Techniques, 410–423.

53 Buttler, W.T., Lamoreaux, S.K., Torgerson, J.R. et al. (2003). Fast, efficient error reconciliation for quantum cryptography. *Physical Review A* 67: 052303.

54 Gallager, R. (1962). Low-density parity-check codes. *IRE Transactions on Information Theory* 8: 21–28.

55 Bennett, C.H., Brassard, G., Crépeau, C., and Maurer, U.M. (1995). Generalized privacy amplification. *IEEE Transactions on Information Theory* 41: 1915–1923.

56 Renner, R. (2008). Security of quantum key distribution. *International Journal of Quantum Information* 6: 1–127.

57 Csiszár, I. and Korner, J. (1998). Broadcast channels with confidential messages. *IEEE Transactions on Information Theory* 24: 339–348.

58 Acín, A., Brunner, N., Gisin, N. et al. (2007). Device-independent security of quantum cryptography against collective attacks. *Physical Review Letters* 98: 230501.

59 Biham, E., Boyer, M., Brassard, G. et al. (2002). Security of quantum key distribution against all collective attacks. *Algorithmica* 34: 372–388.

60 Ghorai, S., Grangier, P., Diamanti, E., and Leverrier, A. (2019). Asymptotic security of continuous-variable quantum key distribution with a discrete modulation. *Physical Review X* 9: 021059.

61 Renner, R. (2007). Symmetry of large physical systems implies independence of subsystems. *Nature Physics* 3: 645–649.

62 Christandl, M., König, R., and Renner, R. (2009). Postselection technique for quantum channels with applications to quantum cryptography. *Physical Review Letters* 102: 020504.

63 Tomamichel, M., Lim, C.C.W., Gisin, N., and Renner, R. (2012). Tight finite-key analysis for quantum cryptography. *Nature Communications* 3: 1–6.

64 Yuen, H.P. (2016). Security of quantum key distribution. *IEEE Access* 4: 724–749.

65 Seshadreesan, K.P., Krovi, H., and Guha, S. (2020). Continuous-variable quantum repeater based on quantum scissors and mode multiplexing. *Physical Review Research* 2: 013310.

66 McMahon, N., Lund, A., and Ralph, T. (2014). Optimal architecture for a nondeterministic noiseless linear amplifier. *Physical Review A* 89: 023846.

67 Cerf, N.J., Levy, M., and Van Assche, G. (2001). Quantum distribution of Gaussian keys using squeezed states. *Physical Review A* 63: 052311.

68 Grosshans, F. and Grangier, P. (2002). Continuous variable quantum cryptography using coherent states. *Physical Review Letters* 88: 057902.

69 García-Patrón, R. and Cerf, N.J. (2009). Continuous-variable quantum key distribution protocols over noisy channels. *Physical Review Letters* 102: 130501.

70 Weedbrook, C., Lance, A.M., Bowen, W.P. et al. (2004). Quantum cryptography without switching. *Physical Review Letters* 93: 170504.

71 Walther, P., Franz, E., and Strufe, T. (2019). Blind synchronization of channel impulse responses for channel reciprocity-based key generation. IEEE 44th Conference on Local Computer Networks (LCN), 76–83.

72 Walther, P., Janda, C., Franz, E. et al. (2018). Improving quantization for channel reciprocity based key generation. 43rd IEEE Conference on Local Computer Networks (LCN), 545–552.

73 Walther, P., Knauer, R., and Strufe, T. (2020). Passive angriffe auf kanalbasierten schlüsselaustausch. *GI SICHERHEIT*.

74 Walther, P., Richter, M., and Strufe, T. (2021). Ray-tracing based inference attacks on physical layer security. *Electronic Communications of the EASST*, Volume 80.

75 Walther, P. and Strufe, T. (2020). Blind twins: Siamese networks for non-interactive information reconciliation. IEEE 31st Annual International Symposium on Personal, Indoor and Mobile Radio Communications (PIMRC). Best Paper Award.

76 Walther, P. and Strufe, T. (2020). Inference attacks on physical layer channel state information. International Conference on Trust, Security and Privacy in Computing and Communications (TrustCom), 935–942.

77 Wyner, A.D. (1975). The wire-tap channel. *Bell System Technical Journal* 54 (8): 1355–1387.

78 Bellare, M., Tessaro, S., and Vardy, A. (2012). A cryptographic treatment of the wiretap channel. arXiv preprint arXiv:1201.2205.

79 Bellare, M. (2012). Semantic security for the wiretap channel. Advances in Cryptology–CRYPTO 2012: 32nd Annual Cryptology Conference (19–23 August 2012), 294–311. Santa Barbara, CA, USA.

80 Bellare, M. and Tessaro, S. (2012). Polynomial-time, semantically-secure encryption achieving the secrecy capacity. arXiv preprint arXiv:1201.3160.

81 Hayashi, M. and Matsumoto, R. (2016). Secure multiplex coding with dependent and non-uniform multiple messages. *IEEE Transactions on Information Theory* 62 (5): 2355–2409.

82 Wiese, M. and Boche, H. (2020). Semantic security via seeded modular coding schemes and Ramanujan graphs. *IEEE Transactions on Information Theory* 67 (1): 52–80.

83 Torres-Figueroa, L., Mönich, U.J., Voichtleitner, J. et al. (2021). Experimental evaluation of a modular coding scheme for physical layer security. In: *2021 IEEE Global Communications Conference (GLOBECOM)*, 1–6.

84 Frank, A., Voichtleitner, J., Wiese, M., and Boche, H. (2022). Implementation of a modular coding scheme for secure communication. ICC 2022 - IEEE International Conference on Communications, 2900–2905.

85 Torres-Figueroa, L., Hörmann, M., Wiese, M. et al. (2022). Implementation of physical layer security into 5G NR systems and E2E latency assessment. GLOBECOM 2022 - 2022 IEEE Global Communications Conference, Rio de Janeiro, Brazil, 4044–4050. https://doi.org/10.1109/GLOBECOM48099 2022.10001457.

86 Voichtleitner, J., Wiese, M., Frank, A., and Boche, H. (2023). Statistical verification of upper and lower bounds for the security performance of wiretap channels. 2023 IEEE Wireless Communications and Networking Conference (WCNC), Glasgow, UK, 1–6. https://doi.org/10.1109/WCNC55385.2023.10119098.

87 Ivan, J.S., Sabapathy, K.K., and Simon, R. (2011). Operator-sum representation for bosonic Gaussian channels. *Physical Review A* 84: 042311.

88 Dehdashti, S., Notzel, J., and van Loock, P. (2022). Quantum capacity of a deformed bosonic dephasing channel.

89 Kunz, L., Paris, M.G.A., and Banaszek, K. (2018). Noisy propagation of coherent states in a lossy Kerr medium. *Journal of the Optical Society of America B: Optical Physics* 35 (2): 214–222.

90 Banaszek, K., Kunz, L., Jachura, M., and Jarzyna, M. (2020). Quantum limits in optical communications. *Journal of Lightwave Technology* 38 (10): 2741–2754.

91 Devetak, I. and Shor, P.W. (2005). The capacity of a quantum channel for simultaneous transmission of classical and quantum information. *Communications in Mathematical Physics* 256 (2): 287–303.

92 Boche, H., Cai, M., Deppe, C. et al. (2022). Semantic security for quantum wiretap channels. *Journal of Mathematical Physics* 63 (9): 092204.

93 Boche, H., Cai, M., and Wiese, M. (2022). Mosaics of combinatorial designs for semantic security on quantum wiretap channels. 2022 IEEE International Symposium on Information Theory (ISIT), 856–861.

94 Frey, M., Bjelaković, I., Nótzel, J., and Stańczak, S. (2022). Semantic security with infinite dimensional quantum eavesdropping channel. 2022 IEEE Information Theory Workshop (ITW), 720–725.

95 Boche, H., Cai, M., Deppe, C., and Nötzel, J. (2017). Classical-quantum arbitrarily varying wiretap channel: secret message transmission under jamming attacks. *Journal of Mathematical Physics* 58 (10): 102203.

96 Nötzel, J., Wiese, M., and Boche, H. (2015). The arbitrarily varying wiretap channel - secret randomness, stability and super-activation. 2015 IEEE International Symposium on Information Theory (ISIT), 2151–2155.

97 Wiese, M., Nötzel, J., and Boche, H. (2015). The arbitrarily varying wiretap channel-communication under uncoordinated attacks. 2015 IEEE International Symposium on Information Theory (ISIT), 2146–2150.

98 Ahlswede, R. (1978). Elimination of correlation in random codes for arbitrarily varying channels. *Zeitschrift ffir Wahrscheinlichkeitstheorie und verwandte Gebiete* 44: 159–175.

99 Ahlswede, R., Bjelaković, I., Boche, H., and Nötzel, J. (2013). Quantum capacity under adversarial quantum noise: arbitrarily varying quantum channels. *Communications in Mathematical Physics* 317: 103–156.

100 Ahlswede, R. and Blinovsky, V. (2007). Classical capacity of classical-quantum arbitrarily varying channels. *IEEE Transactions on Information Theory* 53 (2): 526–533.

101 Boche, H., Cai, M., Nötzel, J., and Deppe, C. (2019). Secret message transmission over quantum channels under adversarial quantum noise: secrecy capacity and super-activation. *Journal of Mathematical Physics* 60 (6): 062202.

102 Aggarwal, D., Brennen, G.K., Lee, T. et al. (2017). Quantum attacks on bitcoin, and how to protect against them. *arXiv preprint arXiv:1710.10377.*

103 Inayat, U., Zia, M.F., Mahmood, S. et al. (2022). Learning-based methods for cyber attacks detection in IoT systems: a survey on methods, analysis, and future prospects. *Electronics* 11 (9): 1502.

104 Monroe, J.T., Halpern, N.Y., Lee, T., and Murch, K.W. (2021). Weak measurement of a superconducting qubit reconciles incompatible operators. *Physical Review Letters* 126 (10): 100403.

105 Huang, A., Li, R., Egorov, V. et al. (2020). Laser-damage attack against optical attenuators in quantum key distribution. *Physical Review Applied* 13: 034017.

106 Huang, A., Navarrete, Á., Sun, S.-H. et al. (2019). Laser-seeding attack in quantum key distribution. *Physical Review Applied* 12: 064043.

107 Jain, N., Anisimova, E., Khan, I. et al. (2014). Trojan-horse attacks threaten the security of practical quantum cryptography. *New Journal of Physics* 16: 123030.

108 Navarrete, Á. and Curty, M. (2022). Improved finite-key security analysis of quantum key distribution against Trojan-Horse attacks. *Quantum Science and Technology* 7 (3): 035021.

109 Wang, W., Tamaki, K., and Curty, M. (2018). Finite-key security analysis for quantum key distribution with leaky sources. *New Journal of Physics* 20: 083027.

110 Zhang, Y., Coles, P.J., Winick, A. et al. (2021). Security proof of practical quantum key distribution with detection-efficiency mismatch. *Physical Review Research* 3: 013076.

111 Zhao, Y., Fung, C.-H.F., Qi, B. et al. (2008). Quantum hacking: experimental demonstration of time-shift attack against practical quantum-key-distribution systems. *Physical Review A* 78: 042333.

112 Lee, J. and Han, D.-G. (2020). Security analysis on dummy based side-channel countermeasures—case study: AES with dummy and shuffling. *Applied Soft Computing* 93: 106352.

113 Takesue, H., Diamanti, E., Honjo, T. et al. (2005). Differential phase shift quantum key distribution experiment over 105 km fibre. *New Journal of Physics* 7: 232.

114 Pang, X.-L., Yang, A.-L., Zhang, C.-N. et al. (2020). Hacking quantum key distribution via injection locking. *Physical Review Applied* 13: 034008.

115 Lucamarini, M., Choi, I., Ward, M.B. et al. (2015). Practical security bounds against the Trojan-Horse attack in quantum key distribution. *Physical Review X* 5: 031030.

116 Makarov, V., Bourgoin, J.-P., Chaiwongkhot, P. et al. (2016). Creation of backdoors in quantum communications via laser damage. *Physical Review A* 94: 030302.

117 Todoroki, S.-i. and Inoue, S. (2004). Observation of blowing out in low loss passive optical fuse formed in silica glass optical fiber circuit. *Japanese Journal of Applied Physics* 43: L728.

118 Lo, H.-K., Curty, M., and Qi, B. (2012). Measurement-device-independent quantum key distribution. *Physical Review Letters* 108: 130503.

119 Lydersen, L., Wiechers, C., Wittmann, C. et al. (2010). Hacking commercial quantum cryptography systems by tailored bright illumination. *Nature Photonics* 4: 686–689.

120 Yuan, Z., Dynes, J., and Shields, A. (2011). Resilience of gated avalanche photodiodes against bright illumination attacks in quantum cryptography. *Applied Physics Letters* 98: 231104.

121 Park, D., Kim, G., Heo, D. et al. (2021). Single trace side-channel attack on key reconciliation in quantum key distribution system and its efficient countermeasures. *ICT Express* 7: 36–40.

122 Epping, M., Kampermann, H., and Bruß, D. (2017). Multi-partite entanglement can speed up quantum key distribution in networks. *New Journal of Physics* 19: 093012.

123 Horodecki, K., Winczewski, M., and Das, S. (2022). Fundamental limitations on the device-independent quantum conference key agreement. *Physical Review A* 105: 022604.

124 Murta, G., Grasselli, F., Kampermann, H., and Bruß, D. (2020). Quantum conference key agreement: a review. *Advanced Quantum Technologies* 3: 2000025.

125 Thalacker, C., Hahn, F., de Jong, J. et al. (2021). Anonymous and secret communication in quantum networks. *New Journal of Physics* 23: 083026.

126 Proietti, M., Ho, J., Grasselli, F. et al. (2021). Experimental quantum conference key agreement. *Science Advances* 7: eabe0395.

127 Even, S., Goldreich, O., and Lempel, A. (1985). A randomized protocol for signing contracts. *Communications of the ACM* 28 (6): 637–647.

128 Jakobi, M., Simon, C., Gisin, N. et al. (2011). Practical private database queries based on a quantum-key-distribution protocol. *Physical Review A* 83: 022301.

129 Rabin, M.O. (2005). How to exchange secrets with oblivious transfer. Cryptology ePrint Archive.

130 Gao, F. (2014). Postprocessing of the oblivious key in quantum private query. *IEEE Journal of Selected Topics in Quantum Electronics* 21: 98–108.

131 Maitra, A., Paul, G., and Roy, S. (2017). Device-independent quantum private query. *Physical Review A* 95: 042344.

132 Rao, M.P. and Jakobi, M. (2013). Towards communication-efficient quantum oblivious key distribution. *Physical Review A* 87: 012331.

133 Kilian, J. (1988). Founding crytpography on oblivious transfer. In: *Proceedings of the 20th Annual ACM Symposium on Theory of Computing*, STOC '88, 20–31. New York, NY, USA: Association for Computing Machinery.

134 Brassard, G., Crépeau, C., Mayers, D., and Salvail, L. (1997). A brief review on the impossibility of quantum bit commitment.

135 Mayers, D. (1996). The trouble with quantum bit commitment.

136 Lo, H.-K. and Chau, H.F. (1997). Is quantum bit commitment really possible? *Physical Review Letters* 78: 3410–3413.

137 Brassard, G., Crepeau, C., Jozsa, R., and Langlois, D. (1993). A quantum bit commitment scheme provably unbreakable by both parties. Proceedings of the 1993 IEEE 34th Annual Foundations of Computer Science, 362–371.

138 Bennett, C.H., Brassard, G., Crépeau, C., and Skubiszewska, M.-H. (1992). Practical quantum oblivious transfer. In: *Advances in Cryptology—CRYPTO '91* (ed. J. Feigenbaum), 351–366. Berlin, Heidelberg: Springer-Verlag.

139 Broadbent, A. and Schaffner, C. (2016). Quantum cryptography beyond quantum key distribution. *Designs, Codes and Cryptography* 78: 351–382.

140 Damgård, I., Fehr, S., Lunemann, C. et al. (2009). Improving the security of quantum protocols via commit-and-open. In: *Advances in Cryptology - CRYPTO 2009* (ed. S. Halevi), 408–427. Berlin, Heidelberg: Springer-Verlag.

141 D'Ariano, G.M., Kretschmann, D., Schlingemann, D., and Werner, R.F. (2007). Reexamination of quantum bit commitment: the possible and the impossible. *Physical Review A* 76: 032328.

8

Physical Layer Security for Terahertz Communications in Massive IoT

Vitaly Petrov[1,*], *Hichem Guerboukha*[2,*], *Zhambyl Shaikhanov*[3,*], *Edward W. Knightly*[3], *Daniel M. Mittleman*[4], *and Josep M. Jornet*[5]

[1] *Division of Communication Systems, KTH Royal Institute of Technology, Stockholm, Sweden*
[2] *School of Science and Engineering, University of Missouri - Kansas City, Kansas, MO, USA*
[3] *Department of Electrical and Computer Engineering, Rice University, Houston, TX, USA*
[4] *School of Engineering, Brown University, Providence, RI, USA*
[5] *Institute for the Wireless Internet of Things, Northeastern University, Boston, MA, USA*

8.1 Introduction

Internet-of-things (IoT) is one of the important elements of state-of-the-art 5G-grade wireless networks, as well as of prospective 6G-grade solutions. With the continuous development of intelligent connected sensors and actuators, IoT networks find important applications in multiple diverse domains, ranging from industrial to medical sectors [1]. The prospective landscape of 6G IoT systems include not only conventional sensing devices, but also control and actuation through the network, leading to innovative applications such as intelligent transportation systems consisting of connected (partially) self-driving vehicles [2], wearable and implantable devices, i.e. eXtended Reality (XR) equipment [3], and intelligent connected robots [4].

The abovementioned trend features the following challenge: On one side, the prospective IoT devices are envisioned to collect, process, store, and exchange notable amounts of sensitive data (i.e. medical data from wearable sensors facilitating digital healthcare) [5]. On the other side, the long target battery lifetime of IoT devices demands low-power solutions to be used, while the computational resources may also be limited compared to full-scale computers, such as desktops, laptops, or even modern smartphones [6]. Hence, designing efficient security mechanisms for 6G-grade IoT networks is currently an important research challenge.

The conventional ways of maintaining the secrecy and authenticity of wireless data exchange – through encryption-based solutions – will likely continue to be widely used in 6G as well. However, the security-centric characteristics of prospective networks may potentially be improved by other means, such as exploiting the nature of the wireless channel itself. This concept, often referred to as "physical layer security (PLS)" [7], suggests protecting the transmitted data already at the physical layer. The approach is based on the

* These authors contributed equally to this chapter.

Security and Privacy for 6G Massive IoT, First Edition.
Edited by Georgios Mantas, Firooz B. Saghezchi, Jonathan Rodriguez, and Victor Sucasas.

fact that if it is possible to design the system, where the target receiver always has better channel conditions than a potential attacking node, the transmitting node can exploit this fact by encoding the data in a way that only the target receiver has high chances to robustly receive and decode.

Here, the envisioned adoption of extremely directional wireless links in terahertz (THz, 300 GHz–3 THz) and sub-THz (100–300 GHz) frequency bands in 6G [8] facilitates the use of PLS mechanisms [9]. It is worth noting that the evolution of wireless communication systems brings not only benefits but also novel risks or attack vectors (i.e. by exploiting metasurfaces), as further discussed in the chapter. Therefore, the studies on this research topic are far from over, with many more technical contributions envisioned in the upcoming years.

8.2 Information Theoretic Analysis of Eavesdropping

The first study of secure communication from an information theoretic perspective dates back to Shannon and his landmark paper in 1949 [10]. In that work, he considered the problem of securely transmitting a message over a noiseless channel where legitimate communicating parties share a secret key. As illustrated in Figure 8.1a, the legitimate transmitter Alice aims to secretly send the message M to the legitimate receiver Bob. For that, Alice and Bob share a secret key K, and Alice encodes the message M with a codeword X. During the transmission, Bob observes X and recovers the message using K. Meanwhile, eavesdropper Eve also overhears X (without any degradation). It was shown that secure communication can be achieved if the mutual information between the message and the codeword is zero, namely, $I(M;X) = 0$ where $I(\cdot;\cdot)$ denotes mutual information. In other words, the message M and the codeword X must be statistically independent to yield perfect secrecy. Subsequently, Shannon demonstrated the perfect secrecy with a one-time pad procedure, computing the codeword based on the XOR operation of each message bit with a separate key bit.

Shannon's cipher system with a noiseless channel has been serving as the fundamental principle of information-theoretic security for decades. Building on that foundation, researchers have explored noisy channels, which is generally the case in practice, and reasonably relaxed some secrecy constraints, such as asymptotic (rather than exact) statistical independence between the message and Eve's observation. In the real world, wireless channels usually undergo various imperfections, such as fading, multi-path, and interference. As shown previously, these imperfections could be exploited to enable PLS [7]. That is, the underlying physical properties of the channel could be leveraged to assist secure and reliable communication without requiring any key exchange between communicating parties. These findings are particularly important for IoT applications, which generally have limited resources, such as low computational power, at the end devices, and data encryption is challenging in such scenarios [11].

Modeling a noisy channel and a keyless communication system, Wyner introduced his renowned wiretap channel studies in 1975 [12]. This is illustrated in Figure 8.1b. Unlike in Shannon's model, in this scheme, Alice encodes her message M into a codeword X^n consisting of n symbols and transmits it over a noisy channel. Intended receiver Bob observes the signal Y^n while unauthorized Eve observes the noisy version of that signal, indicated as Z^n. The objective here is twofold. The first goal is to ensure reliable recovery of the

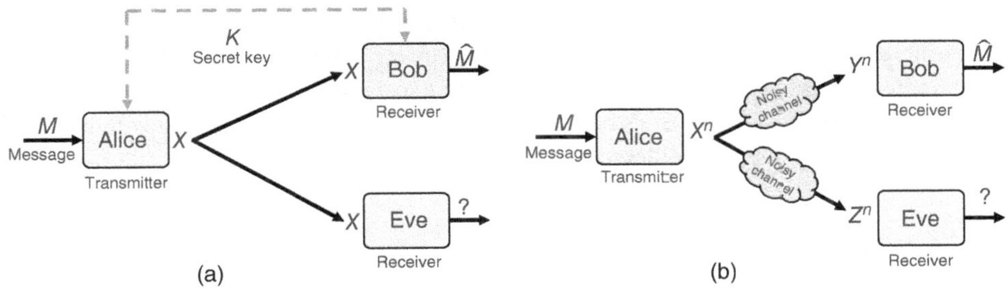

Figure 8.1 (a) Shannon's cipher model and (b) Wyner's wiretap channel.

message at Bob, namely, $P\{\hat{M} \neq M\} \xrightarrow[n\to\infty]{} 0$ where $\hat{M}$ denotes recovered message at Bob. The second goal is to provide secrecy of the message from Eve. There are different definitions of secrecy and the most frequently referred to are strong secrecy and weak secrecy [13, 14]. The strong secrecy criterion requires the amount of information leaked to Eve to vanish as n approaches infinity, namely, $\lim_{n\to\infty} I(M; Z^n)$. However, the weak secrecy relaxes that constraints and requires the rate of information leaked to Eve to vanish, that is, $\lim_{n\to\infty} \frac{1}{n} I(M; Z^n)$ [7].

Wyner jointly optimized the reliability and the confidential requirements described above and introduced the concept of secrecy capacity to characterize it [12]. The secrecy capacity, which is denoted as C_s, describes the maximum rate at which Eve gains no information about the message M after observing Z^n. Formally, it is expressed as [7]:

$$C_s = \max_{p(u,x)} I(U; Y) - I(U; Z) \tag{8.1}$$

which considers a discrete memoryless wiretap channel and an auxiliary random variable U such that $U \to X \to (Y, Z)$. The random variable U is employed to introduce artificial noise into the system and creates virtual channels from U to Y and from U to Z, respectively. The optimization is over all possible joint input distributions $p_{U,X}(u, x)$. Conceptually, the $I(U; Y)$ expression in Eq. (8.1) describes Alice to Bob's channel quality and quantifies the rate at which they can reliably communicate. The expression $I(U; Z)$, however, represents the Alice to Eve channel quality and specifies how much the rate needs to back off to ensure secure communication. Importantly, the secrecy capacity highlights the importance of Alice to Bob's channel advantage in order to ensure positive secrecy.

Perhaps the most fundamental wireless channel frequently employed in the literature is the Gaussian wiretap channel [15]. This model considers an additive white Gaussian noise in the channel. Namely, at each channel use i, Bob's and Eve's observations are defined as $Y_i = h_{Bob} X_i + N_{i,Bob}$ and $Z_i = h_{Eve} X_i + N_{i,Eve}$, respectively. The notations h_{Bob} and h_{Eve} indicate corresponding channel gains while $N_{i,Bob}$ and $N_{i,Eve}$ are zero-mean Gaussian noise with σ^2_{Bob} and σ^2_{Eve} variances. For such a channel, it was shown that the secrecy capacity yields:

$$C_s = \left[\frac{1}{2}\log\left(1 + \frac{P|h_{Bob}|}{\sigma^2_{Bob}}\right) - \frac{1}{2}\log\left(1 + \frac{P|h_{Eve}|}{\sigma^2_{Eve}}\right)\right]^+ \tag{8.2}$$

in which P denotes average transmit power and $[x]^+ = max(x, 0)$. Equation (8.2) indicates that the secrecy capacity of the Gaussian wiretap channel equals the difference between the Shannon capacity of the Alice to Bob channel and the Shanno capacity of the Alice to Eve channel [7]. Consequently, the positive secrecy can be achieved when Bob's signal-to-noise ratio (SNR) exceeds Eves', that is $|h_{Bob}|^2/\sigma_{Bob}^2 > |h_{Eve}|^2/\sigma_{Eve}^2$. In other words, Bob's observed channel must be less noisy than Eve's to yield positive secrecy. Then, by using the appropriate coding and artificial noise, Eve's best guess of the message M becomes not better than a uniformly random guess, while Bob can still reliably decode the data.

8.3 Eavesdropping THz Links

8.3.1 Security of Directional THz Links

Securing THz wireless links from eavesdropping is of paramount importance for the privacy of transmitted sensitive data. At THz frequencies, this added security is a natural consequence of the use of highly directional beams, which are necessary to counter propagation losses that are very large at high frequencies. Friis transmission equation relates the received power on the detector side P_{Rx} to the transmitted power P_{Tx}, and the antenna gains G_{Rx}, G_{Tx} of the receiver and transmitter respectively:

$$P_{Rx}^{dB} = P_{Tx}^{dB} + G_{Tx}^{dBi} + G_{Rx}^{dBi} - 20\log_{10}\left(\frac{4\pi d\nu}{c}\right) \tag{8.3}$$

where c is the speed of light, ν is the frequency, and d is the distance between the transmitter and receiver. The fourth term in Eq. (8.3) is the free-space path loss. This term scales rapidly with frequency; for example, at 300 GHz, it reaches ~80 dB over 1 m. From Eq. (8.3), there are two options to counteract these losses: either increase the radiated power of the transmitter or increase the antenna gains. The former is an ongoing research topic as THz emitters are still lacking in terms of maximal generated power (not to mention that it is not a preferred approach with respect to power management). The other solution is to increase the antenna gains. The gain is simply the product of the antenna efficiency and the antenna directivity. The antenna efficiency describes how much of the power fed to the antenna is effectively radiated in free space, while the directivity relates to how well the energy is radiated in a particular direction. Because of these, antennas used in the THz band are required to have high gain and therefore exhibit highly directional radiation patterns.

This high directivity is, of course, a challenge in terms of how to effectively communicate with a mobile user, for example. However, from the point of view of PLS, this challenge can be transformed into an opportunity. Indeed, because of the narrow nature of the link, eavesdropping is very challenging and close to impossible when the transmitted signal is sufficiently highly directional. The eavesdropping equipment required to detect and demodulate the signal is larger than the detector's aperture, so blockage would always be a concern. However, as shown by [16], this is not the only strategy for eavesdropping. They consider placing passive objects in the beam path to scatter part of the radiation. In their geometry, Alice and Bob are in a line-of-sight configuration, while Eve is placed off-axis (Figure 8.2). Eve's goal is to scatter enough signal to intercept the communication, without attenuating Bob's measured power to not alert him of her presence. In [16], the blockage is defined as:

$$b = 1 - \frac{SNR_{Bob}^{object}}{SNR_{Bob}^{no\ object}} \tag{8.4}$$

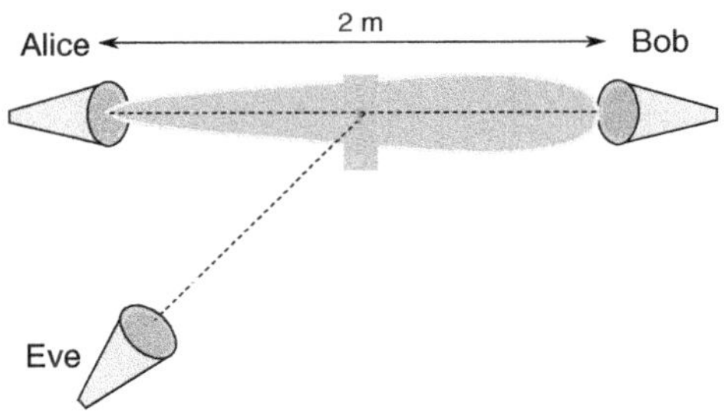

Figure 8.2 Eavesdropping geometry. Alice and Bob are in a direct line-of-sight, while Eve is off-axis and intercepts the signal using a scattering object placed in the beam path.

where *SNR* is the signal-to-noise ratio on a linear scale. For example, $b = 1/2$ if Eve's scattering object blocks half of Bob's signal (3 dB losses), and can be considered a threshold value beyond which Bob can be certain that his signal is being blocked. A normalized secrecy capacity is also defined as:

$$\overline{c}_s = \frac{\log(1 + SNR_{Bob}) - \log(1 + SNR_{Eve})}{\log(1 + SNR_{Bob})} \tag{8.5}$$

This equation relates Bob to Eve signal in such a way that if $\overline{c}_s = 1$, Eve receives no signal, and if $\overline{c}_s = 0$, Eve and Bob receive the same signal. The authors consider an arbitrary value of $\overline{c}_s = 0.5$ below which eavesdropping is possible.

Figure 8.3 shows the field measurement results obtained for different sub-THz and THz frequencies, including 100, 200, and 400 GHz. Figure 8.3a shows measured values of the blockage and the secrecy capacity as a function of the diameter of metallic cylinders placed in the center of the line-of-sight. As can be expected, when large cylinders are used, more signal is directed toward Eve. However, this also introduces enough blockage of Bob's signal, that may alert him of Eve's presence. As an alternative strategy, Eve can position her scattering object slightly off-axis as shown in Figure 8.3b. Then, in certain cases, the blockage is significantly reduced ($b < 0.5$) while allowing successful eavesdropping ($\overline{c}_s < 0.5$). These results indicate that the use of narrow, pencil-like beams is not a guarantee against eavesdropping. Attacks in the THz band are still possible, and therefore require the development of countermeasures.

8.3.2 Metasurface-in-the-Middle Attack

With the advancement toward 6G communications, new sophisticated structures such as metasurfaces are emerging to enable the highly controllable capability of high-frequency (THz) and high data rate (Tb/s) wireless links. Designed as artificially engineered 2D structures, these metasurfaces can provide customizable electromagnetic properties over the impinging waves, even beyond what is available in nature. They have been shown to improve communication performance in various ways, e.g. relaying directional signals via transparent window-embedded metasurfaces [17] and extending signal coverage via metamorphic surfaces on curtains and blinds [18]. Similarly, metasurfaces have been investigated to secure wireless links, e.g. generating Radio Frequency (RF) fingerprints [19] or physical layer keys [20], and are envisioned to be a ubiquitous part of the environment in the 6G era [21, 22].

Recently, [23] explored the notion of adversarial agents exploiting these sophisticated metasurfaces for malicious purposes in THz links. Particularly, they exposed a novel class of attacks, dubbed the "Metasurface-in-the-Middle (MSITM)" attack, and investigated the

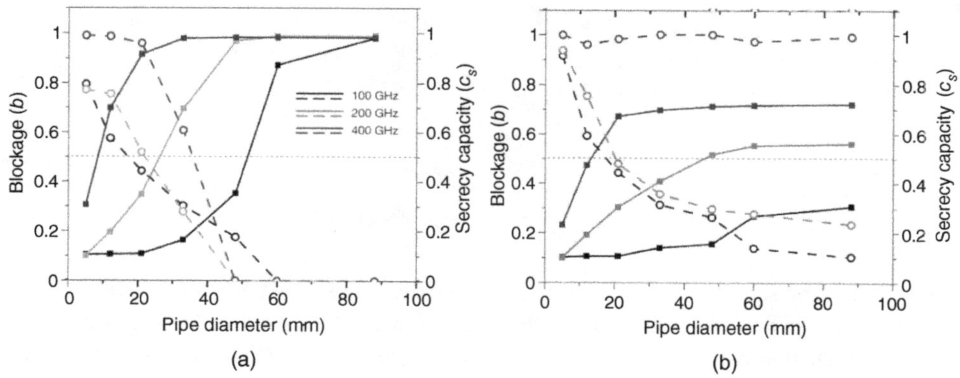

Figure 8.3 Blockage b (filled squares, left axis) and secrecy capacity $\bar{c}_s$ (open circles, right axis) as a function of the diameter of the metallic cylinder when (a) placed in the center of the line-of-sight, and (b) when moved off the center by a distance equal to radius.

threat from theoretical foundations to experimental evaluations with demonstrations in $\nu_c = 150$ GHz links. In the attack, eavesdropper Eve first places a metasurface in the path of the directional transmission between transmitter Alice and receiver Bob. The MSITM could be hidden in the environment as a "bug," e.g. concealed between (THz transparent) objects or disguised as a part of the decoration in that area. Eve then intercepts the THz transmission and stealthily manipulates the wavefront, altering the transmission radiation pattern. Specifically, she induces abrupt and targeted phase shifts at the transmissive MSITM interface. In doing so, Eve secretly establishes an eavesdropping diffractive link directed toward herself at an angle θ relative to Bob's position.

As the authors in [23] describe, the key element of the MSITM attack is the induction of the targeted phase discontinuities at the metasurface interface. Considering a surface periodic in x-axis with periodicity Γ and the phase gradient designated as $\frac{d\Phi(x)}{dx}$, the attacker's eavesdropping angle θ can be modeled, based on the generalized Snell's law [24] as:

$$\theta = \sin^{-1}\left(\frac{\frac{c}{2\pi\nu_c}\frac{d\Phi(x)}{dx} + n_\gamma \sin(\gamma)}{n_\theta}\right) \tag{8.6}$$

where c denotes the speed of light, γ is the angle of incidence of Alice's transmission relative to the surface norm, and the n_θ and n_γ designate the refractive index of the propagation environment. With zero phase gradient at the surface interface, $\frac{d\Phi(x)}{dx} = 0$, the expression in Eq. (8.6) reduces to the standard Snell's law and describes the change in the direction of the incidence wave due to different mediums. However, as highlighted in [23], Eve purposefully generates linear phase profile $\frac{d\Phi(x)}{dx} = \frac{2\pi}{\Gamma}$ and uniform amplitude transmission to establish an eavesdropping diffraction link at θ. She can also control her targeted θ by adjusting periodicity Γ in her design.

MSITM attacker has different design choices in physically realizing the aforementioned phase profile changes. Yet, the overall principle of the attacker remains similar; she (i) designs sub-wavelength scale meta-atom resonators and (ii) systematically arranges a group of meta-atoms, referred to as supercells, across the THz transparent substrate. It is demonstrated based on C-shape split ring meta-atoms design by [25]. As the authors describe, the attacker generates targeted amplitude transmission and phase shifts by controllably adjusting radius r, slit opening α (the angle of the break in the ring making it C-shape, not O-shape), and orientation β of the C-shape meta-atom. In doing so, she can obtain the entire 2π phase shift control. In addition, the attacker arranges eight different meta-atoms to form a supercell (each meta-atom inducing $\pi/4$ shift relative to the neighboring one and maintaining approximately uniform amplitude transmission). The paper [23] has demonstrated a prototype of MSITM with $r = 240$, 284, 296, and 320 μm and with $\alpha = 136°$, 82°, 32°, and 12°, respectively, and considered exemplary value of $\Gamma = 6.1$ mm.

Moreover, it has been shown in [23] that Eve could fabricate the adversarial metasurface in less than 5 min and at the cost of several cents. For that, she takes advantage of the hot-stamping technique described by [26]. As shown in Figure 8.4b–d, the fabrication process involves only few steps. First, Eve prints the metasurface on a piece of glossy paper. Next, she puts the printed design paper and the foil together and passes them through the laminator heated to 263 °F (temperature is configured based on the paper thickness and the type of the foil sheet). During the hot-stamping process, the metallic powder from the foil sheet gets transferred to the ink-printed paper, metalizing the pattern. Then, she peels off extra parts and cleans the surface with a simple tape. As noted, such a rapid fabrication method requires only standard office supplies such as a printer, laminator, and a foil sheet, reducing the attack cost to very minimal. Yet, there is a trade-off of the achievable resolution; thus, multiple trials and iterations are necessary. The MSITM prototype is shown in Figure 8.4a, while its microscopic view is illustrated in Figure 8.4e.

The effectiveness of the attack was later explored in [23] by performing an experimental evaluation with the T-Ray 5000 THz system and investigating how much signal power Eve can obtain by establishing an eavesdropping diffraction link. To do so, Alice and Bob were positioned 1m apart, while the MSITM was placed just in between them. Eve was eavesdropping from her empirically optimal location 22°. Eve's observation without the

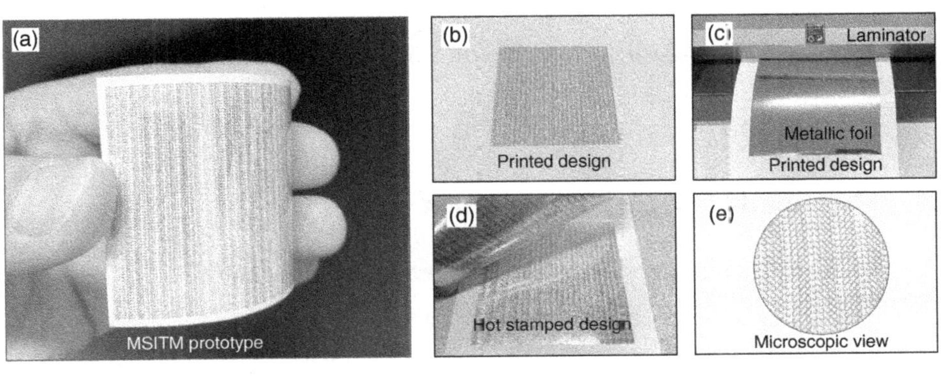

Figure 8.4 (a) Is a prototype of the MSITM. Eve employs a low-cost and rapid fabrication method that involves (b) printing the designed pattern and (c) passing printed paper patterns along with metallic foil through a laminator. The surface (d) can also be cleaned with tape to remove excessive foil powder. (e) Is a microscopic view of the fabricated metasurface.

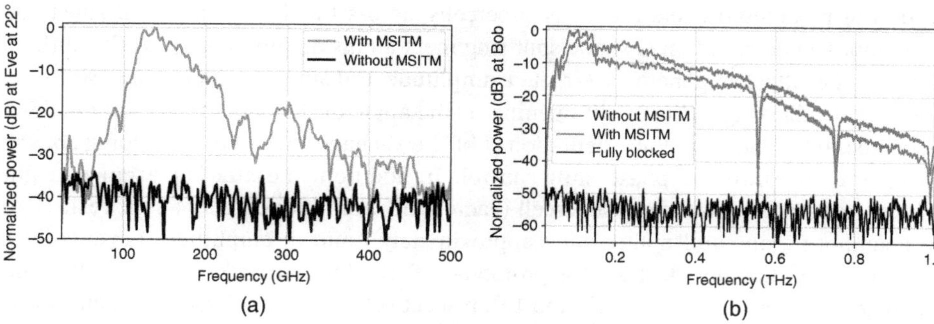

Figure 8.5 (a) MSITM attack from perspective of Eve and (b) from perspective of Bob.

metasurface was analyzed as a baseline. As depicted in Figure 8.5a, without the MSITM, Eve largely observes a noise illustrated in the black color. However, she can deflect significant signal power toward herself with the MSITM shown in the gray color. That is, she can obtain over 30 dB more signal at her targeted center frequency. The authors also investigated how specifically these diffracted signal powers contribute to the link's secrecy capacity and demonstrated that Eve can induce a large insecure (Alice–Bob link secrecy compromised) zone by carrying out such an attack.

Moreover, detecting an ongoing MSITM attack is also challenging [27]. That is, significant disruption of Bob's communication link (such as blockage) is likely to alert Bob of the attack, however, Eve's MSITM attack leaves a minimal energy footprint. The authors experimentally investigated this impact, and the results are depicted in Figure 8.5b. Observe that the power spectrum with the MSITM (the middle curve) is quite similar to the one without the MSITM (the top curve), albeit with some downward shift. That is, fluctuations in the middle curve accurately resemble the ones in the top curve, Eve inducing only a few dB (3–4 dB) power loss at Bob. The authors note that the attack impact on Bob is hardly distinguishable because a few dB power loss is characteristic of wireless channels, especially at these high frequencies. For instance, small-scale mobility or antenna misalignment also yields a similar effect. Thus, the exposed MSITM attack is devastating, as Eve can not only compromise the secrecy of the Alice–Bob link but also accurately maintain it, avoiding revealing the attack.

8.4 Multi-path THz Communications

8.4.1 Multi-path THz Communications to Improve Security Against Eavesdropping

As discussed above, exploiting the inherently directional nature of THz communication links allows us to notably improve the level of data confidentiality already at the physical layer. Still, the achievable gains are not sufficient to, for instance, completely replace the state-of-the-art encryption techniques. While the chances of successfully eavesdropping on the data exchange over a highly directional THz beam are notably lower than those over an omni or quasi-omni transmission (i.e. over microwave frequencies), the question remains open if any further improvements can be made in this area.

One of the possible solutions to enhance the security level of directional THz communications is based on the idea of utilizing several directional beams instead of a single (i.e. LoS path) beam. The rationale here is the following: in many prospective deployments (i.e. indoor wireless local networks or outdoor cellular networks), there are often multiple paths available between the nodes. For instance, the state-of-the-art 3GPP stochastic channel model for the frequencies up to 100 GHz detailed in TR 38.901 [28], assumes that there are usually in-between 5 and 12 paths termed as "ray clusters" available in a typical urban outdoor deployment. In practice, this means that (regardless of the presence of the LoS path), there are other alternative paths available i.e. the 1st-order and the second-order reflections from the surrounding objects in the environment, such as building walls and car bodies. By design, these "alternative" paths are characterized by higher losses than the preferred "best available" path (i.e. LoS path in the presence of the LoS between the nodes). However, as per [28–31], the difference is not that drastic. For the THz systems, this leads to the possibility of reliable data exchange at lower rates than the ones offered by the "best" path.

Hence, the approach introduced in [32] and termed as *multi-path THz communications* works as follows: In the presence of M alternative communication paths, the Tx node first splits the message into M *shares* and encodes these parts in such a way that the full message cannot be decoded without all M shares, i.e. by using a simple method from [33] or another applicable technique. Then, the Tx node transmits all the M shares one after another (or simultaneously, i.e. by using Multiple-Input Multiple-Output (MIMO)) over different communication paths, one share per path. The Rx node, informed in advance, of such a way to communicate sensitive data, receives all the M shares over different paths and combines them together to decode the initial message. Here, the main idea is that, while it is possible to eavesdrop on one message shared over one path, a single attacker cannot be physically present within all the paths, so it cannot eavesdrop on them all. In practice, even a coordinated group of attackers will face severe challenges when aiming to eavesdrop on all M shares over M alternative paths. However, as discussed above, this multi-path communication scheme naturally leads to lower capacity (and consequently, data rates) of the data exchange. Therefore, there is a clear security-performance trade-off introduced by this scheme that needs to be evaluated, as outlined below in this section.

8.4.2 Evaluation Framework

For the sake of analysis, we model an outdoor cellular access THz communication link between a THz access point (AP) and a THz user equipment (UE), i.e. a smartphone or an extended reality (XR) device. The nodes are assumed to be stationary (for a short duration of the data exchange, as per [34]), located d meters apart from each other (2D distance, projection) and placed at heights h_A and h_U, respectively, as illustrated in Figure 8.6. In addition to these two nodes, there are two more categories of objects present in the modeled scenario. The first group are other humans acting as potential blockers of THz radiation [35, 36]. These humans are deployed following the Poisson point process (PPP), with the density of μ humans per square meter. Following [35, 37], the humans are modeled as cylinders with a base radius of r_B and a height of height h_B. For the sake of tractability, interference is neglected in the analysis, which is a reasonable assumption in highly directional centralized control cellular systems.

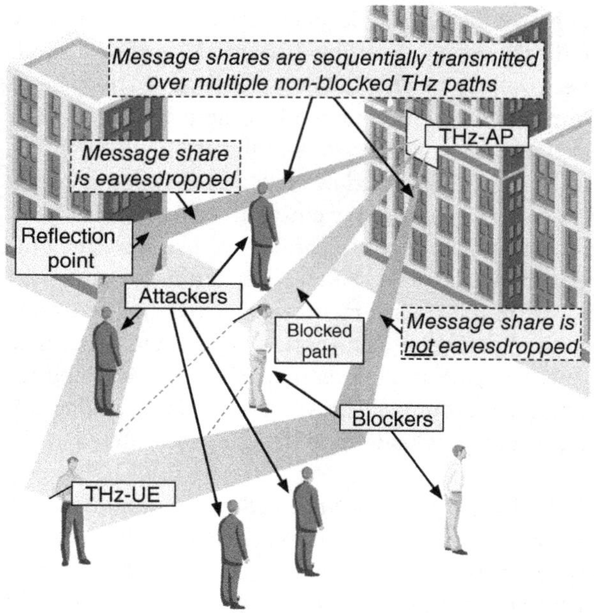

Figure 8.6 Secure multi-path THz communications.

The second group in the deployment are attackers. As the geometrical configuration of the paths between the UE and the AP may not be known in advance, the attackers are also assumed to be deployed randomly (following the PPP) around the UE with the density λ. For simplicity, the attackers are assumed not to block or reflect any THz radiation. The latter assumption is in favor of the attackers, as it has been shown i.e. above and in [16] that blocking the THz radiation by the attacker's body is possible in some cases and also reduces the chances of making the attack unnoticeable. For the antenna model, the cone-shaped circular model from [38] is used. The propagation between the nodes is modeled using the approximations presented and verified in [39]. While it is understood that the model in [39] is verified mainly for mmWave frequencies, this is the simplest stochastic model to date applicable to the discussed study.

The two schemes are compared in terms of their security and performance. The first scheme is the state-of-the-art communication approach termed as *single-path* scheme. In this scheme, the Tx node selects the best out of M available paths, namely the path currently associated with the highest SNR, and uses this path for the data exchange. The introduced *multi-path* scheme splits the message into several shares and transmits them individually over different paths, as further detailed above in Section 8.4.1. Here, a time-domain duplex is used to first transmit the first share over the first path, then the second share over the second path, and so on. The attackers cooperate, so the message is considered eavesdropped if and only if each of the message shares has been eavesdropped by any of the attackers.

A detailed mathematical analysis of the described scenario is introduced in [32] and delivers the following equations for the security and performance levels of the single-path and multi-path schemes.

The Shannon capacity for the single-path scheme is given by

$$E[C(x)] = \sum_{i=1}^{M} \left[\prod_{j=1}^{i-1} p_{B,j} \right] (1 - p_{B,i}) \int_{0}^{\infty} f_{C_i}(y; x) y dy \tag{8.7}$$

where $p_{B,i}$ is the probability of the path i (out of M) to be blocked and C_i is the individual capacity of the THz wireless link over the path i.

Meanwhile, the capacity for the multi-path scheme is determined by:

$$E[C(x)] = \frac{1 - p_B^M}{M} \sum_{i=1}^{M} \int_{0}^{\infty} f_{C_i}(y; x) y dy \tag{8.8}$$

where the notation is the same as above.

Proceeding further with the security analysis, for the considered scenario, the eavesdropping probability for the single-path scheme is given by:

$$p_E = \frac{\sum_{i=1}^{M} \left((1 - p_{B,i}) p_{E,i} \prod_{j=1}^{i-1} p_{B,j} \right)}{1 - \prod_{i=1}^{M} p_{B,i}} \tag{8.9}$$

where $p_{E,i}$ is the probability of eavesdropping the message share sent over the path i that is further computed in [32].

For the multi-path scheme, the eavesdropping probability is approximated by

$$p_E = \frac{\sum_{i=1}^{M} q_{M,i} \left(\frac{M-i}{M} p_{E,nL}^{i} + \frac{i}{M} p_{E,L} p_{E,nL}^{i-1} \right)}{1 - \prod_{i=1}^{M} p_{B,i}} \tag{8.10}$$

where $q_{M,i}$ is the probability of having i out of M clusters blocked, while $p_{E,L}$ and $p_{E,nL}$ are the eavesdropping probabilities over LoS and nLoS paths, respectively, as further detailed in [32].

8.4.3 Security vs. Capacity Trade-Off

In order to explore the trade-offs between the eavesdropping probability and link capacity, associated with the multi-path THz communications scheme, a numerical study has been conducted. The main parameters for this study are summarized in [32]. The eavesdropping probability is studied in Figure 8.7a as a function of the density of attackers (eavesdroppers). Particularly, the discussed multi-path scheme is compared to the baseline state-of-the-art single-path scheme for a different number of antenna elements at the THz AP. Here, the greater number of antenna elements leads to higher antenna gain and, consequently, narrower beamwidth.

It is first observed from Figure 8.7a that the higher number of antenna elements at the THz AP, as expected, leads to lower eavesdropping probability, as narrower beamwidth results in smaller sizes of the corresponding "eavesdropping zones" – the physical areas, where the message over a given THz path can be eavesdropped. Hence, the principal observation made earlier in this chapter holds: the narrower the THz beam, the better, on

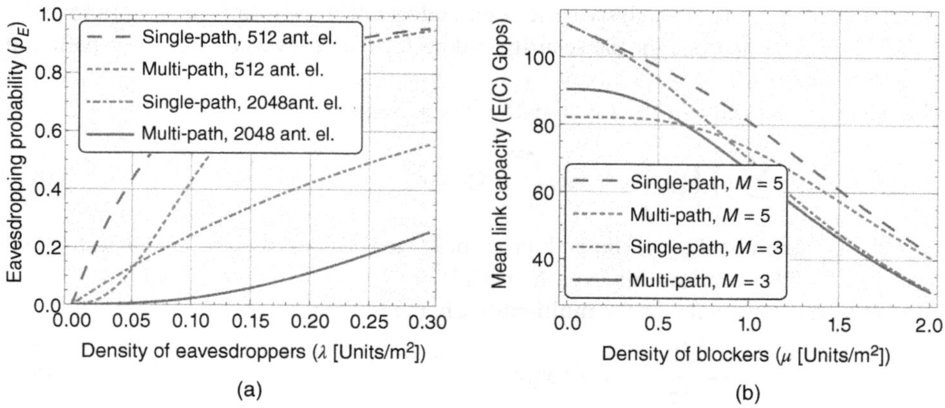

Figure 8.7 (a) Eavesdropping probability and (b) capacity of THz communications.

average, the system is protected against eavesdropping. Second, it is noted that the use of the multi-path scheme leads to a notably reduced eavesdropping probability. The latter is especially visible at low and moderate densities of eavesdroppers (i.e. $\lambda \approx 0.05$), as in these regimes the density of eavesdroppers is already sufficient to have high chances of intercepting a single THz beam, but is far from enough to intercept all the spatially diverse THz beams, each carrying a message share. Finally, it should be noted that at extremely high densities of eavesdroppers (i.e. $\lambda \approx 0.3$ for 512×512 THz AP antenna elements, resulting in approximately one eavesdropper per every three square meters), the density of eavesdroppers is sufficient to likely intercept every spatially diverse path. Hence, there is always a certain critical density of eavesdroppers, where the gain of multi-path scheme vs. single-path scheme becomes negligible. The corresponding value for 2048 antenna elements is much higher, in the order of one eavesdropper per square meter.

The improvement in security (much lower eavesdropping probability) introduced by the multi-path scheme does not come for free. Particularly, the corresponding link capacity can be compromised as illustrated in Figure 8.7b. In this figure, the mean link capacity in gigabits-per-second (Gbps) is plotted against the density of blockers, as blockers are one of the major challenges for reliable and efficient data exchange over THz frequencies. Here, it is first observed that in the absence of blockers ($\mu = 0$), the capacity of the single-path scheme is around 25–30% higher than that of the multi-path scheme. This is explained by the fact that the single-path scheme operates exclusively over the best available path (highest SNR), while the multi-path scheme communicates over several paths, hence the "average" channel is notably worse. Analyzing another extreme point, it is visible from Figure 8.7b that the capacity is decreasing rapidly with the density of blockers and, importantly, the capacity values of multi-path and single-path schemes are comparable and very low for an extremely dense crowd surrounding the UE (i.e. $\mu \approx 2$ or 2 persons per square meter). Finally, for the intermediate realistic values of μ (in-between $\mu = 0.1$ and $\mu = 0.5$), the capacity of the multi-path scheme is approximately 15% to 20% lower than the capacity of the single-path scheme. Here, it is important to note that while this reduction is non-negligible, the corresponding loss is not extreme (i.e. not several orders of magnitude). Hence, reliable and efficient data exchange is still possible with multi-path THz communications at lower data rates.

In summary, the discussed multi-path THz communication scheme allows to: (i) rapidly reduce the eavesdropping probability in certain regimes (up to dozens of times, as illustrated in Figure 8.7a); (ii) at a cost of 5% to 30% reduction in the channel capacity. Therefore, while such a performance reduction is not desired for regular data exchange, the introduced scheme still looks promising when applied selectively, i.e. for the exchange of sensitive data, including but not limited to encryption keys, online banking, medical data, and other types of personal communications.

In the following Section 8.5, another approach is described to further increase the security level of THz communications by jointly exploiting the spatial diversity and frequency-selective properties of specific THz antenna systems.

8.5 Absolute Security

Standard approaches to security generally consider some statistical assumptions about the noise. For example, Wyner's wiretap model considers the link to be secure, in an averaged sense, when Bob's signal-to-noise ratio exceeds that of Eve [12]. In [40], a new approach was proposed to security termed *absolute* security: security with *probability one* regardless of the assumption of the noise.

Their approach relies on the inherent radiation properties of Alice's antenna along with a secure coding scheme. Many antennas have multiple minima at different angular locations, effectively dividing the radiation pattern into several lobes. These minima define a blind region where an eavesdropper cannot detect the transmission with a probability of one. The fact that these minima change their angular position with frequency is key to the absolute security concept.

Indeed, the secure coding scheme relies on a frequency division of the message. In brief, it divides the message into multiple frequency channels. The message is encoded in such a way that if Eve misses one of the subchannels, she cannot decipher the message. Because different frequencies have minima at different positions, this defines a blind region in space where at least one of the channels is always missing. Then, the blind region for the total transmission band is the union of the blind regions of each subchannel. This region grows with the number of subchannels, meaning that it grows with the bandwidth. This remarkable feature contrasts with most encoding schemes, where having more security also means having less bandwidth.

To understand the secure encoding scheme, consider the simple case of three subchannels at frequencies $f_1 f_2$, and f_3. Alice maps her message into blocks and then maps each block into a symbol selected from a finite field. For example, for three subchannels, we can consider the finite field $\mathbb{F}_{11}$. Alice divides her message into blocks with length $[\log_2(11)]$, each mapped to a symbol in $\mathbb{F}_{11}$. To transmit a single message $M \in \mathbb{F}_{11}$, Alice generates two random symbols $T_1, T_2 \in \mathbb{F}_{11}$. Then, three encoded symbols $X_1, X_2, X_3 \in \mathbb{F}_{11}$ from the linear combination:

$$X_1 = M + T_1 + T_2 \tag{8.11}$$

$$X_2 = M + 2T_1 + 4T_2 \tag{8.12}$$

$$X_3 = M + 3T_1 + 9T_2 \tag{8.13}$$

The symbols X_1, X_2, and X_3 are then transmitted via the three frequency bands f_1, f_2, and f_3. Assuming that Bob is not in the blind region, he can receive the three symbols and therefore reconstruct the message M by the inversion:

$$\begin{pmatrix} M \\ T_1 \\ T_2 \end{pmatrix} = \begin{pmatrix} 3 & 8 & 1 \\ 3 & 4 & 4 \\ 6 & 10 & 6 \end{pmatrix} \begin{pmatrix} X_1 \\ X_2 \\ X_3 \end{pmatrix} \tag{8.14}$$

On the other hand, if Eve is in the blind region, she misses at least one of the symbols in the set $\{X_1, X_2, X_3\}$. Then, it can be shown that Eve cannot reconstruct M, regardless of the symbol that she is missing.

This secure encoding relies on the presence of frequency-dependent minima (or zeros) in the radiation pattern. Fundamentally, this behavior is a consequence of diffraction theory. A square aperture, for example, has an intensity in the far field that depends on the aperture width W and the frequency ν:

$$I(\theta) \propto \mathrm{sinc}^2\left[\frac{\pi W \sin(\theta)\nu}{c}\right] \tag{8.15}$$

where c is the speed of light and $\mathrm{sinc}(x) = \sin(x)/x$. From this, it is clear that minima (zeros) appear at the angular positions θ_m defined by:

$$\sin(\theta_m) = \frac{mc}{W\nu} \tag{8.16}$$

with $m = 1,2,3,\ldots$. Clearly, the position of these zeros varies as a function of frequency.

Many antennas have frequency-dependent minima, including phased array antennas and parabolic dishes. Other antennas are designed to avoid the presence of sidelobes and, therefore, the presence of zeros in the radiation pattern. In those cases, it is still possible to introduce minima with a clever design at the antenna output. Figure 8.8a shows finite element method simulations of the radiation pattern for different frequencies with a geometry consisting of a horn antenna, a lens antenna, and a beam block (shown in the inset). The horn antenna is an example of an antenna that exhibits no sidelobes. Therefore, to introduce

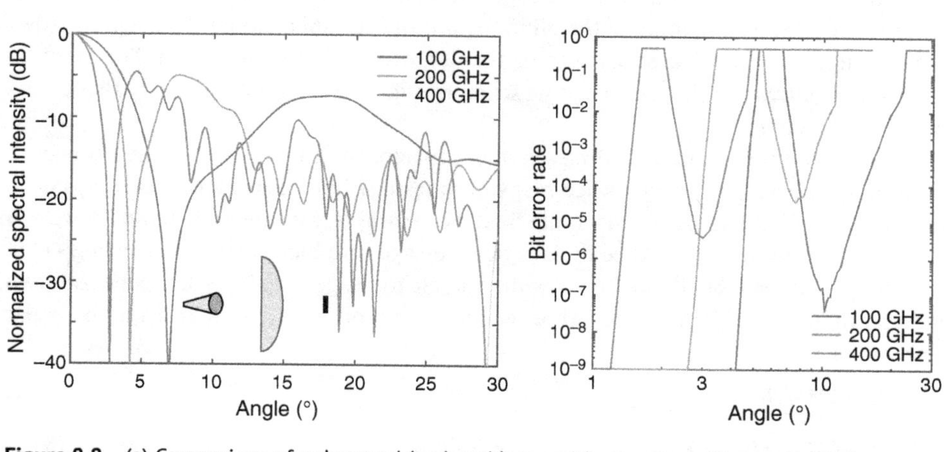

Figure 8.8 (a) Comparison of unjammed (top) and jammed (bottom) eye diagrams. (b) The geometry of the jamming (top) and radiation pattern of Bob's horn antenna (bottom).

sidelobes, a beamblock (a metallic bar) is put after the lens. This results in adding a pronounced minima (<40 dB compared to the main lobe) whose position varies as a function of frequency.

In their work, Cohen et al. experimentally realized this geometry using a communication system. They sent a 1 Gbps data stream with an on-off keying (OOK) modulation format at frequencies of 100, 200, and 400 GHz. Then, they measured the bit error rate as a function of the angle for the three frequencies as shown in Figure 8.8b. In their particular configuration, blind regions were created in the regions of $1.6° < \theta < 2.0°$ and $\theta > 3\,4°$. There, at least one of the channels was absent, which can be experimentally measured by a bit error rate of 0.5 (i.e. the 0 bit and the 1 bit cannot be distinguished). Hence, using this antenna geometry with a frequency division of the message, as discussed earlier can be a realization of absolute security.

8.6 Jamming THz Links

Up until now, the works we have presented focused on the threat of eavesdropping. Jamming is another important threat. It can be seen as a physical denial-of-service attack where the receiver (Bob) is unable to receive the information sent by the transmitter (Alice). Generally, in a wireless communication scenario, jamming refers to malicious flooding of the intended receiver by means of strong electromagnetic interference. At lower frequencies, jamming attacks are well known and have been an important concern since the Second World War. As the systems progress toward the THz band, new challenges and opportunities arise. As in the case of eavesdropping, jamming at THz frequencies is more difficult due to the narrow nature of the channel. However having large bandwidths – one of the primary benefits of THz – also leads to a new challenge since a jammer can use this bandwidth for malicious interference.

Recently, the first study of the vulnerability of THz links to jamming attacks was performed in [41]. They considered an OOK modulation scheme where the information is transmitted through amplitude modulation of the electric field, i.e. bit 1 when the pulse is present, bit 0 when it is absent. This simple type of modulation is one of the two identified physical layer modes of the IEEE 802.15.3d standard [42], and is intended for low-cost low-complexity devices that require rates up to 10 Gbps. Surprisingly, this simple communication system presents some peculiar characteristics when jammed.

In this system, the envelope detector (a Schottky diode in the case described by Shrestha et al.) is sensitive to the square of the incident electric field E_B. Assuming a single-tone jammer that emits at a frequency of ν_M, Bob then measures:

$$|E_B(t)|^2 = |A_A u_A(t)\cos(2\pi\nu_A t + \phi_A) + A_M \cos(2\pi\nu_M t + \phi_M)|^2 + N(t) \tag{8.17}$$

Here, we assumed that Alice transmits an OOK signal at a carrier frequency ν_A modulated by u_A (t) that corresponds to the data signal. A_A and A_M are the amplitudes of Alice and Mallory (the jammer) respectively, while ϕ_A and ϕ_M are their respective phases. $N(t)$ is the power expressed in power. On the detector side, Bob has a circuit that filters the high frequency and the DC components, which reduces the expression to:

$$|E_B(t)|^2 = \frac{A_A^2 u_A(t)}{2} + A_A A_M u_A(t)\cos(2\pi\Delta\nu t + \Delta\phi) + N(t) \tag{8.18}$$

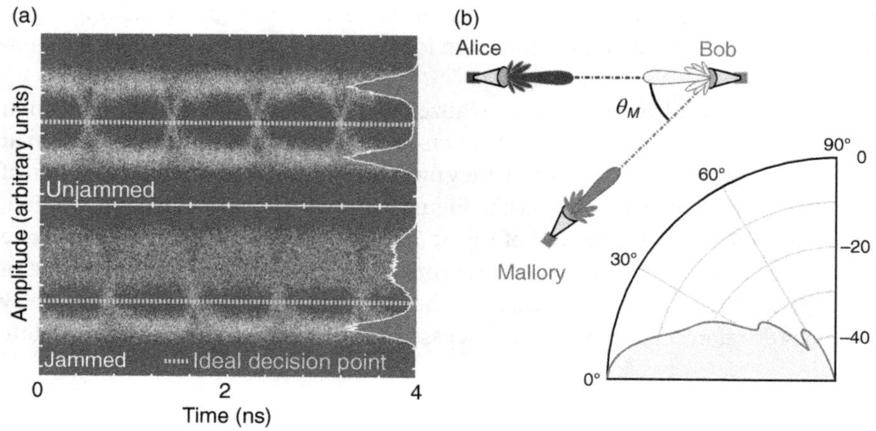

Figure 8.9 (a) Comparison of unjammed (top) and jammed (bottom) eye diagrams. The dashed lines show the optimal decision thresholds. (b) The geometry of the jamming (top) and radiation pattern of Bob's horn antenna (bottom).

where $\Delta\nu = |\nu_A - \nu_B|$ and $\Delta\phi = |\phi_A - \phi_M|$ are the differences in frequencies and phases between Alice and Mallory. In this equation, the first term is the original data transmitted by Alice (baseband signal centered at the 0 frequency), while the second term is the effect of Mallory's jamming. The latter is the product of Alice's data and a sinusoidal term oscillating at the beat frequency $\Delta\phi$. Figure 8.9a shows an example of the measured eye diagram in the presence or absence of Mallory's jamming. Remarkably, the jamming occurs only on bit 1, broadening its response, while leaving bit 0 untouched. This intriguing behavior can be understood in light of the expression (8.18). There, we can observe that the jamming term (the second term) appears only when the pulse is present, i.e. when $u_A = 1$.

This asymmetric behavior may have some other consequences on Bob's decoding of the bits. In an unjammed eye diagram where the noise affects both bits in a symmetric way, Bob usually sets his decision point at the midpoint between the two bits, meaning that when the measured signal is larger than this midpoint, the bit is considered a 1, and when it is lower it is considered as a 0. The asymmetric behavior posed by the jamming can be an additional form of attack, as Mallory could constantly change the level of jamming (by changing her power, for example) and consequently force Bob to change his decision threshold for error-free detection.

In general, the efficiency of the jamming depends on how well Mallory couples into Bob's antenna relative to how well Alice couples into the same antenna. Consider the jamming geometry shown in Figure 8.9b. Alice is located at a distance R_{AB} from Bob's detector, which she faces at 0°. Mallory is at a distance R_{MB} and attacks Bob's antenna at an angle θ_M. From Friis's equation that describes the power budget in this geometry, we write the ratio of Mallory to Alice's contribution on Bob's detection as:

$$\frac{P_B^M}{P_B^A} = \frac{P_M G_M^{\theta=0} G_B^{\theta=\theta_M}}{P_A G_A^{\theta=0} G_B^{\theta=0}} \left(\frac{R_{AB}}{R_{MB}}\right)^2 \tag{8.19}$$

where P_A and P_M are the nominal powers emitted by Alice and Mallory respectively. $G_A^{(\theta=0)}$ and $G_M^{(\theta=0)}$ are the maximal gains of Alice and Mallory antennas (usually in the front

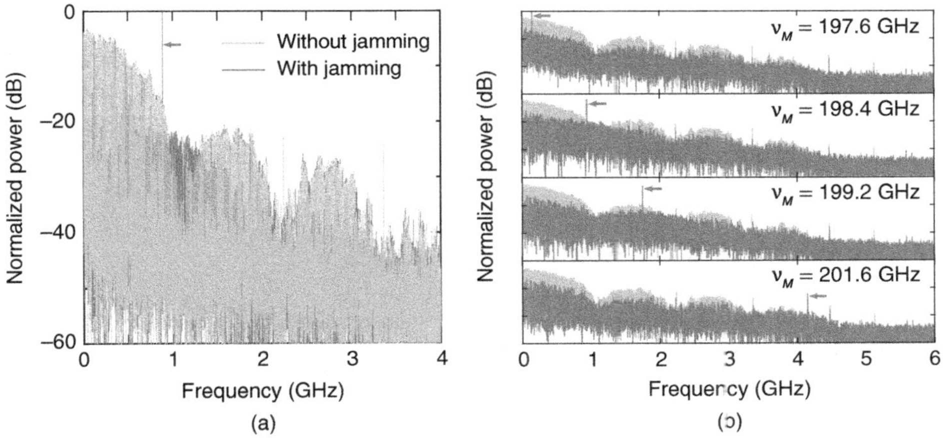

Figure 8.10 (a) Unjammed (light gray) and jammed (dark gray) spectra. (b) Jammed spectrum (light gray) and the difference between the jammed spectrum and the unjammed spectrum (dark gray). The dark gray arrow indicates the single-tone frequency.

direction $\theta = 0$), while $G_B^{(\theta=\theta_M)}$ is the gain of Bob's antenna in the direction of Mallory. This expression corresponds to how efficient the jamming is. To increase the efficacy of her attack, Mallory must increase the ratio; she can do this by either (i) increasing her power, (ii) increasing her gain, and/or (iii) reducing her distance from Bob. At the same time, to protect the link from jamming, Alice can (i) increase her power, (ii) increase her gain and/or reduce her distance from Bob. As for Bob, he can secure the link by increasing the directivity of his receiving antenna such that the gain in the front direction is larger than the gain at the side angle $G_B^{(\theta=0)} > G_B^{(\theta=\theta_M)}$. For example, consider the radiation pattern shown in Figure 8.9b that corresponds to the horn antenna used by Bob that shows a drop of 30 dB at 30°. For Mallory to have a significant effect on the jamming while at $\theta_M = 30°$, she must increase her power by 30 dB, considering that she is located at the same distance from Bob as Alice.

Another remarkable feature of jamming OOK signals is the effect on the spectrum. Figure 8.10a compares the spectra when jammed and unjammed when Mallory has her single-tone offset by 890 MHz from Alice. Mallory's jamming introduces a broad response centered at the beat frequency (shown by the arrows). This behavior can be understood by taking the Fourier transform of Eq. (8.18):

$$\mathfrak{F}\{|E_B(t)|^2\} = \frac{A_A^2}{2}\mathfrak{F}\{u_A^2(t)\} + A_A A_M \mathfrak{F}\{u_A(t)\} * \mathfrak{F}\{\cos(2\pi\Delta\nu t)\} + N(\nu) \tag{8.20}$$

where $\mathfrak{F}\{...\}$ is the Fourier transform. The second term is the convolution of Alice's data with a delta function centered at the beat frequency $\Delta\nu$. By the sifting property, this results in an additional interference term where Alice's data spectrum is moved to the position of the beat frequency and overlaps Alice's original spectrum. Figure 8.10b shows the resulting spectra when changing Mallory's frequency while Alice's frequency is fixed. As can be seen, the delta function moves toward higher frequency when Mallory increases her frequency as predicted by Eq. (8.20).

This type of jamming, referred to as beat jamming, is a consequence of Bob having an envelope detector. It appears even if Mallory only uses a single-tone signal. The size of the jamming term depends on the bandwidth B of Alice's data (which ultimately depends on the modulation format, data rate, etc.). If the beat frequency is smaller than the bandwidth ($\Delta\nu < B$), then the jamming term can significantly overlap with Alice's data. Therefore, it cannot be simply filtered out by Bob without also filtering important parts of Alice's data.

8.7 Summary and the Road Ahead

In summary, the use of highly directional THz beams allows for a notably improved, achievable level of security already at the physical layer. The observed advantages lead to novel opportunities for securing various IoT services in 6G-grade networks and systems. Specifically, battery-less IoT sensors installed in buildings and implantable small-scale IoT meters with miniature sub-THz and THz antennas may benefit from the PLS solutions to complement and even sometimes replace the encryption-based techniques involving heavy computations on the IoT devices to encrypt the transmitted message and then decrypt the received one.

At the same time, the chances of eavesdropping on the data from a single THz link are still non-negligible. There are two main outcomes from this observation. First, at the moment, it seems unlikely that the entire set of encryption-centric security mechanisms will be completely replaced with PLS methods for the entire set of IoT devices and services utilizing 6G-grade THz communications. Second, this fact opens the door to many promising enhancements to rapidly increase the achievable levels of security, including but not limited to the ones discussed in this chapter. The use of new security methods also always leads to the further development of novel attack techniques. Therefore, it is likely that new attack vectors may appear in the foreseeable future targeting specifically THz wireless communications, i.e. related to jamming or new types of man-in-the-middle attacks using metasurfaces. Still, the PLS of THz communications is a promising research direction that requires further attention within the scope of 6G-grade and beyond IoT systems.

References

1 Nguyen, D.C., Ding, M., Pathirana, P.N. et al. (2022). 6G Internet of Things: a comprehensive survey. *IEEE Internet of Things Journal* 9 (1): 359–383. https://doi.org/10.1109/JIOT.2021.3103320.

2 Andreev, S., Petrov, V., Dohler, M., and Yanikomeroglu, H. (2019). Future of ultra-dense networks beyond 5G: harnessing heterogeneous moving cells. *IEEE Communications Magazine* 57 (6): 86–92. https://doi.org/10.1109/MCOM.2019.1800056.

3 Gapeyenko, M., Petrov, V., Paris, S. et al. (2022). Standardization of extended reality (XR) over 5G and 5G-advanced 3GPP new radio. *IEEE Network* 37 (4): 22–28 https://doi.org/10.1109/MNET.003.2300062.

4 Andreev, S., Petrov, V., Huang, K. et al. (2019). Dense moving fog for intelligent IoT: key challenges and opportunities. *IEEE Communications Magazine* 57 (5): 34–41. https://doi.org/10.1109/MCOM.2019.1800226.

5 Ometov, A., Petrov, V., Bezzateev, S. et al. (2019). Challenges of multi-factor authentication for securing advanced IoT applications. *IEEE Network* 33 (2): 82–88. https://doi.org/10.1109/MNET.2019.1800240.

6 Mikhaylov, K., Stusek, M., Masek, P. et al. (2018). Multi-rat LPWAN in smart cities: trial of LoRaWAN and NB-IoT integration. 2018 IEEE International Conference on Communications (ICC), 1–6. https://doi.org/10.1109/ICC.2018.8422979.

7 Bloch, M. and Barros, J. (2011). *Physical-Layer Security: From Information Theory to Security Engineering*. Cambridge University Press.

8 Shafie, A., Yang, G.N., Han, C. et al. (2022). Terahertz communications for 6G and beyond wireless networks: challenges, key advancements, and opportunities. *IEEE Network* 37 (3): 162–169. https://doi.org/10.1109/MNET.118.2200057.

9 Gao, W., Han, C., and Chen, Z. (2020). Receiver artificial noise aided terahertz secure communications with eavesdropper in close proximity. GLOBECOM 2020 - 2020 IEEE Global Communications Conference, 1–6. https://doi.org/10.1109/GLOBECOM42002.2020.9322487.

10 Shannon, C.E. (1949). Communication theory of secrecy systems. *The Bell System Technical Journal* 28 (4): 656–715.

11 Mukherjee, A. (2015). Physical-layer security in the Internet of Things: sensing and communication confidentiality under resource constraints. *Proceedings of the IEEE* 103 (10): 1747–1761.

12 Wyner, A.D. (1975). The wire-tap channel. *The Bell System Technical Journal* 54 (8): 1355–1387. https://doi.org/10.1002/j.1538-7305.1975.tb02040.x.

13 Poor, H.V. and Schaefer, R.F. (2017). Wireless physical layer security. *Proceedings of the National Academy of Sciences of the United States of America* 114 (1): 19–26.

14 Yener, A. and Ulukus, S. (2015). Wireless physical-layer security: lessons learned from information theory. *Proceedings of the IEEE* 103 (10): 1814–1825.

15 Leung-Yan-Cheong, S. and Hellman, M. (1978). The Gaussian wire-tap channel. *IEEE Transactions on Information Theory* 24 (4): 451–456.

16 Ma, J., Shrestha, R., Adelberg, J. et al. (2018). Security and eavesdropping in terahertz wireless links. *Nature* 563 (7729): 89–93.

17 Kitayama, D., Hama, Y., Goto, K. et al. (2021). Transparent dynamic metasurface for a visually unaffected reconfigurable intelligent surface: controlling transmission/reflection and making a window into an RF lens. *Optics Express* 29 (18): 29292–29307.

18 Zelaya, R.I., Ma, R., and Hu, W. (2021). Towards 6G and beyond: smarten everything with metamorphic surfaces. Proceedings of the 20th ACM Workshop on Hot Topics in Networks, 155–162.

19 Rajendran, S., Sun, Z., Lin, F., and Ren, K. (2020). Injecting reliable radio frequency fingerprints using metasurface for the Internet of Things. *IEEE Transactions on Information Forensics and Security* 16: 1896–1911.

20 Jiao, L., Sun, G., Le, J., and Zeng, K. (2021). Machine learning-assisted wireless PHY key generation with reconfigurable intelligent surfaces. Proceedings of the 3rd ACM Workshop on Wireless Security and Machine Learning, 61–66.

21 Di Renzo, M., Zappone, A., Debbah, M. et al. (2020). Smart radio environments empowered by reconfigurable intelligent surfaces: how it works, state of research, and the road ahead. *IEEE Journal on Selected Areas in Communications* 38 (11): 2450–2525.

22 Sengupta, K., Nagatsuma, T., and Mittleman, D.M. (2018). Terahertz integrated electronic and hybrid electronic–photonic systems. *Nature Electronics* 1 (12): 622–635.

23 Shaikhanov, Z., Hassan, F., Guerboukha, H. et al. (2022). Metasurface-in-the-middle attack: from theory to experiment. Proceedings of the 15th ACM Conference on Security and Privacy in Wireless and Mobile Networks, 257–267.

24 Yu, N., Genevet, P., Kats, M.A. et al. (2011). Light propagation with phase discontinuities: generalized laws of reflection and refraction. *Science* 334 (6054): 333–337.

25 Zhang, X., Tian, Z., Yue, W. et al. (2013). Broadband terahertz wave deflection based on C-shape complex metamaterials with phase discontinuities. *Advanced Materials* 25 (33): 4567–4572.

26 Guerboukha, H., Amarasinghe, Y., Shrestha, R. et al. (2021). High-volume rapid prototyping technique for terahertz metallic metasurfaces. *Optics Express* 29 (9): 13806–13814.

27 Shaikhanov, Z., Hassan, F., Guerboukha, H. et al. (2022). Adversarial metasurfaces: metasurface-in-the-middle attack. Proceedings of the 15th ACM Conference on Security and Privacy in Wireless and Mobile Networks, 274–276.

28 3GPP (2020). Study on channel model for frequencies from 0.5 to 100 GHz (Release 16). 3GPP TR 38.901 V16.1.0, January 2020.

29 Khalid, N. and Akan, O.B. (2016). Wideband THz communication channel measurements for 5G indoor wireless networks. Proceedings of IEEE International Conference on Communications (IEEE ICC), 1–6.

30 Ko, J., Cho, Y., Hur, S. et al. (2017). Millimeter-wave channel measurements and analysis for statistical spatial channel model in in-building and urban environments at 28 GHz. *IEEE Transactions on Wireless Communications* 16 (9): 5853–5868.

31 Kokkoniemi, J., Petrov, V., Moltchanov, D. et al. (2016). Wideband terahertz band reflection and diffuse scattering measurements for beyond 5G indoor wireless networks. Proceedings of the 22th European Wireless Conference, 1–6.

32 Petrov, V., Moltchanov, D., Jornet, J.M., and Koucheryavy, Y. (2019). Exploiting multipath terahertz communications for physical layer security in beyond 5G networks. Proceedings of IEEE Conference on Computer Communications Workshops (IEEE INFOCOM WKSHPS), 1–8.

33 Wu, Y., Khisti, A., Xiao, C. et al. (2018). A survey of physical layer security techniques for 5G wireless networks and challenges ahead. *IEEE Journal on Selected Areas in Communications* 36 (4): 679–695.

34 Akyildiz, I.F., Jornet, J.M., and Han, C. (2014). TeraNets: Ultra-broadband communication networks in the terahertz band. *IEEE Wireless Communications* 21 (4): 130–135.

35 Petrov, V., Solomitckii, D., Samuylov, A. et al. (2017). Dynamic multi-connectivity performance in ultra-dense urban mmWave deployments. *IEEE Journal on Selected Areas in Communications* 35 (9): 2038–2055.

36 Petrov, V., Kokkoniemi, J., Moltchanov, D. et al. (2018). Last meter indoor terahertz wireless access: performance insights and implementation roadmap. *IEEE Communications Magazine* 56 (6): 158–165.

37 Peng, B., Rey, S., Rose, D.M. et al. (2018). Statistical characteristics study of human blockage effect in future indoor millimeter and sub-millimeter wave wireless communications. Proceedings of the IEEE VTC Spring, 1–5.

38 Petrov, V., Komarov, M., Moltchanov, D. et al. (2017). Interference and SINR in millimeter wave and terahertz communication systems with blocking and directional antennas. *IEEE Transactions on Wireless Communications* 16 (3): 1791–1808.

39 Gapeyenko, M., Petrov, V., Moltchanov, D. et al. (2018). An analytical representation of the 3GPP 3D channel model parameters for mmWave bands. Proceedings of the 2nd ACM Workshop on Millimeter Wave Networks and Sensing Systems (mmNets), 33–38.

40 Cohen, A., D'Oliveira, R.G.L., Yeh, C.-Y. et al. (2022). Absolute security in high-frequency wireless links. https://arxiv.org/abs/2208.05907 (accessed 15 March 2024).

41 Shrestha, R., Guerboukha, H., Fang, Z. et al. (2022). Jamming a terahertz wireless link. *Nature* 13: 3045.

42 Petrov, V., Kurner, T., and Hosako, I. (2020). IEEE 802.15.3d: First standardization efforts for sub-terahertz band communications toward 6G. *IEEE Communications Magazine* 58 (11): 28–33. https://doi.org/10.1109/MCOM.001.2000273.

Index

a

absolute security 241–243
ABY^3 74–77
active security 84
AI-assisted Vehicle-to-Everything (V2X) 21–22, 27
anomaly based detection techniques 103
asymmetric cryptography 42
authenticity 136
availability 135

b

Bayes' theorem 104
BB84 QKD protocol 207–209
Bell state 202
Bitcoin 132, 133, 147
bit generation rate (BGR) 159
bit mismatch rate (BMR) 159
blockchain 132
- applications 132–133
- based IoT framework 131
- consensus mechanisms 133–134
- scalability 147–148

boosting algorithm 80

c

channel-based keyless physical layer authentication
- CIR-based authentication 167–168
- frequency-based authentication 167
- machine learning-based authentication 168
- pilot-based authentication 168
- RSS-based authentication 167

channel impulse response (CIR)-based authentication 167–168
channel reciprocity-based key establishment, PLKE
- evaluation metrics 159
- key establishment model 159–161
- principles and assumptions 157–159

chirp spread spectrum (CSS)-based PLE 173
Clauser–Horne–Shimony–Holt (CHSH) test 206
cloud computing
- categories 36–37
- to edge computing 37

computational offloading 40
confidentiality 135
conjugate coding (CC) 201
continuous variable (CV) quantum system 199–200
convolutional neural networks (CNN) 73
Crypten 77–78
cryptocurrency 132, 133
cryptography 155, 203
- post-quantum 42

CryptoNets 69–70

d

data injection attack detection 177–178
data security 203–204
- formal cryptographic frameworks 204
- practical source and security 204

Security and Privacy for 6G Massive IoT, First Edition.
Edited by Georgios Mantas, Firooz B. Saghezchi, Jonathan Rodriguez, and Victor Sucasas.

Decision Trees (DTs)
  computational setups 80–81
  function 105
  inference 81–83
  relevant contributions 83–84
  structure of 79
  training 79–80
DeepSecure 72–73
Delegated Proof of Stake (DPoS) 134
Delphi 74
denial of sleep attack 179
device-based keyless physical layer authentication
  physically unclonable function 169–170
  radio frequency fingerprinting 168–169
device independent (DI-QKD) protocol 206
Diffie-Hellman key exchange 204
Digital Twins (DT) for manufacturing 9–10
discrete variable protocols 205–207
  BB84 207–209
  security argument 209–211
distributed ledger technology (DLT) 144
distribution model 107

e

eavesdropping
  information theoretic analysis of 230–232
  THz links
    directional security 232–233
    metasurface-in-the-middle attack 233–236
edge computing 36–37
edge intelligence (EI)
  edge computing, importance 36–37
  emergence 38–39
  fusion of 39–40
  future development 35–36
  need for secure 40
  network slicing 147
  privacy challenges 42–44
  security challenges in
    computational offloading 40
    machine learning security 41
    post-quantum cryptography 42
  security standardization for 46–47
  trust challenges in 44–46
Einstein–Podolsky–Rosen (EPR) pairs 202
elliptic curve Diffie–Hellman (ECDH) problem 42
enabling sustainability
  autonomous supply chains 6–7
  earth monitor 6
  E-health for all 5–6
  institutional coverage 6
  network functionality for crisis resilience 8
  network trade-offs for minimized environmental impact 7–8
  sustainable food production 7
Ensemble Learning (EL) 106
entanglement swapping 202–203
European 6G Flagship Project Hexa-X 2, 3

f

FANNG–MPC 78
Fantastic Four 77
federated learning (FL) 43
  categories of 108
  definition of 107–108
  federated transfer learning 111–113
  horizontal federated learning 110–111
  model aggregation algorithms in 119–121
  model aggregation approaches in 117–119
  vertical federated learning 111
federated learning-based IDSs for IoT
  challenges and future directions in
    communication efficiency and scalability 122
    data heterogeneity and non-IID data 121
    explainability and interpretability 122
    security and privacy enhancements 121–122
    validation and standardization 121
  lightweight and privacy-preserving, cross-layer 116–117
  use cases
    agricultural IoT 115–116
    industrial IoT 114–115

in smart homes 113–114
vehicular IoT networks 116
federated transfer learning (FTL) 109, 111–113
finite dimensional (FD) quantum systems 197
five-sense communications 13–14
Fock space model systems 199
frequency-based authentication 167
frequency hopping spread spectrum (FHSS)-based PLE 173
fully homomorphic encryption (FHE)
privacy-preserving machine learning (PPML) 63–64
bottlenecks and advantages 64–65
hybrid 87–89
vs. MPC 65
schemes 66
private data process 55
functional encryption (FE) 43

g

Garbled Circuits (GC) 56–57, 84
Garbled Decision Tree (GDT) 84
Gauss function 200–201
Gaussian state 200–201
Gazelle 73–74
General Data Protection Regulation (GDPR) 43
generic signal modulation-based PLE 172
Gradient-Boosting Decision Trees (GBDT) 80
greedy algorithms 80

h

haptic communications 193
haptic synchronization 196
Hexa-X 6G vision 2
Hierarchical Federated Averaging (HFA) 119
Hilbert spaces 197–199, 201
holographic communications 13
homomorphic encryption (HE) 40, 83
horizontal federated learning (HFL) 108–111
hybrid-based detection techniques 103
hyperconnected resilient network infrastructures
AI-assisted Vehicle-to-Everything (V2X) 21–22
enhanced public protection 23
interconnected IoT micro-networks 22–23
sensor infrastructure web 21

i

Industrial Internet of Things (IIoT) 114
inference, stages 81, 83
Information and Communications Technology (ICT) 37
information theoretic analysis 230–232
integrated physical layer security 211
conventional approaches 212–213
quantum approaches
notation 213
quantum physical layer security 214
integrity 135
Internet of Things (IoT)
cloud computing to edge computing 37
data flow and trust 54
6G 36
PPML for 54–56
intrusion detection systems (IDSs) in IoT
fundamentals of 103–104
limitations of ML-based 106
machine learning techniques for 104–106
IoT. *see* Internet of Things (IoT)
IoT communication network 135

j

jamming THz links 243–246

k

key-based data confidentiality 171–173
key-based physical layer authentication 165–166
key entropy 159
keyless data confidentiality
artificial noise-based extensions 175

keyless data confidentiality (*contd.*)
 channel coding-based data confidentiality 174–175
 directional modulation-based extensions 176
 overview of 173–174
 security metrics 174
K-means clustering 105
K-nearest neighbor (KNN) algorithm 104–105

l
leveled fully homomorphic scheme (LHE) 64

m
machine learning (ML)
 based authentication 168
 IDSs in IoT 104–106
 PPML for IoT 54–55
 6G EI 41
massive IoT 131
 location privacy challenges 138
 privacy requirements in 136–137
 security requirements in 135–136
massive twinning
 digital twins for manufacturing 9–10
 immersive smart cities 10–11
 internet of tags 11–12
measurement device independence (MDI) 206
Metasurface-in-the-Middle (MSITM) attack 233–236
MiniONN 70–71
multiparty computation (MPC) 43–44
 privacy-preserving machine learning (PPML) 56–60
  arithmetic 61–62
  outsourcing 86–87
  preprocessing 63
  privacy and output guarantees 61
  security models and setups 60
 private data process 55
 protocol types 61
multi-path THz communications
 evaluation framework 237–239
 security against eavesdropping 236–237
 security *vs.* capacity trade-off 239–241

n
Naive Bayes (NB) classifier 104
network domain 194
network function virtualization (NFV) 129, 141, 142
network slicing
 blockchain-enabled secure
  distributed ledger technology 144
  multiple participant coordination and trust management 145–146
  in verticals 144
 blockchain scalability 147–148
 edge intelligence in 147
 lifecycle of 140
 post-quantum security 147
 RAN slicing 148
 security, privacy, and trust threats for 5G 142–143
 standardization 138–141

o
oblivious neural networks (ONN)
 n-party (MPC) 69
 three-party (3PC) and four-party (4PC) 69
 two-party (2PC) 67
 two-party with helper ($2PC^+$) 67–68
oblivious transfer (OT) 217–218
 bit commitment 219–220
 probabilistic formulation of 218
 relevance of 219
one-time pad encryption 203
order-revealing encryption (ORE) 144
orthogonal frequency division multiplexing (OFDM)-based PLE 173

p
Packet Additive Homomorphic Encryption (PAHE) 73
passive security 83–84

personally identifiable information (PII) 42–43
phase shift keying (PSK)-based PLE 172–173
physical layer-based data confidentiality
  concept of 171
  key-based data confidentiality–physical layer encryption 171–173
  keyless data confidentiality
    artificial noise-based extensions 175
    channel coding-based data confidentiality 174–175
    directional modulation-based extensions 176
    overview of 173–174
    security metrics 174
physical layer-based detection of malicious nodes
  concept of 177
  data injection attack detection 177–178
  sleep deprivation attack detection 179
  Sybil attack detection 178–179
physical layer-based key establishment (PLKE)
  channel reciprocity-based key establishment
    evaluation metrics 159
    key establishment model 159–161
    principles and assumptions 157–159
  concept of 156–157
  signal source indistinguishability-based key establishment
    evaluation metrics 162
    key establishment model 162–163
    principles and assumptions 161–162
    security and performance evaluation 163–164
physical layer-based node authentication
  channel-based keyless physical layer authentication
    CIR-based authentication 167–168
    frequency-based authentication 167
    machine learning-based authentication 168
    pilot-based authentication 168
    RSS-based authentication 167
  concept of 164–165
  device-based keyless physical layer authentication
    physically unclonable function 169–170
    radio frequency fingerprinting 168–169
  key-based physical layer authentication 165–166
physical layer encryption (PLE)
  channel coding-based 172
  chirp spread spectrum 173
  frequency hopping spread spectrum 173
  generic signal modulation-based 172
  orthogonal frequency division multiplexing 173
  phase shift keying-based 172–173
  quadrature amplitude modulation 173
physical layer security (PLS) 229
physically unclonable function (PUFs) 169–170
pilot-based authentication 168
Poisson point process (PPP) 237
post-quantum cryptography 42
PPML. *see* privacy-preserving machine learning (PPML)
privacy homomorphism 63. *see also* fully homomorphic encryption (FHE)
Privacy-Preserving Decision Trees (PDT) 78
privacy-preserving machine learning (PPML)
  alternative solutions 90
  decision trees
    computational setups 80–81
    inference 81–83
    relevant contributions 83–84
    structure of 79
    training 79–80
  frameworks
    ABY^3 74–77
    Crypten 77–78
    CryptoNets 69–70
    DeepSecure 72–73
    Delphi 74
    FANNG-MPC 78

privacy-preserving machine learning (PPML) (*contd.*)
Fantastic Four 77
Gazelle 73–74
MiniONN 70–71
SecureML 70
fully homomorphic encryption 63–64
bottlenecks and advantages 64–65
comparison with MPC 65
hybrid 87–89
schemes 66
for IoT 54–56
multiparty computation 56–60
arithmetic 61–62
outsourcing 86–87
preprocessing 63
privacy and output guarantees 61
security models and setups 60
oblivious neural networks
n-party (MPC) 69
three-party (3PC) and four-party (4PC) 69
two-party (2PC) 67
two-party with helper (2PC+) 67–68
software and frameworks 85–86
Proof of Capacity (PoC) 134
Proof of Stake (PoS) 134
public-key cryptography 203

q

quadrature amplitude modulation (QAM)-based PLE 173
quantum bit error rate (QBER) 209
quantum computing 42
quantum conference key agreement (QCKA) 218
quantum entanglement 201–202
quantum gates 197–198
quantum key distribution (QKD)
continuous variable protocols 211
discrete variable protocols 205–207
BB84 207–209
security argument 209–211
quantum measurement 198–199
quantum mechanics
conjugate coding 201
continuous variable quantum states 199–200
entanglement swapping 202–203
finite dimensional quantum systems 197
Fock space 199
Gaussian states 200–201
quantum entanglement 201–202
quantum gates 197–198
quantum measurement 198–199
quantum physical layer security 214
quantum private queries (QPQ) 218
quantum security 214–217
Eve's attack 215
Trojan-horse attack 216

r

radio access network (RAN) 39
radio frequency fingerprinting (RFF) 168–169
random algorithms 80
random forests 80
Random Forests (RFs) 105
RAN slicing 148
Rivest–Shamir–Adleman (RSA) 204
RSS-based authentication 167

s

sea/underwater network 26
SecureML 70
security argument 209–211
security threats in massive IoT in 6G
authentication 31
authorization 31–32
availability 30–31
data confidentiality 29
data integrity 29–30
security *vs.* capacity trade-off 239–241
Shannon's cipher system 230–231
signature-based detection techniques 103
6G era, secure edge intelligence
edge computing, importance 36–37
emergence 38–39
fusion of 39–40
future development 35–36

need for secure 40
privacy challenges 42–44
security challenges in
computational offloading 40
machine learning security 41
post-quantum cryptography 42
security standardization for 46–47
trust challenges in 44–46
6th Generation (6G)-enabled massive IoT
air network 25
enabling sustainability
autonomous supply chains 6–7
earth monitor 6
E-health for all 5–6
institutional coverage 6
network functionality for crisis resilience 8
network trade-offs for minimized environmental impact 7–8
sustainable food production 7
ground network 25
hyperconnected resilient network infrastructures
AI-assisted Vehicle-to-Everything (V2X) 21–22
enhanced public protection 23
interconnected IoT micro-networks 22–23
sensor infrastructure web 21
massive twinning
digital twins for manufacturing 9–10
immersive smart cities 10–11
internet of tags 11–12
from robots to cobots
AI partners 16
consumer robots 15–16
flexible manufacturing 17
interacting and cooperative mobile robots 16–17
situation-aware device reconfiguration 17–18
SAGSIN 24–26
security objectives in
authentication 28
authorization 28
availability 27–28
confidentiality 26–27
integrity 27
security threats in
authentication 31
authorization 31–32
availability 30–31
data confidentiality 29
data integrity 29–30
space network 24–25
surface network 26
telepresence
fully merged cyber–physical worlds 12–14
immersive sport event 14–15
merged reality game/work 15
Mixed Reality co-design 14
trusted embedded networks
human-centric communications 18–19
infrastructure-less network extensions and embedded networks 19
local coverage for temporary usage 20
small coverage, low-power micro-network in networks for production and manufacturing 20
underwater network 26
vision and core values 2–3
sleep deprivation attack detection 179
Software Defined Network (SDN) 129
Space-Air-Ground-Sea/Underwater Integrated Network (SAGSIN) 23, 24
air network 25
ground network 25
sea/underwater network 26
security objectives
authentication 28
authorization 28
availability 27–28
confidentiality 26–27
space network 24–25
surface network 26
Support Vector Machines (SVMs) 105
Sybil attack detection 178–179
symmetric cryptography 42

t

Tactile Edge 194, 196
Tactile Internet (TI) 193–195
telepresence
  fully merged cyber–physical worlds 12–14
  immersive sport event 14–15
  merged reality game/work 15
  Mixed Reality co-design 14
third generation partnership project (3GPP) 138–140
THz links, eavesdropping
  directional security 232–233
  metasurface-in-the-middle attack 233–236
traceability 136
Trojan-horse attack 216
trusted embedded networks
  human-centric communications 18–19
  infrastructure-less network extensions and embedded networks 19
  local coverage for temporary usage 20
  small coverage, low-power micro-network in networks for production and manufacturing 20
Trusted Third Party (TTP) model 86

u

ultra-reliable low-latency communication (URLLC) 194

v

Vehicle-to-Everything (V2X) 21–22
Vehicular Ad-hoc Networks (VANETs) 116
verifiable computation (VC) 40, 65
Vernam cipher 203
vertical federated learning (VFL) 108, 109, 111

w

Wireless Brain–Computer Interfaces (WBCI) 3, 14
Wireless Mind–Machine Interfaces (WMMI) 3
Wyner's wiretap channel 231

www.ingramcontent.com/pod-product-compliance
Lightning Source LLC
Chambersburg PA
CBHW080834310825
31883CB00004B/12
9781119987970